Phytoplasmas

Genomes, Plant Hosts and Vectors

Phytoplasmas

Genomes, Plant Hosts and Vectors

Edited by

Phyllis G. Weintraub

Agricultural Research Organization
Gilat Research Center
Israel

and

Phil Jones

Rothamsted Research
UK

www.cabi.org

CABI is a trading name of CAB International

CABI Head Office
Nosworthy Way
Wallingford
Oxfordshire OX10 8DE
UK

Tel: +44 (0)1491 832111
Fax: +44 (0)1491 833508
E-mail: cabi@cabi.org
Website: www.cabi.org

CABI North American Office
875 Massachusetts Avenue
7th Floor
Cambridge, MA 02139
USA

Tel: +1 617 395 4056
Fax: +1 617 354 6875
E-mail: cabi-nao@cabi.org

A catalogue record for this book is available from the British Library, London, UK.

Library of Congress Cataloging-in-Publication Data

Phytoplasmas : genomes, plant hosts, and vectors / editors, Phyllis G. Weintraub & Phil Jones.

p. cm.

Includes bibliographical references and index.

ISBN 978-1-84593-530-6 (alk. paper)

1. Phytoplasmas. I. Weintraub, Phyllis G. II. Jones, Phil, 1947- III. Title.

SB738.P59 2010

632'.32--dc22

2009019575

ISBN-13: 978 1 84593 530 6

Typeset by AMA Dataset, Preston, UK.
Printed and bound in the UK by MPG Books Group.

The paper used for the text pages in this book is FSC certified. The FSC (Forest Stewardship Council) is an international network to promote responsible management of the world's forests.

Contents

The colour plate section can be found following p. 146

Contributors

Alberto Alma, *DIVAPRA – Entomologia e Zoologia applicate all'Ambiente, Università degli Studi di Torino, via Leonardo da Vinci 44, 10095 Grugliasco, Torino, Italy. E-mail: alberto.alma@unito.it*

Domenico Bosco, *DIVAPRA – Entomologia e Zoologia applicate all'Ambiente, 'Carlo Vidano' Università degli Studi di Torino, via Leonardo da Vinci 44, 10095, Grugliasco, Torino, Italy. E-mail: domenico.bosco@unito.it*

Fiona E. Constable, *Department of Primary Industries, Private Bag 15, Ferntree, Gully Delivery Centre, Victoria 3156, Australia. E-mail: fiona.constable@dpi.vic.gov.au*

Daniele Daffonchio, *DISTRAM, Università degli Studi di Milano, via Celoria 2, 20133 Milano Italy. E-mail: daniele.daffonchio@unimi.it*

Romina D'Amelio, *DIVAPRA – Entomologia e Zoologia applicate all'Ambiente, 'Carlo Vidano' Università degli Studi di Torino, via Leonardo da Vinci 44, 10095, Grugliasco, Torino, Italy. E-mail: romina.damelio@unito.it*

Robert E. Davis, *Molecular Plant Pathology Laboratory, USDA-ARS, Rm 118, Bldg. 004, 10300 Baltimore Avenue, Beltsville, MD 2070, USA. E-mail: robert.davis@ars.usda.gov*

Matthew Dickinson, *University of Nottingham, School of Biosciences, Plant Sciences Division, Loughborough LE12 5RD, UK. E-mail: Matthew.Dickinson@nottingham.ac.uk*

Xavier Foissac, *UMR-1090 Génomique Diversité Pouvoir Pathogène, INRA et Université Victor Ségalen Bordeaux 2, 71 avenue Edouard Bourlaux – BP 81, 33883 Villenave d'Ornon, France. E-mail: foissac@bordeaux.inra.fr*

Luciana Galetto, *Istituto di Virologia Vegetale, CNR, Strada delle Cacce, 73, I-10135 Torino, Italy. E-mail: l.galetto@ivv.cnr.it*

Elena Gonella, *DIVAPRA – Entomologia e Zoologia applícate all'Ambiente, Università degli Studi di Torino, Italy, via Leonardo da Vinci 44, 10095 Grugliasco (Torino), Italy. E-mail: Elena.gonella@unito.it*

Hugh Harries, *Centro de Investigación Científica de Yucatán, Mérida, Mexico and 2 Beech Road, Broadway, Weymouth, Dorset DT3 5NP, UK. E-mail: harrieshc@yahoo.com*

Jennifer Hodgetts, *University of Nottingham, School of Biosciences, Plant Sciences Division, Sutton Bonington campus, Loughborough LE12 5RD, UK. E-mail: sbxjh2@nottingham.ac.uk*

Saskia A. Hogenhout, *Department of Disease and Stress Biology, The John Innes Centre, Norwich Research Park, Colney Lane, Colney, Norwich, NR4 7UH, UK. E-mail: saskia.hogenhout@bbsrc.ac.uk*

Barbara Jarausch, *RLP AgroScience GmbH, AlPlanta, Fruit Crop Diseases, Breitenweg 71, 67435 Neustadt, Germany. E-mail: barbara.jarausch@agroscience.rlp.de*

Wolfgang Jarausch, *RLP AgroScience GmbH, AlPlanta, Fruit Crop Diseases, Breitenweg 71, 67435 Neustadt, Germany. E-mail: wolfgang.jarausch@agroscience.rlp.de*

Phil Jones, *Rothamsted Research, UK. E-mail: phytophil@gmail.com*

Shigeyuki Kakizawa, *Department of Agricultural and Environmental Biology, University of Tokyo, 1-1-11 Yayoi, Bunkyo-ku, 113-8657, Tokyo, Japan. E-mail: kakizawa@mail.ecc.u-tokyo.ac.jp*

Ing-Ming Lee, *Molecular Plant Pathology Laboratory, USDA-ARS, Rm 218, Bldg. 004, 10300 Baltimore Avenue, Beltsville, MD 20705, USA. E-mail: ingming.lee@ars.usda.gov*

Michael Maixner, *Julius Kühn Institute (JKI), Federal Research Centre for Cultivated Plants Institute for Plant Protection in Fruit Crops and Viticulture, Brüningstraße 84, D-54470 Bernkastel-Kues, Germany. E-mail: Michael.Maixner@jki.bund.de*

Carmine Marcone, *Dipartimento di Scienze Farmaceutiche, Università degli Studi di Salerno, via Ponte Don Melillo, I-84084 Fisciano (Salerno), Italy. E-mail: cmarcone@unisa.it*

Cristina Marzachì, *Istituto di Virologia Vegetale, CNR, Strada delle Cacce, 73, I-10135 Torino, Italy. E-mail: C.Marzachi@ivv.cnr.it*

Rita Musetti, *Dipartimento di Biologia e Protezione delle Piante, Università di Udine, via delle Scienze, 208, 33100 Udine, Italy. E-mail: Rita.Musetti@uniud.it*

Shigetou Namba, *Laboratory of Plant Pathology, Department of Agricultural and Environmental Biology, University of Tokyo, 1-1-11 Yayoi, Bunkyo-ku, 113-8657, Tokyo, Japan. E-mail: anamba@mail.ecc.u-tokyo.ac.jp*

Kenro Oshima, *Laboratory of Plant Pathology, Department of Agricultural and Environmental Biology, University of Tokyo, 1-1-11 Yayoi, Bunkyo-ku, 113-8657, Tokyo, Japan. E-mail: kenro@ims.u-tokyo.ac.jp*

Noura Raddadi, *DIVAPRA – Entomologia e Zoologia applicate all'Ambiente, Università degli Studi di Torino, Italy, via Leonardo da Vinci 44, 10095 Grugliasco (Torino), Italy. E-mail: noura.raddadi@unito.it*

Yaima Arocha Rosete, *Rothamsted Research, Harpenden, Hertfordshire, AL5 2JQ, United Kingdom and: CABI-Europe, Bakeham Lane, Egham TW20 9TY, UK. E-mail: arocharosete57@googlemail.com*

Erich Seemüller, *Julius Kühn Institute (JKI), Federal Research Centre for Cultivated Plants Institute for Plant Protection in Fruit Crops and Viticulture, Schwabenheimer Str. 101, 69221 Dossenheim, Germany. E-mail: Erich.Seemueller@jki.bund.de*

Martina Šeruga Music, *Department of Biology, Faculty of Science, University of Zagreb, Marulicev trg 9A, HR-10000 Zagreb, Croatia. E-mail: martina@botanic.hr*

Wei Wei, *Molecular Plant Pathology Laboratory, USDA-ARS, Rm 213, Bldg. 004, 10300 Baltimore Avenue, Beltsville, MD 20705, USA. E-mail: wei.wei@ars.usda.gov*

Phyllis G. Weintraub, *Agricultural Research Organization, Gilat Research Center, D.N. Negev, 85280, Israel. E-mail: phyllisw@volcani.agri.gov.il*

Michael R. Wilson, *Entomology Section, Department of Biodiversity & Systematic Biology, National Museum of Wales, Cardiff, CF10 3NP, UK. E-mail: mike.wilson@museumwales.ac.uk*

Yan Zhao, *Molecular Plant Pathology Laboratory, USDA-ARS, Rm 213, Bldg. 004, 10300 Baltimore Avenue, Beltsville, MD 20705, USA. E-mail: yan.zhao@ars.usda.gov*

Preface

Since their discovery in 1967 as 'mycoplasma-like organisms', the phytoplasmas have quickly become established as a unique group of plant pathogens. Diseases, frequently called 'yellows', have been known since the late 1800s; originally thought to be associated with viruses, many are now known to be caused by phytoplasmas. During the 1970s, research centred on diagnosis using symptoms and electron microscopy to visualize the phytoplasmas in the phloem sieve cells of their hosts, transmission by insect vector and studies on the spread of the diseases they caused. The biology and taxonomy of these obligate pathogens were still shrouded in mystery. It was the advent of the molecular biological revolution in the 1980s that saw the introduction of techniques such as nucleic acid purification, DNA hybridization and the polymerase chain reaction, which with the secrets of these fastidious bacteria begin to emerge. In the 1990s the term phytoplasma had been proposed, and by 2004 a distinct taxonomic group, '*Candidatus* Phytoplasma', was defined. The evolution of molecular techniques has led to more information and, paradoxically, less clarity in grouping different phytoplasma 'taxa'. As of today there are hundreds of diseases caused by phytoplasmas and about 100 known insect vectors.

In this book we have tried to examine all aspects related to phytoplasmas, their plant hosts and insect vectors and so present the reader with the state of the art in a logical, coherent fashion. Since phytoplasmas are fastidious, diagnostic methods and quantification assume greater importance because one is limited in the scope of available techniques. For example, serological methods are very limited with an organism that cannot be artificially cultured. The opening chapter is followed by chapters on sequencing and functional genomics, which relies heavily on comparing phytoplasma genomics with that of other known bacteria. As mentioned, there are hundreds of diseases caused by phytoplasmas, and visual methods (transmission electron microscopy, DAPI, etc.) do not allow for identification of these bacteria, so means of differentiation are primarily based on genome sequences. We have three chapters that take different approaches to differentiation, classification and taxonomy. Having thoroughly examined the phytoplasma from the inside out, we then turn to organismal aspects, the first group of chapters being related to aspects of phytoplasmas in plants. The first of these chapters examines the movement of phytoplasmas within the plant and the development of disease. We then look at the biochemical changes precipitated by the replication of the phytoplasma in plants, and finally at aspects of plant resistance. Chapters on the epidemiology of disease in grasses and grapevines delve into the disease process in plants. The last of the plant-related chapters examines epidemiological systems with multiple host plants. Turning to the insect vectors, we start with a chapter describing general aspects of vectors and their control, followed by an in-depth examination of the psyllid vectors and their control. Unique control methods are evolving, particularly symbiotic control of phytoplasmoses. As with plants, phytoplasmas can have multiple insect vectors, and the ramifications of this are examined. We conclude with an examination of the distribution and potential spread of phytoplasma diseases and vectors worldwide.

We believe that we have brought together an ensemble of authors from all regions of the world that are at the forefront of their respective disciplines. We hope this book will be useful to researchers/professionals at all levels and will help illuminate and stimulate thoughts and interest in this challenging and difficult host–pathogen–vector system.

Acknowledgements

The editors would like to extend a special thanks to the many people who contributed their time to review the chapters presented herein.

Athanassios S. Alivizatos, *Benaki Phytopathological Institute, Kifissia, Greece*
Mark T. Andersen, *The New Zealand Institute for Plant & Food Research Limited, Auckland, New Zealand*
Dez Barbara, *University of Warwick, Warwick, UK*
André Bervillé, *INRA UMR 1097 DIAPC, Montpellier, France*
Neil Boonham, *The Food and Environment Research Agency, York, UK*
Toby J.A. Bruce, *Rothamsted Research, Harpenden, UK*
Michael Deadman, *College of Agricultural and Marine Sciences, Sultan Qaboos University, Sultanate of Oman*
Murray J. Fletcher, *Orange Agricultural Institute, New South Wales, Australia*
Jürgen Gross, *Julius Kuehn Institute (JKI), Federal Research Centre for Cultivated Plants Institute for Plant Protection in Fruit Crops and Viticulture, Dossenheim, Germany*
Michael Kube, *Max Planck Institute for Molecular Genetics, Berlin, Germany*
Daniel Pascal Klaehre, *Gartenbauzentrum Bayern Nord, Amt für Ernaehrung Landwirtschaft und Forsten Mainbernheimerstr, Kitzingen, Germany*
Lia W. Liefting, *MAF Biosecurity New Zealand, Auckland, New Zealand*
Roberto Michelutti, *Agriculture and Agri-Food Canada, Greenhouse and Processing Crops Research Centre, Ontario, Canada*
Thomas Albert Miller, *University of California, California, USA*
Helena Guglielmi Montano, *Universidade Federal Rural do Rio de Janeiro (UFRRJ), Rio de Janeiro, Brasil*
Mogens Nicolaisen, *Aarhus University, Slagelse, Denmark*
Janice Proud, *Anglican Church in Ethiopia and the Horn of Africa, Addis Ababa, Ethiopia*

Joël Renaudin, *INRA-Université Bordeaux2, UMR 1090 Génomique Diversité Pouvoir Pathogène, IBVM, Centre INRA de Bordeaux, Villenave d'Ornon, France*

Monika Riedle-Bauer, *Höhere Bundeslehranstalt und Bundesamt für Wein-und, Langenzersdorf, Austria*

Gianfranco Romanazzi, *Marche Polytechnic University, Ancona, Italy*

Bernd Schneider, *Julius Kuehn Institute (JKI), Federal Research Centre for Cultivated Plants Institute for Plant Protection in Fruit Crops and Viticulture, Dossenheim, Germany*

Michael Shaw, *University of Reading, Reading, UK*

Claire Streten, *Charles Darwin University, Darwin, Australia*

Rosemarie Tedeschi, *DIVAPRA – Entomologia e Zoologia applicate all'Ambiente 'C. Vidano', Universita di Torino, Torino, Italy*

Ester Torres, *Laboratori de Sanitat Vegetal (Generalitat de Catalunya), Barcelona, Spain*

Ivo Toševski, *CABI Europe – Switzerland, Delémont, Switzerland*

Lucy Tran-Nguyen, *Department of Regional Development, Primary Industry, Fisheries and Resources, NT, Australia*

Chandrashekara A. Viraktamath, *University of Agricultural Sciences, Bangalore, India*

Astri Wayadanda, *Oklahoma State University, Stillwater, USA*

Einat Zchori-Fein, *Agricultural Research Organization, Ministry of Agriculture, Ramat Yishay, Israel*

1 Real-time PCR Diagnosis and Quantification of Phytoplasmas

LUCIANA GALETTO AND CRISTINA MARZACHÌ

Istituto di Virologia Vegetale, CNR, Torino, Italy

Introduction

Since their identification (Doi *et al.*, 1967), phytoplasmas have been identified as pathogens in numerous plant genera and in some cases have caused severe epidemics in major crops such as grapevine, sugarcane and coconut. Phytoplasmas are vectored by phloem-feeding leafhoppers, planthoppers and psyllids (Weintraub and Beanland, 2006), and the contemporaneous presence of phytoplasma, weed reservoir and vector has often been the cause of severe losses, especially in countries with weak rural economies. As phytoplasmas have still not been cultured *in vitro*, their diagnosis relies mainly on molecular techniques such as PCR, usually followed by RFLP for assignation to a '*Candidatus* (*Ca.*) Phytoplasma' species or to a 16S rDNA group. The complete diagnostic procedure is laborious and requires several post-amplification steps. To overcome these problems, several approaches have been developed, amongst which universal and group-specific real-time PCR protocols have been proposed since 2004.

We review here the real-time PCR systems that have been developed for the diagnosis of phytoplasmas and highlight problems and possible solutions for the use of this technique in routine diagnosis. We also discuss possible future applications and modifications.

Real-time PCR

Real-time PCR has recently replaced the traditional PCR in efforts to increase the speed and sensitivity of detection and to improve techniques for mass screening. Real-time PCR reagents are now readily available and offer specific, sensitive and quantitative detection. During a real-time PCR run, accumulation of newly generated amplicons is monitored at each cycle by fluorescent detection methods, and so there is no need for post-PCR manipulation such as electrophoresis, which is required at the end of regular PCR. The amount of fluorescence, monitored at each amplification cycle, is proportional to the log of concentration of the PCR target, and for this reason real-time PCR is also a powerful technique for the quantification of specific DNA. Real-time PCR amplicons are visualized through several labelling techniques, most of which specifically bind to a target sequence on the amplicon, while others aspecifically stain double-stranded (ds) DNA amplicons. TaqMan® probes are the most commonly used ones for the diagnosis of phytoplasmas, although several other sequence-specific detection tools are available (Monis and Giglio, 2006), which, in theory, can be adapted to detect these phytopathogens. TaqMan probes are labelled at the 5′ end with a reporter dye and at the 3′ end with a quenching molecule; during each PCR cycle in the presence of a specific target DNA, the TaqMan probe, bound to its target sequence, is degraded by the 5′–3′ exonuclease activity of the Taq polymerase as it extends the primer. The fluorescence moiety of the probe is therefore freed from its quencher-labelled portion and the fluorescence is detected by the optical system of the apparatus.

SYBR Green I®, a highly specific, double-stranded DNA binding dye, is also used to detect phytoplasma-specific PCR product as it accumulates during real-time PCR cycles. The most important difference between the TaqMan and SYBR Green I dye chemistries is that the SYBR Green I dye chemistry will detect all double-stranded DNA, including non-specific reaction products. A well-optimized reaction is therefore essential for accurate results, which must be further analysed by running a melting curve analysis.

Real-time PCR Applications for Phytoplasma Diagnostics

Phytoplasmas have resisted all attempts to cultivate them *in vitro* in cell-free media, although the recent complete sequencing of several '*Ca.* Phytoplasma' species genomes and the discovery of their lack of some metabolic pathways (Hogenhout *et al.*, 2008) will add new impetus to this task. Moreover, the uneven distribution of phytoplasmas in the phloem of the infected plant, their low concentration (especially in woody hosts) and variations in titre according to the season and plant organ (reviewed in Firrao *et al.*, 2007) are also important obstacles for efficient diagnosis. For these reasons, although phytoplasmas seriously threaten the cultivation of some very important crop species, their diagnosis is not simple. The development of DNA-based tools such as PCR has been a major step in detection, identification and classification, and the 16S ribosomal gene has been the major target for designing phytoplasma-universal as well as group-specific primers (Bertaccini, 2007).

The success of these protocols relies on obtaining nucleic acid preparations of good quality, if possible enriched in phytoplasma DNA. Thus several methods have been developed to extract total phytoplasma DNA, aiming to concentrate it while reducing plant phenolics and polysaccharides that can inhibit the Taq DNA polymerase used in the PCR assay (Marzachì *et al.*, 2004). Most of these protocols require a phytoplasma enrichment step and this adds to their complexity, reducing the number of samples that can be processed. To speed up the procedure and increase throughput in routine diagnostic facilities, several quicker protocols have been developed (Firrao *et al.*, 2007). These are specially suited when decreased sensitivity can be compensated for by increasing sample number. Moreover, most universal as well as specific diagnostic protocols rely on nested PCR, which, although extremely sensitive, is also time-consuming and poses risks in terms of carry-over contamination between the two rounds of amplification.

Universal phytoplasma detection

Recently three protocols for the universal diagnosis of phytoplasmas using direct real-time PCR amplification of the 16S rDNA gene have been developed (Christensen *et al.*, 2004; Galetto *et al.*, 2005; Hren *et al.*, 2007). All of them exploited a TaqMan probe for detection. Universal phytoplasma DNA amplification is usually achieved with primers based on the ribosomal operon sequence, but these can also amplify DNA from other bacteria such as closely related *Acholeplasma* spp., which may be present on the surface of some plants (Tully *et al.*, 1994). To avoid false positives from contaminating bacterial DNA, a specific probe can be included. In all protocols, following optimization of the starting amount of total template DNA healthy controls were always absent. 16S rDNA-based primer/probe systems can be used to detect phytoplasmas belonging to several ribosomal subgroups with sensitivity similar to that of conventional nested PCR. Such sensitivity can be achieved even in detecting pathogens from field-collected woody hosts and insect vectors (Galetto *et al.*, 2005) with an obvious improvement in the speed of the assay.

Group-specific phytoplasma identification

Most diagnostic protocols for phytoplasmas include a first PCR amplification driven with universal primers, followed by a nested PCR with group-specific primers (Bertaccini, 2007). RFLP analysis of the group-specific amplicon may then be required for final identification of the phytoplasma strain. The entire procedure requires time-consuming post-amplification steps and is thus laborious. This problem is overcome by the sensitivity of real-time PCR, coupled to the possibility of designing group-specific primers and even probes to further enhance the specificity of detection.

Flavescence dorée (FD) and bois noir (BN) phytoplasmas infecting grapevines cause important losses in several countries. It is not surprising therefore that several protocols have been developed in recent years for diagnosis of these agents using real-time PCR (Bianco *et al.*, 2004; Galetto *et al.*, 2005; Angelini *et al.*, 2007; Gori *et al.*, 2007; Hren *et al.*, 2007). '*Ca.* Phytoplasma mali' (apple proliferation, AP), '*Ca.* Phytoplasma pyri' (pear decline, PD) and '*Ca.* Phytoplasma prunorum' (European stone fruit yellows, ESFY) are important pathogens of fruit trees, and several laboratories have proposed rapid, specific and sensitive diagnostic protocols for these pathogens (Baric and Dalla-Via, 2004; Jarausch *et al.*, 2004; Galetto *et al.*, 2005; Torres *et al.*, 2005; Aldaghi *et al.*, 2007; Martini *et al.*, 2007; Bisognin *et al.*, 2008). Table 1.1 lists the currently available reagents for specific diagnosis of several phytoplasmas using real-time PCR.

In most cases the 16S rDNA gene is the preferred target, but other genes or even randomly cloned DNA fragments to which no specific function is assigned have been used. SYBR Green I detection of 16S rDNA amplicons has been applied for the diagnosis of AP, PD, ESFY and FD, all quarantine phytoplasmas affecting fruit trees in Europe. SYBR Green I is the most economical chemistry for real-time PCR detection, but the specificity of the reaction is extremely important and needs to be carefully checked. Phytoplasma concentration in the host plant may be extremely low, and this results in high threshold cycles (CT), corresponding to late phases during which amplification of non-specific DNA may occur. This often happens in field-collected woody plants and even more with assays of individual insect vectors, despite the fact that phytoplasmas are usually present in high concentration in the vector body. In both hosts, bacterial contamination is common and unpredictable, and bacterial sequences may interfere with the diagnosis, especially when primers derived from the ribosomal operon are used (Wally *et al.*, 2008). When the real-time PCR amplification of field-collected plants or vectors results in high CT values (usually around or above 30), analysis of the melting curve of the amplicon is indispensable, since only those templates yielding amplicons with the expected melting temperature (MT) are phytoplasma-infected (Galetto *et al.*, 2005; Torres *et al.*, 2005). Amplicon detection with a specific TaqMan probe increases the specificity of the reaction and eliminates the need to run a melting analysis.

Baric and Dalla-Via (2004) developed a real-time PCR assay using a TaqMan minor groove binding (MGB) probe to detect AP in plant material. The TaqManMGB probe has an MGB ligand and a non-fluorescent quencher conjugated to the 3′ end, plus a fluorescent reporter dye at the 5′ end. The MGB ligand allows the use of shorter and more specific probes by increasing the stability of the probe–target bond. This property allows the use of shorter probes, with higher specificity than conventional TaqMan ones, and the discrimination of even single nucleotide mismatches (Kostina *et al.*, 2007). The same approach has been applied for specific detection of '*Ca.* Phytoplasma mali' amongst members of the 16SrX taxonomic group (Aldaghi *et al.*, 2007) as well as FD, BN and other phytoplasmas less frequently infecting grapevines (Hren *et al.*, 2007).

Rapid and sensitive detection of infected individuals in field populations of known phytoplasma vectors is extremely important for disease management and to study the characteristics of transmission. It is also decisive in the search for other potential vectors. In theory, any of the protocols described can be applied to real-time PCR detection of phytoplasma DNA in the insect,

Table 1.1. Name and sequence of primers and probes designed for the universal or group-specific detection of phytoplasma DNA by real-time PCR.

Specificity	Target gene	Forward primer 5´–3´	Reverse primer 5´–3´	Probe 5´–3´	Reference
Universal	16S rDNA	CGTACGCAAGTATGAAACTTAAAGGA	TCTTCGAATTAAACAACATGATCCA	TGACGGGACTCCGCACAAGCG	Christensen *et al.*, 2004
Universal	16S rDNA	CYS2Fw AGGTTGAACGGCCACATTG	CYS2Rv TTGCTCGGTCAGAGTTTCCTC	CYS2 Probe ACACGGCCCAAACTCCTACGGGA	Galetto *et al.*, 2005
Universal	16S rDNA	UniRNA Forward AAATATAGTGGAGGTTATCAGGGATACAG	UniRNA Reverse AACCTAACATCTCACGACACGAACT	UniRNA Probe ACGACAACCATGCACCA	Hren *et al.*, 2007
FD	16S rDNA	fAY GCACGTAATGGTGGGGACTT	rEY GCTTCAATTCGGTGACGAAAG	/	Galetto *et al.*, 2005
FD	16S rDNA	Flavescence dorée Forward AAGTCGAACGGAGACCCTTC	Flavescence dorée Reverse TAGCAACCGTTTCCGATTGT	Flavescence dorée Probe AAAAGGTCTTAGTGGCGAACGGGT	Angelini *et al.*, 2007
FD	sec Y	FDgen Forward TTATGCCTTATGTTACTGCTTCTATTGTTA	FDgen Reverse TCTCCTTGTTCTTGCCATTCTTT	FDgen Probe ACCTTTTGACTCAATTGA	Hren *et al.*, 2007
FD	16S rDNA	F1024 GTGAGATGTTAGGTTAAGTCCTAAAACGA	R1112 TTGGCAGTCTCGCTAAAGTCC	iProbe AACCCCTGTCGCTAGTTGCCAGC	Bianco *et al.*, 2004
BN	Genomic fragment	StolFw AACCGCTCGCAAACAGC	StolRev ATTAGCGCCTTAGCTGTG	/	Galetto *et al.*, 2005
BN	16S rDNA	Bois noir Forward GGTTAAGTCCCGCAACGAG	Bois noir Reverse CCCACCTTCCTCCAATTTATCA	Bois noir Probe AACCCTTGTTGTTAATTGCCATCATTAAG	Angelini *et al.*, 2007
BN	Genomic fragment	BNgen Forward AAGCAGGTTTAGCGATGGTTGT	BNgen Reverse TGGTACCGTTGCTTCATCATTT	BNgen Probe TTAATACCACCTTCAGGAAA	Hren *et al.*, 2007

(*continued*)

Table 1.1. *continued*

Specificity	Target gene	Forward primer 5′–3′	Reverse primer 5′–3′	Probe 5′–3′	Reference
AP	Nitro-reductase	fAP_2 AAGAGCAATTCGTACTTTCG	rAP_2 GCCGAACTAGTTTCTAATTGAC	/	Galetto *et al.*, 2005
AP	Genomic fragment	AP3 GAAACATGTCCTATTGGTGG	AP4 CCAATGTGTGAAATCTGTAG	/	Jarausch *et al.*, 2004
AP	16S rDNA	qAP-16S-F CGAACGGGTGAGTAACACGTAA	qAP-16S-R CCAGTCTTAGCAGTCGTTTCCA	qAP-16S TAACCTGCCTCTTAGACG	Baric and Dalla-Via, 2004
AP	16S rDNA	qAP-16S-F CGAACGGGTGAGTAACACGTAA	qAP-16S-R CCAGTCTTAGCAGTCGTTTCCA	AP-MGB CTGCCTCTTAGACGAGG	Aldaghi *et al.*, 2007
AP	16S rDNA	fAT CATCATTTAGTTGGGCACTT	rATRT CGCTTCAGCTACTCTTTGTG	TaqMan Probe CCCTTATGACCTGGGCTACA	Bisognin *et al.*, 2008
ESFY	Ribosomal protein	rpLNS2f GTGCTGAAGCTAATTTATTG	rpLNS2r2 CAATATGGCTAGTTCTTTTT	/	Martini *et al.*, 2007
16SrX	16S rDNA	P1 AAGAGTTTGATCCTGGCTCAGATT	R16(X)F1r CATCTCTCAGCATACTTGCGGGTC	/	Torres *et al.*, 2005
‘*Ca*. P. asteris’ (onion yellows)	tuf	Tuf1 GCTAAAACTTGTCCACGTTGTACG	Tuf2 CGGAAATAGAATTGAGGACGGT	TGTTTTAACTAAAAGAAGAAGGAGGACGTCACACTGCCTTTTTCTCTC	Wei *et al.*, 2004
‘*Ca*. P. asteris’ (aster yellows)	16S rDNA	Aster yellows Forward TTGGGTTAAGTCCCGCAAC	Aster yellows Reverse CCCACCTTCCTCCAATTTATCA	Aster yellows Probe CCAGCACGTAATGGTGGGGACTT	Angelini *et al.*, 2007
‘*Ca*. P. asteris’ (aster yellows)	16S rDNA	AACCCTCACCAGGTCTTGACA	CACGAGCTGACGACAACCAT	/	Hollingsworth *et al.*, 2008
Beet leafhopper-transmitted virescence agent	16S rDNA	16Sp303F AGGGCCTATAGCTCAGTTGGTTAGA	16Sp378R GTGGGCCTAAATGGACTTGAAC	16TM329 CACACGCCTGATAAGCGTGAGGTCG	Crosslin *et al.*, 2006

and ESFY has been detected with success in batches of *Cacopsylla pruni* Scopoli vectors (Martini *et al.*, 2007), while FD, BN and AP have been detected even in single field-collected *Scaphoideus titanus* Ball, *Hyalesthes obsoletus* Signoret, *Reptalus panzeri* (Löw), *Euscelis incisus* (Kirschbaum) and *Ca. melanoneura* (Förster) vectors (Jarausch *et al.*, 2004; Galetto *et al.*, 2005; Hren *et al.*, 2007).

Additional real-time PCR controls

Diagnosis of pathogen in woody plants is often hampered by the presence of PCR inhibitors such as polyphenolics, polysaccharides and other molecules that may produce false-negative results even from heavily infected samples. To prove that the absence of signal is not due to such causes, protocols for control amplification and detection of the host DNA have been developed. The chloroplast chaperonin 21 gene (Angelini *et al.*, 2007) and cytochrome oxidase gene (Hren *et al.*, 2007), the chloroplast gene for tRNA leucine (Baric and Dalla-Via, 2004) and the 18S rDNA gene (Christensen *et al.*, 2004; Marzachi and Bosco, 2005; Martini *et al.*, 2007) have been addressed as targets to control for the quality of total DNA extracted from grapevine, apple, *Prunus* spp. and other plant species, as indicated in Table 1.2. A similar approach can be applied to check the quality of total DNA extracted from field-collected phytoplasma vectors, and the 18S rDNA gene has been suggested as target for this purpose (Marzachì and Bosco, 2005).

Reverse transcription real-time PCR

It is conceivable that, if phytoplasma DNA concentration is low in the host plant, mRNAs, especially the highly expressed ribosomal ones, may offer a better target for diagnosis (Firrao *et al.*, 2007). A simple protocol for crude sap preparation from leaves (Osman and Rowhani, 2006), followed by diagnosis through reverse transcription (RT) coupled to PCR in a single tube has been proposed for specific detection of FD in field-collected grapevines (Margaria *et al.*, 2007). The protocol has recently been modified to include direct detection of the pathogen-specific amplicon in RT real-time PCR driven with group-specific primers and TaqMan probes (Margaria *et al.*, 2008). It is also important to consider that grapevine can be infected by several viruses with RNA genomes, and so a single total nucleic acid extract can be used for RT-PCR assays, driven with reagents specific for several grapevine viruses (Osman *et al.*, 2008) as well as for the most important grapevine phytoplasmas. In theory, multiplex reverse transcriptase RT-PCR protocols can be developed for the simultaneous detection of the most important virus and virus-like diseases of grapevine, although the different concentrations of each pathogen in the infected plant may seriously interfere with the linearity of detection of the least concentrated ones. Extraction of leaf sap is rapid and straightforward, and less prone to contamination between samples, so

Table 1.2. Name and sequence of primers and probes designed for the quality control of the total DNA extracted from several phytoplasma hosts.

Host	Target gene	Forward primer 5´–3´	Reverse primer 5´–3´	Probe 5´–3´	Reference
Periwinkle, Poinsettia, *Prunus* spp.	18S rDNA	GACTACGTCCCTGCCCTTTG	AACACTTCACCGGAC CATTCA	ACACACCGCCCG TCGCTCC	Christensen *et al.*, 2004; Martini *et al.*, 2007
Grapevine, potato	Cytochrome oxidase	COX-F CGTCGCATTCCAGATT ATCCA	COX-R CAACTACGGATATATAAGA GCCAAAACTG	COXP TGCTTACGCTGG ATGGAATGCCCT	Hren *et al.*, 2007
Grapevine	Chaperonin	Chaperonin grapevine gene Forward GGTCCTTTGGATGAGG ATGG	Chaperonin grapevine gene Reverse GAAGTCATTCCCTGCAT ACTTGG	Chaperonin grapevine gene Probe GAAACCACTGTCT GTGAGCCCAGGA	Angelini *et al.*, 2007
Apple	tRNA leucine	qMd-cpLeu-F CCTTCATCCTTTCTGAAG TTTCG	qMd-cpLeu-R AACAAATGGAGTTG GCTGCAT	qMd-cpLeu TGGAAGGATTCCTTT ACTAAC	Baric and Dalla-Via, 2004
Marguerite	18S rDNA (ITS1)	ChrysFw AAGGAAAACTAAACTTAAGA AGCTT–GTT	ChrysRv GTGGCTTCTTTATAATCAC	Chrys Probe CCCCGTTCGCGGT GTGCTCATG	Marzachì and Bosco, 2005
Leafhopper species	18S rDNA	MqFw AACGGCTACCACATCCAAGG	MqRv GCCTCGGATGAGTCCCG	Mq Probe AGGCAGCAGGCA CGCAAATTACCC	Marzachì and Bosco, 2005

reverse transcription of phytoplasma rRNAs from leaf sap extracts is a good choice when screening a large number of samples. Moreover, the crude leaf extract can be stored for some months at −20°C without affecting results. Since phytoplasmas are not always present in every part of an infected plant (Firrao *et al.*, 2007), correct sampling procedure is crucial to obtain reliable and reproducible results.

Real-time PCR Quantification of Phytoplasma DNA

Competitive PCR was first used to monitor the multiplication of a '*Ca.* Phytoplasma asteris' strain in males and females of its vector *Macrosteles quadrinlineatus* (Forbes) (=*fascifrons* Stål) by Liu *et al.* (1994). Quantification was achieved following co-amplification of phytoplasma DNA and several dilutions of an appropriate internal standard. This approach was complex; several steps, such as electrophoresis, image analysis of the gels, compensating for differences in intensity due to the different sizes of the products from the pathogen target and the internal standard, were required before the band intensities could be plotted for linear regression analysis. Nevertheless, this approach did demonstrate that different quantities of '*Ca.* Phytoplasma asteris' accumulated in female and male vectors. A few years later, a similar protocol, based on the construction of an internally deleted phytoplasma sequence transformed into a plasmid vector, was used to quantify phytoplasma cells in various plant hosts (Berges *et al.*, 2000), and to compare four methods for extraction of phytoplasma DNA from infected plant tissue (Palmano, 2001).

Real-time PCR is the most suitable method to quantify the nucleic acids of many plant pathogens, although the lack of growth in pure culture means that quantification of phytoplasmas can only be achieved in the presence of high levels of host DNA. Several laboratories have studied this problem and different approaches have been pursued. In some cases, absolute quantification of phytoplasma DNA was achieved per gram of extracted tissue (Wei *et al.*, 2004; Bisognin *et al.*, 2008) or per insect vector (Jarausch *et al.*, 2004). As already mentioned, recovery of DNA is strongly influenced by the extraction method, by different extraction runs and by different species of plant and insect host, and therefore quantification of phytoplasma DNA in relation to host DNA yield has been suggested (Baric and Dalla-Via, 2004; Marzachì and Bosco, 2005; Martini *et al.*, 2007). Results obtained in this way are easily comparable. For example, grapevines and apricots can host between 1.5×10^2 and 2×10^4 cells of FD and ESFY phytoplasmas per ng of plant DNA, and in both cases vectors have a much higher phytoplasma concentration, ranging between 10^6 and 10^7 phytoplasma cells per ng of insect DNA (D'Amelio *et al.*, 2007; Martini *et al.*, 2007).

The phytoplasma genome-sequencing era, which is just beginning, should provide much information relative to the molecular pathways followed by these bacteria in their parasitic lives and will also give clues on how to culture them *in vitro*. Until then, quantitative real-time PCR (Q-PCR) will be a powerful technique to study several aspects of their biology and epidemiology, such as their different multiplication rates in their plant and vector hosts. The multiplication of phytoplasmas in different compartments of the host plant has already been studied by measuring phytoplasma concentration at different times after vector inoculation at a localized point. Fast multiplication rates and high concentration of two '*Ca.* Phytoplasma asteris' strains have been reported in young developing leaves and roots of infected daisy plants (Wei *et al.*, 2004; Saracco *et al.*, 2006), although conflicting results were described for a branch-inducing phytoplasma in *Euphorbia pulcherrima* (Christensen *et al.*, 2004). Quantification of AP in the roots and shoots of several apple cultivars has also been applied to correlate phytoplasma concentration and disease development, in order to unveil mechanisms of resistance to AP infection (Bisognin *et al.*, 2008).

Quantitative analysis has also confirmed that phytoplasma concentration varies in individual plants of the same species, even following inoculation under controlled conditions (Saracco *et al.*, 2006), and differences of several thousand-fold are common in extracts from field-collected potato plants showing phytoplasma-associated purple top symptoms (Crosslin *et al.*, 2006). Q-PCR may also be applied in development of resistant varieties, a hot topic for economically important woody crops such as palms and grapevines. It is not clear if plants harbouring different phytoplasma concentrations behave differently as sources of inoculum for vectors, but it is known that FD-infected grapes of a sensitive cultivar are a better source of inoculum for *S. titanus* than infected vines of a less sensitive cultivar (Bressan *et al.*, 2005).

Q-PCR technology has also been applied to calculate the concentration of several phytoplasmas in their insect vectors (Jarausch *et al.*, 2004; Bosco *et al.*, 2007; D'Amelio *et al.*, 2007; Martini *et al.*, 2007) and to examine phytoplasma multiplication in vectors following acquisition under controlled conditions (Bosco *et al.*, 2007; D'Amelio *et al.*, 2007). Epidemiological studies of phytoplasmas can also benefit from the introduction of a user-friendly measurement technique. For example, active multiplication of the pathogen and increasing times elapsed after inoculation could explain the decrease of phytoplasma concentration from the edge of wheat fields towards the centre (Hollingsworth *et al.*, 2008). This information supported the hypothesis that '*Ca.* Phytoplasma asteris' vectors move inwards from the perimeter of these fields and has resulted in improved control strategies.

Q-PCR also has a possible application in studying the interactions of different phytoplasma species or strains present in mixed infections in the host plant or the vector. Mixed infections have been reported under field conditions in several important crops, such as grapevine (Alma *et al.*, 1996), palm (Harrison *et al.*, 2008), potato (Leyva-Lopez *et al.*, 2002) and ornamentals (Leyva-Lopez *et al.*, 2002; Bertaccini *et al.*, 2005); Q-PCR would enable detailed study of the interactions of different phytoplasmas in the same host.

Phytoplasma vectors may be generalist or specialist, depending on their feeding habits, and this has a tremendous influence on their ability to transmit one or more phytoplasma species or strains (Marzachì *et al.*, 2004). Phytoplasmas belonging to some species, such as '*Ca.* Phytoplasma asteris', can be vectored to plants of different genera by more than one insect species

(Marzachì *et al.*, 2004; Weintraub and Beanland, 2006; Bosco *et al.*, 2007). Interactions between different phytoplasmas and the effects of mixed infections on pathogen multiplication in the vector body and on vectoring abilities may be followed by Q-PCR; an example is the study of the fate of FD and '*Ca.* Phytoplasma asteris' phytoplasmas following double acquisition by the vector *Euscelidius variegatus* Kirschbaum (D'Amelio *et al.*, 2007).

Future Developments for Phytoplasma Detection

A protocol based on the terminal restriction length polymorphism (T-RFLP) analysis of a 23S rDNA sequence using a DNA sequence analysis system has been developed to provide the simultaneous detection and taxonomic grouping of phytoplasmas (Hodgetts *et al.*, 2007; see Hodgetts and Dickinson, Chapter 6, this volume). The technique requires a PCR amplification of the target gene with a couple of specific primers, one of which is fluorescently labelled, followed by restriction digestion and analysis of the labelled fragments. The possibility of amplifying both phytoplasma and plant chloroplast DNA in each single PCR tube is an interesting feature of this protocol, allowing the identification of false-negative results. Moreover, the presence of other bacteria in the analysed sample can be easily detected, since they produce distinct terminal restriction fragments.

Co-operational PCR (co-PCR) has been proposed by Bertolini *et al.* (2007), to detect 16SrX phytoplasmas, avoiding nested PCR. The method uses a tri-primer reaction coupled with dot-blot hybridization with universal and specific probes. Different concentrations of one external and two internal primers were used in the co-PCR, with 60 amplification cycles. The method was tested on infected periwinkle and field tree samples, and was as sensitive as conventional nested PCR.

Microarray technology, essentially a reverse dot-blotting technique, has the ability to simultaneously display the expression of thousands of genes at a time, and its potential application for phytoplasma detection has been described (Hadidi *et al.*, 2004). Frosini *et al.* (2002) combined a ligase detection reaction (LDR) with hybridization on universal arrays as a quick and reliable tool for universal phytoplasma diagnosis and group-specific detection of FD and BN. Nicolaisen and Bertaccini (2007) proposed the universal and specific detection of phytoplasmas by printing on a microarray two universal and several group-specific oligonucleotides designed on the 16S rDNA gene. PCR products, obtained with universal primers, were labelled and used as hybridization probes on the array. The technique could efficiently identify periwinkle samples infected with phytoplasmas belonging to several groups, producing no cross-reaction among groups and no signal from out-group negative controls.

Low-density PCR arrays (LDAs) have recently been introduced as a novel approach to gene expression profiling and molecular detection. Arrays are prepared by drying real-time TaqMan PCR reagents (primers and probes) specific for several target genes into different wells of a plate. Primers and probes are reconstituted by adding cDNA together with PCR master mix, and the reaction takes place in a real-time thermal cycler. LDAs, while retaining the sensitivity of TaqMan RT real-time PCR, allow the simultaneous quantification of large numbers of target genes present in single samples. Key features of the LDA assessment include convenience, ease of use, rapidity, sensitivity and reproducibility (Osman *et al.*, 2008). Thus, LDAs may be an interesting technique for phytoplasma diagnosis and quantification, especially because several TaqMan real-time PCR assays are now available. Thirteen grapevine viruses were successfully detected with LDAs and the method was found to be more sensitive than RT-PCR and TaqMan RT real-time PCR (Osman *et al.*, 2008).

New perspectives in molecular diagnosis of pathogens have been opened by the miniaturization of biological and chemical analytical devices. PCR microchips are studied extensively, and thus great progress has been made on aspects of fabrication, materials and detection chemistry. There are several versions and prototypes, which vary according to the materials used for the chip (silicon, glass and polymers) or according to the microfluidic design of the chip (stationary chamber, very similar to a small version of a conventional thermal cycler, or flow-through PCR, in which the sample repeatedly flows through the three PCR temperatures). The microfluidic PCR can be integrated with other analytical functional units, such as sample preparation, capillary electrophoresis, DNA microarray hybridization or real-time detection (by TaqMan or SYBR Green I chemistry). PCR microfluidics are currently used for biomedical and bioanalytical applications such as detection of bacteria and viruses (Zhang *et al.*, 2006).

Isothermal amplification of nucleic acids has recently been described as an alternative to PCR. Some of these protocols, such as nucleic acid sequence-based amplification (NASBA) (Compton, 1991), loop-mediated isothermal amplification (LAMP) (Notomi *et al.*, 2000) and rolling-circle amplification (RCA) (Fire and Xu, 1995), have been applied to plant pathogens.

NASBA, a method to amplify RNA, has been used in diagnostic bacteriology for clinical, environmental and food applications (Gracias and McKillip, 2007). NASBA involves an isothermal series of reactions using avian myeloblastosis virus reverse transcriptase (Rtase), RNase H, T7 RNA polymerase, transcript-specific primers and associated cofactors to amplify large amounts of target RNA. NASBA allows target RNA detection by real-time chemistries, such as SYBR dyes or molecular beacon probes, and several kits are currently available for RNA amplification and analyses. Moreover, NASBA diagnostic methods usually provide an automated nucleic acid extraction procedure (NucliSens® EasyMAG®), based upon silica extraction technology and magnetic particles (Gracias and McKillip, 2007).

LAMP uses a set of four or six primers and a DNA polymerase with strand displacement activity (Bst DNA polymerase) to amplify DNA with high specificity under isothermal conditions in less than 1 h. This method has been used to detect several bacteria, among them *Mycoplasma pneumoniae* (Saito *et al.*, 2005), as well as plant pathogens, such as the non-culturable, phloem-limited '*Ca.* Liberobacter spp.', the causal agent of citrus greening

(Okuda *et al.*, 2005). LAMP products are visualized either by gel electrophoresis or with other methods suitable for use in the field. The synthesis of large amounts of DNA in a LAMP reaction yields a white precipitate of magnesium pyrophosphate, which can be detected either with the naked eye or by real-time turbidimeter. Alternatively, in the presence of SYBR Green I, a positive LAMP reaction produces a colour change that can again be detected by the naked eye or under a UV lamp. This method has potential for testing in the field or in under-equipped laboratories, although most LAMP assays have been performed using nucleic acid extraction methods not suitable for field use (Mumford *et al.*, 2006).

Physiological rolling-circle amplification (replication) (RCA) is used by many plasmids and viruses, generally involves a double-stranded template with a helicase or single-strand DNA binding activity preceding the polymerase, and operates on templates in the order of kilobases and larger (Fire and Xu, 1995). RCA enzymatic activity of bacteriophage ϕ 29 DNA polymerase has been exploited as a highly efficient method of DNA synthesis and further as a diagnostic and genomic tool for many animal and plant viruses, such as those in the *Geminiviridae* and *Nanoviridae* families (Haible *et al.*, 2006). Recently RCA has been applied together with circular probes (padlock probes) in diagnostic genomics. Briefly, a padlock probe is a long oligonucleotide (approximately 100 bp) containing target-complementary regions at the 3′ and 5′ ends; these regions are complementary to adjacent sequences on the target pathogen to be detected. When the target nucleic acid and padlock probe are hybridized together, they form a circularized molecule that can be amplified by RCA. The presence of amplified probes can be detected using an array (Mumford *et al.*, 2006). A new diagnostic assay based on RCA and padlock probes has been described for the multiplex detection of five *Mycoplasma* spp. (Wang *et al.*, 2008). However, none of these isothermal methods for amplification of nucleic acids has yet been developed for phytoplasma detection.

Looking Forward

Despite the development of protocols which overcome most of the difficulties of phytoplasma diagnosis, the detection of these pathogens is still quite laborious. Real-time PCR offers the opportunity to detect these pathogens in a sensitive and specific manner, bypassing all post-PCR manipulations. Improvement of diagnostic procedures is still hampered by the absence of reliable sampling protocols. In fact, due to the irregular distribution of phytoplasmas in the phloem of their host plants, obtaining reproducible results is still not trivial.

The recent developments of microarray technology, low-density PCR arrays and miniaturization of PCR into microfluidic devices have the potential to increase the number of phytoplasmas detected in a given sample, as well as to minimize detection time, although all these techniques rely on the extraction of good-quality nucleic acid, which is still a bottleneck in the diagnosis. Future work is needed to develop quicker procedures to extract phytoplasma-enriched nucleic acids. Automation may play a part, perhaps involving silica or magnetic beads.

Reverse transcription of mRNA followed by group-specific amplification of transcripts from crude nucleic acid extracts could potentially be developed into a sensitive, rapid and reliable diagnostic method, which can also be broadened to include the diagnosis of other pathogens such as viruses.

Real-time quantification of phytoplasma DNA offers the chance to study several aspects of phytoplasma infection in both plant and vector. Biological aspects of the relationships between these pathogens and their hosts, from susceptibility of different plant genotypes to competition of different '*Ca.* Phytoplasma' species infecting the same host plant or insect vector, may benefit from a quantitative approach. The same technique can be used to follow the fate of the phytoplasma in the host plant during development of the disease, which is a prerequisite to studying the biology of these pathogens in the host.

Since control of phytoplasma diseases with insecticide treatments to minimize vector populations raises ecological concerns, for example honey bee survival (Desneux *et al.*, 2007), several laboratories are looking for alternative strategies to control phytoplasma diseases. Phytoplasma-infected woody plants in some cases may show decreased symptom expression up to complete recovery (Morone *et al.*, 2007). The presence and concentration of phytoplasmas in such plants are important data to consider before recovery is valued among the possible strategies to manage phytoplasma diseases.

Quantification of phytoplasma concentration in different parts of the vector may also help to elucidate the fate of the phytoplasma in the vector and therefore to understand what makes an insect a vector of phytoplasmas. This also has clear implications in the development of non-insecticidal strategies to control vector populations.

The development of protocols to quantify phytoplasma mRNAs may also prove useful to evaluate pathogen viability at different times after inoculation in the host plant or after acquisition by the vector (Lahtinen *et al.*, 2008). Coupling of microarrays and Q-PCR will provide gene expression profiles in different hosts, in order to understand the strategies adopted by phytoplasmas to survive inside the cells of plants and insects.

In conclusion, phytoplasmas represent a challenging experimental model to study complex biological and molecular interactions among the three partners involved. Use of high-throughput, sensitive, rapid and quantitative techniques will help to understand how phytoplasmas exploit their unique ecological niches.

Summary

Real-time PCR is a versatile tool which improves the speed, sensitivity and efficiency of phytoplasma diagnosis. The design of appropriate reagents and reaction conditions for each specific purpose will improve on existing analytical methods. The real-time PCR assays developed so far have clearly shown that this technique facilitates high throughput and provides a high level of fidelity. The probability of obtaining false-positive results is extremely low, due to the quality of the primers used, the possibility of analysing melting curves or including a probe in the reaction mix to improve specificity, and the elimination of post-PCR manipulation. For these reasons real-time PCR-based diagnostics show great promise in all certification and control programmes relating to the more destructive phytoplasmas or those subject to quarantine regulations. Due to its high-throughput potential, and perhaps coupled to automated nucleic acid extraction, real-time PCR also has a future in the development of sanitary programmes in nursery selection. Furthermore, the quantitative property of real-time PCR-based assays can be useful to monitor phytoplasma kinetics such as the progress of an infection, to estimate the number of phytoplasma cells carried by vectors, or to evaluate the phytoplasma tolerance levels in selection for resistant host genotypes.

References

Aldaghi, M., Massart, S., Roussel, S. and Jijakli, M.H. (2007) Development of a new probe for specific and sensitive detection of '*Candidatus* Phytoplasma mali' in inoculated apple trees. *Annals of Applied Biology* 151, 251–258.

Alma, A., Davis, R.E., Vibio, M., Danielli, A., Bosco, D., Arzone, A. and Bertaccini, A. (1996) Mixed infection of grapevines in northern Italy by phytoplasmas including 16S rRNA RFLP subgroup 16SrI-B strains previously unreported in this host. *Plant Disease* 80, 418–421.

Angelini, E., Bianchi, G.L., Filippin, L., Morassutti, C. and Borgo, M. (2007) A new TaqMan method for the identification of phytoplasmas associated with grapevine yellows by real-time PCR assay. *Journal of Microbiological Methods* 68, 613–622.

Baric, S. and Dalla-Via, J. (2004) A new approach to apple proliferation detection: a highly sensitive real-time PCR assay. *Journal of Microbiological Methods* 57, 135–145.

Berges, R., Rott, M. and Seemüller, E. (2000) Range of phytoplasma concentrations in various plant hosts as determined by competitive polymerase chain reaction. *Phytopathology* 90, 1145–1152.

Bertaccini, A. (2007) Phytoplasmas: diversity, taxonomy, and epidemiology. *Frontiers in Bioscience* 12, 673–689.

Bertaccini, A., Franova, J., Botti, S. and Tabanelli, D. (2005) Molecular characterization of phytoplasmas in lilies with fasciation in the Czech Republic. *Fems Microbiology Letters* 249, 79–85.

Bertolini, E., Torres, E., Olmos, A., Martín, M.P., Bertaccini, A. and Cambra, M. (2007) Co-operational PCR coupled with dot blot hybridization for detection and 16SrX grouping of phytoplasmas. *Plant Pathology* 56, 677–682.

Bianco, P.A., Casati, P. and Marziliano, N. (2004) Detection of phytoplasmas associated with grapevine flavescence dorée disease using real-time PCR. *Journal of Plant Pathology* 86, 257–261.

Bisognin, C., Schneider, B., Salm, H., Grando, M.S., Jarausch, W., Moll, E. and Seemüller, E. (2008) Apple proliferation resistance in apomictic rootstocks and its relationship to phytoplasma concentration and simple sequence repeat genotypes. *Phytopathology* 98, 153–158.

Bosco, D., Galetto, L., Leoncini, P., Saracco, P., Raccah, B. and Marzachì, C. (2007) Interrelationships between '*Candidatus* Phytoplasma asteris' and its leafhopper vectors. *Journal of Economic Entomology* 100, 1504–1511.

Bressan, A., Spiazzi, S., Girolami, V. and Boudon-Padieu, E. (2005) Acquisition efficiency of Flavescence dorée phytoplasma by *Scaphoideus titanus* Ball from infected tolerant or susceptible grapevine cultivars or experimental host plants. *Vitis* 44, 143–146.

Christensen, N.M., Nicolaisen, M., Hansen, M. and Schulz, A. (2004) Distribution of phytoplasmas in infected plants as revealed by real-time PCR and bioimaging. *Molecular Plant–Microbe Interactions* 17, 1175–1184.

Compton, J. (1991) Nucleic acid sequence-based amplification. *Nature* 350, 91–92.

Crosslin, J.M., Vandemark, G.J. and Munyaneza, J.E. (2006) Development of a real-time, quantitative PCR for detection of potato purple top phytoplasma in plants and beet leafhoppers. *Phytopathology* 97, S166.

D'Amelio, R., Marzachì, C. and Bosco, D. (2007) Double infection of '*Candidatus* Phytoplasma asteris' and 'flavescence dorée' phytoplasma in the vector *Euscenlidius variegatus*. *Bulletin of Insectology* 60(2), 223–224.

Desneux, N., Decourtye, A. and Delpuech, J.-M. (2007) The sublethal effects of pesticides on beneficial arthropods. *Annual Review of Entomology* 52, 81–106.

Doi, M., Tetranaka, M., Yora, K. and Asuyama, H. (1967) Mycoplasma or PLT-group-like organism found in the phloem elements of plants infected with mulberry dwarf, potato witches' broom, aster yellow, or paolownia witches' broom. *Annals of the Phytopathological Society of Japan* 33, 259–266.

Fire, A. and Xu, S.Q. (1995) Rolling replication of short DNA circles. *Proceedings of the National Academy of Sciences of the USA* 92, 4641–4645.

Firrao, G., Garcia-Chapa, M. and Marzachì, C. (2007) Phytoplasmas: genetics, diagnosis and relationships with the plant and insect host. *Frontiers in Bioscience* 12, 1353–1375.

Frosini, A., Casati, P., Bianco, P.A., Bordoni, R., Consolandi, C., Castiglioni, B., Mezzelani, A., Rizzi, E., Battaglia, C., Belli, G., Rossi Bernardi, L. and De Bellis, G. (2002) Ligase detection reaction and universal array as a tool to detect grapevine infecting phytoplasmas. *Minerva Biotecnologica* 14, 265–267.

Galetto, L., Bosco, D. and Marzachì, C. (2005) Universal and group-specific real-time PCR diagnosis of flavescence dorée (16Sr-V), bois noir (16Sr-XII) and apple proliferation (16Sr-X) phytoplasmas from field-collected plant hosts and insect vectors. *Annals of Applied Biology* 147, 191–201.

Gori, M., Monnanni, R., Bulatti, M., Goti, E., Carnevale, S., da Prato, L., Bertaccini, A. and Biricolti, S. (2007) Establishing a real-time PCR detection procedure of 'flavescence dorée' and 'bois noir' phytoplasmas for mass screening. *Bulletin of Insectology* 60(2), 255–256.

Gracias, K.S. and McKillip, J.L. (2007) Nucleic acid sequence-based amplification (NASBA) in molecular bacteriology: a procedural guide. *Journal of Rapid Methods and Automation in Microbiology* 15, 295–309.

Hadidi, A., Czosnek, H. and Barba, M. (2004) DNA microarrays and their potential applications for the detection of plant viruses, viroids, and phytoplasmas. *Journal of Plant Pathology* 86, 97–104.

Haible, D., Kober, S. and Jeske, H. (2006) Rolling circle amplification revolutionizes diagnosis and genomics of geminiviruses. *Journal of Virological Methods* 135, 9–16.

Harrison, N.A., Helmick, E.E. and Elliott, M.L. (2008) Lethal yellowing-type diseases of palms associated with phytoplasmas newly identified in Florida, USA. *Annals of Applied Biology* 153, 85–94.

Hodgetts, J., Ball, T., Boonham, N., Mumford, R. and Dickinson, M. (2007) Use of terminal restriction fragment length polymorphism (T-RFLP) for identification of phytoplasmas in plants. *Plant Pathology* 56, 357–365.

Hogenhout, S.A., Oshima, K., Ammar, E.D., Kakizawa, S., Kingdom, H.N. and Namba, S. (2008) Phytoplasmas: bacteria that manipulate plants and insects. *Molecular Plant Pathology* 9, 403–423.

Hollingsworth, C.R., Atkinson, L.M., Samac, D.A., Larsen, J.E., Motteberg, C.D., Abrahamson, M.D., Glogoza, P. and MacRae, I.V. (2008) Region and field level distributions of aster yellows phytoplasma in small grain crops. *Plant Disease* 92, 623–630.

Hren, M., Boben, J., Rotter, A., Kralj, P., Gruden, K. and Ravnikar, M. (2007) Real-time PCR detection systems for Flavescence dorée and Bois noir phytoplasmas in grapevine: comparison with conventional PCR detection and application in diagnostics. *Plant Pathology* 56, 785–796.

Jarausch, W., Peccerella, T., Schwind, N., Jarausch, B. and Krczal, G. (2004) Establishment of a quantitative real-time PCR assay for the quantification of apple proliferation phytoplasmas in plants and insects. *Acta Horticulturae* 657, 415–420.

Kostina, E.V., Ryabinin, V.A., Maksakova, G.A. and Sinyakov, A.N. (2007) TaqMan probes based on oligonucleotide–hairpin minor groove binder conjugates. *Russian Journal of Bioorganic Chemistry* 33, 614–616.

Lahtinen, S.J., Ahokoski, H., Reinikainen, J.P., Gueimonde, M., Nurmi, J., Ouwehand, A.C. and Salminen, S.J. (2008) Degradation of 16S rRNA and attributes of viability of viable but nonculturable probiotic bacteria. *Letters in Applied Microbiology* 46, 693–698.

Leyva-Lopez, N.E., Ochoa-Sanchez, J.C., Leal-Klevezas, D.S. and Martinez-Soriano, J.P. (2002) Multiple phytoplasmas associated with potato diseases in Mexico. *Canadian Journal of Microbiology* 48, 1062–1068.

Liu, H.W., Goodwin, P.H. and Kuske, C.R. (1994) Quantification of DNA from aster yellows mycoplasma-like organism in aster leafhoppers (*Macrosteles fascifrons* Stål) by competitive polymerase chain reaction. *Systematic and Applied Microbiology* 17, 274–280.

Margaria, P., Rosa, C., Marzachì, C., Turina, M. and Palmano, S. (2007) Detection of Flavescence dorée phytoplasma in grapevine by reverse-transcription PCR. *Plant Disease* 91, 1496–1501.

Margaria, P., Turina, M. and Palmano, S. (2008) Multiple detection of several grapevine pathogens by Taqman ® real-time RT-PCR assays from crude-sap extracts. *Petria* 18(2), 141–149.

Martini, M., Loi, N., Ermacora, P., Carraro, L. and Pastore, M. (2007) A real-time PCR method for detection and quantification of '*Candidatus* Phytoplasma prunorum' in its natural hosts. *Bulletin of Insectology* 60(2), 251–252.

Marzachì, C. and Bosco, D. (2005) Relative quantification of chrysanthemum yellows (16Sr I) phytoplasma in its plant and insect host using real-time polymerase chain reaction. *Molecular Biotechnology* 30, 117–127.

Marzachì, C., Milne, R.G. and Bosco, D. (2004) Phytoplasma–plant–vector relationships. In: Pandalai, S.G. and Gayathri, A. (eds) *Recent Research Development in Plant Pathology*, Vol. 3. Research Signpost, Trivandrum, Kerala, India, pp. 211–241.

Monis, P.T. and Giglio, S. (2006) Nucleic acid amplification-based techniques for pathogen detection and identification. *Infection Genetics and Evolution* 6, 2–12.

Morone, C., Boveri, M., Giosuè, S., Gotta, P., Rossi, V., Scapin, I. and Marzachì, C. (2007) Epidemiology of Flavescence Dorée in vineyards in northwestern Italy. *Phytopathology* 97, 1422–1427.

Mumford, R., Boonham, N., Tomlinson, J. and Barker, I. (2006) Advances in molecular phytodiagnostics – new solutions for old problems. *European Journal of Plant Pathology* 116, 1–19.

Nicolaisen, M. and Bertaccini, A. (2007) An oligonucleotide microarray-based assay for identification of phytoplasma 16S ribosomal groups. *Plant Pathology* 56, 332–336.

Notomi, T., Okayama, H., Masubuchi, H., Yonekawa, T., Watanabe, K., Amino, N. and Hase, T. (2000) Loop-mediated isothermal amplification of DNA. *Nucleic Acids Research* 28, e63.

Okuda, M., Matsumoto, M., Tanaka, Y., Subandiyah, S. and Iwanami, T. (2005) Characterization of the tufB-secE-nusGrplKAJL-rpoB gene cluster of the citrus greening organism and detection by loop-mediated isothermal amplification. *Plant Disease* 89, 705–711.

Osman, F. and Rowhani, A. (2006) Application of a spotting sample preparation technique for the detection of pathogens in woody plants by RT-PCR and real-time PCR (TaqMan). *Journal of Virological Methods* 133, 130–136.

Osman, F., Leutenegger, C., Golino, D. and Rowhani, A. (2008) Comparison of low-density arrays, RT-PCR and real-time TaqMan RT-PCR in detection of grapevine viruses. *Journal of Virological Methods* 149, 292–299.

Palmano, S. (2001) A comparison of different phytoplasma DNA extraction methods using competitive PCR. *Phytopathologia Mediterranea* 40, 99–107.

Saito, R., Misawa, Y., Moriya, K., Koike, K., Ubukata, K. and Okamura, N. (2005) Development and evaluation of a loop-mediated isothermal amplification assay for rapid detection of *Mycoplasma pneumoniae*. *Journal of Medical Microbiology* 54, 1037–1041.

Saracco, P., Bosco, D., Veratti, F. and Marzachì, C. (2006) Quantification over time of chrysanthemum yellows phytoplasma (16Sr-I) in leaves and roots of the host plant *Chrysanthemum carinatum* (Schousboe) following inoculation with its insect vector. *Physiological and Molecular Plant Pathology* 67, 212–219.

Torres, E., Bertolini, E., Cambra, M., Monton, C. and Martin, M.P. (2005) Real-time PCR for simultaneous and quantitative detection of quarantine phytoplasmas from apple proliferation (16SrX) group. *Molecular and Cellular Probes* 19, 334–340.

Tully, J.G., Whitcomb, R.F., Rose, D.L., Bovè, J.M., Carle, P., Somerson, N.L., Williamson, D.L. and Edengreen, S. (1994) *Acholeplasma brassicae* sp. nov. and *Acholeplasma palmae* sp. nov., two non-sterol-requiring mollicutes from plant surfaces. *International Journal of Systematic Bacteriology* 44, 680–684.

Wally, O., El Hadrami, A., Khadhair, A.H., Adam, L.R., Shinners-Carnelley, T., Elliott, B. and Daayf, F. (2008) DNA sequencing reveals false positives during the detection of aster yellows phytoplasmas in leafhoppers. *Scientia Horticulturae* 116, 130–137.

Wang, H., Kong, F., Zhou, X. and Gilbert, G.L. (2008) Detection of common cell culture contaminant mollicutes using rolling circle amplification. In: *Proceedings of the 17th IOM Congress*, 6–11 July, Tianjin, China, Abstract 38, p. 47.

Wei, W., Kakizawa, S., Suzuki, S., Jung, H.Y., Nishigawa, H., Miyata, S., Oshima, K., Ugaki, M., Hibi, T. and Namba, S. (2004) *In planta* dynamic analysis of onion yellows phytoplasma using localized inoculation by insect transmission. *Phytopathology* 94(3), 244–250.

Weintraub, P.G. and Beanland, L. (2006) Insect vectors of phytoplasmas. *Annual Review of Entomology* 51, 91–111.

Zhang, C., Xu, J., Wenli, M. and Wenling, Z. (2006) PCR microfluidic devices for DNA amplification. *Biotechnology Advances* 24, 243–284.

2 Phytoplasma Genomics, from Sequencing to Comparative and Functional Genomics – What Have We Learnt?

Saskia A. Hogenhout[1] and Martina Šeruga Musić[2]

[1]*The John Innes Centre, UK;* [2]*University of Zagreb, Croatia*

Introduction

Phytoplasmas are insect-transmitted, phloem-limited bacterial pathogens that can cause devastating losses in crops and natural ecosystems worldwide (Lee *et al.*, 2000; Bertaccini, 2007). Phytoplasmas are small bacteria (±500 nm in diameter) surrounded by a single cell membrane, which appear to have suffered extreme genome reductions compared with their Gram-positive relatives. At present, phytoplasmas are not cultivable in cell-free media. Four phytoplasma genomes have been sequenced to completion and tremendous progress has been made with the genome sequencing and understanding of phytoplasma interactions with plants and insects at the molecular level.

Comparative analyses among the four sequenced phytoplasma genomes unveiled diversity in genome size, composition, metabolic pathways and number of repeats. It has now become clear that phytoplasmas undergo rapid genome evolution, which may be a consequence of their life cycle. Phytoplasmas continuously cycle between plants and insects and, in nature, require both organisms for survival and dispersal. This requires adaptation to a broad range of environments, including the phloem of their plant hosts and the gut lumen, haemolymph, saliva and endocellular niches in various organs of their insect hosts.

Mining of the phytoplasma genome sequence data led to the identification of a number of candidate virulence proteins, of which some were further functionally characterized and have confirmed roles in plant and insect invasion. These include proteins located on the phytoplasma cell surface involved in binding of insect gut microfilaments and secreted proteins that target plant cell nuclei. The large repeats present in the majority of phytoplasma genomes appear to have a function in phytoplasma virulence.

The goal of this chapter is to review the technical procedures that led to the completion of the four phytoplasma genomes. We will review similarities and differences in genome content and composition among the four phytoplasmas. Finally, we will describe how mining of phytoplasma genome sequence data can lead to the identification of virulence factors and other aspects of phytoplasma biology.

Phytoplasma Genome Sequencing

Because phytoplasmas cannot be cultured in cell-free media, the isolation of high-quality phytoplasma genomic DNA has been challenging. Nevertheless, to date, four phytoplasma genomes have been sequenced to completion. These are strains Onion Yellows M (OY-M) (Oshima *et al.*, 2004) and Aster yellows witches'-broom (AY-WB) (Bai *et al.*, 2006), which belong to ribosomal subgroups 16SrIB and IA, respectively, of *'Candidatus* (*Ca.*) Phytoplasma asteris', a strain of *'Ca.* Phytoplasma australiense' (subgroup *tuf*-Australia; *rp-A*) (Tran-Nguyen *et al.*, 2008), and strain AT of *'Ca.* Phytoplasma mali' (Kube *et al.*, 2008).

Comparison of the methods used for isolation of genomic DNA of these four phytoplasma genomes revealed that similar approaches were used (Fig. 2.1).

Phytoplasma DNA was extracted from phloem (Oshima *et al.*, 2004; Bai *et al.*, 2006; Kube *et al.*, 2008) or flowers (Tran-Nguyen *et al.*, 2008) of infected plants. Subsequently, the phytoplasma genomic DNA was separated from host DNA by pulsed-field gel electrophoresis. The band corresponding to the phytoplasma genomic DNA was excised from the gel and purified. Additionally, in the *'Ca.* Phytoplasma mali' sequencing project, phytoplasma genomic DNA was obtained by extracting lyophilized tissue using a cetyltrimethyl ammonium bromide (CTAB) procedure, followed by repeated bisbenzimide–CsCl buoyant-density-gradient centrifugation (Kube *et al.*, 2008).

The construction of genomic DNA libraries and sequencing strategies were also similar for all phytoplasmas (Fig. 2.1). For all phytoplasma genome-sequencing projects, shotgun libraries were generated directly from isolated phytoplasma DNA. The phytoplasma genomic DNA was sheared, size separated by agarose gel electrophoresis and fragments between 1.2 and 2.5 kb were gel purified and cloned into plasmids. The inserts were then high-throughput sequenced at six- to 20-fold coverage using dye terminator sequencing, and bioinformatics was used to assemble the sequences into contigs. The various contigs were connected by, for example, primer walking, to obtain a continuous chromosome or plasmid sequence. In addition, phage libraries were constructed for the sequencing of OY-M (Oshima *et al.*, 2004). Phage clones containing DNA fragments of about 15 kb were sonicated, size-fractionated and subcloned into a plasmid vector for subsequent sequencing. A fosmid library was constructed and sequenced for completion of the *'Ca.* Phytoplasma australiense' genome (Tran-Nguyen *et al.*, 2008). To determine the terminal inverted repeats (TIRs) of *'Ca.* Phytoplasma mali', Kube *et al.* (2008) also cloned the AP DNA in lambda libraries.

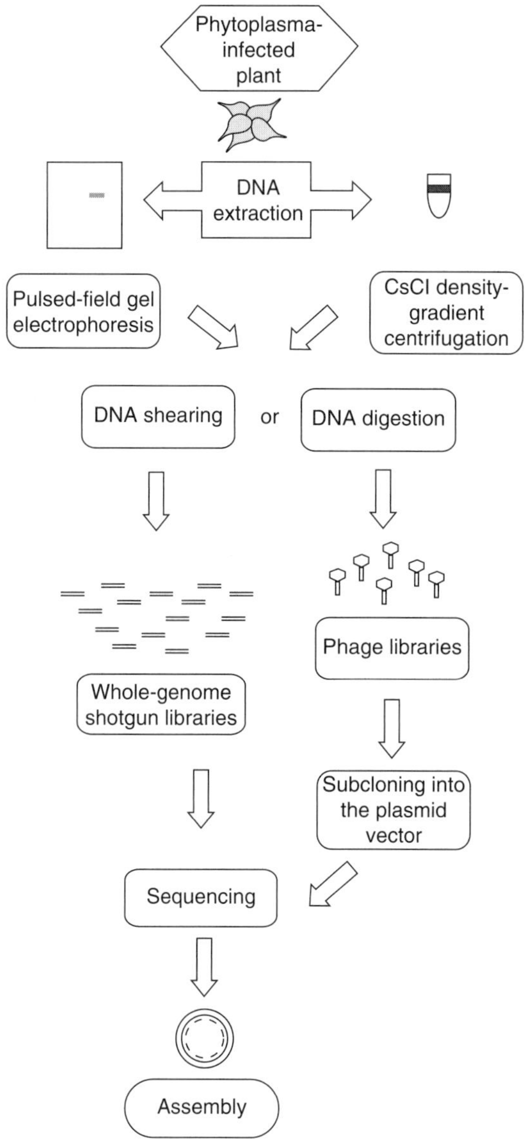

Fig. 2.1. Scheme of sequencing strategies used in phytoplasma genome projects.

Whereas the whole-genome shotgun sequencing strategy resulted in the completion of four phytoplasma genomes, there are certain difficulties with this strategy. First, phytoplasma DNA is AT rich (Table 2.1), which causes cloning problems especially in the presence of contaminating DNA from the host that is less AT rich. The host DNA is then preferentially cloned,

Table 2.1. General features of phytoplasma genomes that are sequenced to completion.

Strain	OY-M	AY-WB	*tuf*-Australia; *rp*-A	AT
'*Ca.* Phytoplasma' species	asteris	asteris	australiense	mali
16S rDNA group	IB	IA	XIIB	X
Cluster	I	I	I	II
Chromosome size (bp)	860,631	706,569	879,324	601,943
Chromosome composition	Circular	Circular	Circular	Linear
G+C content (%)	27.7	26.9	27	21.4
Protein-coding regions (%)	73	72	74	79
Coding sequences	793	708	839	536
Genes encoding proteins	754	671	684	497
Protein-coding genes with assigned functions	446	450	414	338
(Conserved) hypothetical proteins[a]	308	221	270	159
Single-copy proteins	486	482	482	408
Multiple-copy proteins	268	191	202	89
Multiple-copy proteins in PMUs	175	134	143	4
Transposase similar to tra5[b]	7 (6)	6* (20)	5 (1)	1
Fragmented genes	46	102	159	16
Avg ORF size (bp)	833	776	778	955
tRNA genes	32	31	35	32
rRNA operons	2	2	2	2
Extrachromosomal DNAs	2	4	1	0

Data were obtained from Oshima *et al.* (2004), Bai *et al.* (2006), Tran-Nguyen *et al.* (2008) and Kube *et al.* (2008) unless stated otherwise.

[a]Protein-coding genes with annotations 'hypothetical protein' and 'conserved hypothetical protein' were selected and counted from GenBank accessions NC_005303, NC_007716, NC_010544 and NC_011047, which contained the whole chromosome sequences of OY-M, AY-WB, '*Ca.* Phytoplasma australiense' and '*Ca.* Phytoplasma mali', respectively.

[b]The deduced protein sequence of one full-length *tra5* gene of AYWB (GenBank accession YP_456371.1) was searched against the protein sequences of GenBank accessions NC_005303, NC_007716, NC_010544 and NC_011047 using blastp. The output file was analysed for full-length transposases and truncated transposases (the latter are indicated in parentheses).

*Two out of six *tra5* genes consist of two open reading frames that can make one transposase upon a frameshift event.

resulting in a high number of clones containing host DNA relative to phytoplasma DNA. These clones carrying host DNA can dramatically increase the sequencing costs. Hence, scientists spend a substantial amount of time obtaining pure phytoplasma DNA that contains as little as possible host DNA. The other problem is that at least three of the four phytoplasma genomes are repeat rich. Plasmids that contain repeat-rich inserts recombine readily in *Escherichia coli*, thereby generating a bias towards the cloning and sequencing of phytoplasma DNA regions that do not contain repeats. Next-generation sequence technologies, such as 454 Sequencing™ (Roche Diagnostics Corporation), or Illumina® sequencing technology (formerly Solexa®

sequencing), offer new possibilities for phytoplasma whole-genome sequencing or mapping of phytoplasma genomes. These technologies are based on pyrosequencing ('sequencing by synthesis' principle) and do not require cloning into plasmid vectors.

Comparative Analyses of Phytoplasma Genomes

Phylogeny

Phylogeny based on 16S ribosomal DNA (16Sr) sequences divides the phytoplasmas into three distinct clusters (Hogenhout *et al.*, 2008). The first cluster (Cluster I) contains the Aster yellows (AY) 16SrI group and the stolbur (STOL) 16SrXII group phytoplasmas. These two groups have diverged but are clearly more closely related to each other than the other phytoplasma groups (Hogenhout *et al.*, 2008). The second cluster (Cluster II) contains the apple proliferation (AP) 16SrX group phytoplasmas, and the third cluster (Cluster III) contains the largest number of phytoplasma groups, including Western X (WX, 16SrIII), Palm Lethal Yellowing (LY, 16SrIV) and Elm Yellows (EY, 16SrV) (Hogenhout *et al.*, 2008). Three of the four phytoplasmas with fully sequenced genomes belong to Cluster I. These are OY-M, AY-WB and AUSGY. The fourth genome belongs to Cluster II. There are no complete genome sequences available for phytoplasmas in Cluster III, despite this cluster containing the majority of the different '*Ca.* Phytoplasma' species.

Because the four phytoplasmas for which genomes are sequenced to completion belong to different subgroups, groups and clusters in the phytoplasma phylogeny (Table 2.1), useful insights into phytoplasma diversity at the molecular level was obtained. '*Ca.* Phytoplasma asteris' OY-M and AY-WB belong to the AY group, in which OY-M is a 16SrIB and AY-WB a 16SrIA subgroup phytoplasma. Comparison of these two phytoplasmas has led to a better understanding of how closely related phytoplasmas differ in genome content and structure (Oshima *et al.*, 2004; Bai *et al.*, 2006). '*Ca.* Phytoplasma australiense' belongs to the STOL phytoplasma group, which also includes a new STOL (to be described as '*Ca.* Phytoplasma solani'), an economically important pathogen of tomatoes, potatoes and grapevines, and '*Ca.* Phytoplasma fragariae', a pathogen of strawberry (Hogenhout *et al.*, 2008). Comparative genome analyses of members of the AY and STOL phytoplasmas provided insights into similarities and differences in genome content and structure of phytoplasmas belonging to two groups within Cluster I (Tran-Nguyen *et al.*, 2008). Finally, '*Ca.* Phytoplasma mali' is an AP (16SrX) group phytoplasma that belongs to Cluster II, which also includes '*Ca.* Phytoplasma pyri' (pear decline (PD) phytoplasma) and '*Ca.* Phytoplasma prunorum' (European stone fruit yellows (ESFY) phytoplasma). The sequencing of '*Ca.* Phytoplasma mali' allowed for comparison of more distantly related phytoplasmas of Cluster I and II (Kube *et al.*, 2008).

Metabolism

Comparisons of the genome contents of the four sequenced phytoplasma genomes revealed similarities. All phytoplasmas have AT-rich genomes; the GC content ranges from 21.4% for '*Ca.* Phytoplasma mali' to 27.7% for '*Ca.* Phytoplasma asteris' OY-M (Table 2.1). They have similar numbers of tRNA genes and two copies of the rRNA operon. Their genomes are small, ranging from 602 kb for '*Ca.* Phytoplasma mali' to 879 kb for '*Ca.* Phytoplasma australiense'. The genomes are greatly reduced in the number of genes encoding basic metabolic pathways. The phytoplasmas lack genes for the pentose phosphate cycle, salvage pathways for recovery of nucleotides, the urea cycle, and pathways for synthesis of phenylalanine, D-glutamine, D-glutamate, D-arginine, D-ornithine, D-alanine and D-glutathione (Tran-Nguyen *et al.*, 2008). Thus, the phytoplasmas have to scavenge basic metabolites from their plant and insect hosts. Indeed, all four phytoplasmas have a relatively large number of transporter systems compared with their genome size. These include the permeases and solute-binding proteins of ATP-binding cassette (ABC) transporter systems for dipeptides (Dpp) and oligopeptides (Opp), D-methionine (Met), spermidine/putrescine (Pot), uptake of the inorganic ions cobalt (Cbi) and Mn/Zn (Znu), and maltose, trehalose, sucrose and palatinose (Mal) (Bai *et al.*, 2006; Kube *et al.*, 2008). '*Ca.* Phytoplasma mali' lacks ABC transporter systems for several amino acids, including GlnP, GlnQ and the Art system, which are present in the other three phytoplasmas (Kube *et al.*, 2008). Nevertheless, the set of identified ABC transport systems does not fully complement the essential metabolites that are missing. Therefore, it is likely that the ABC transporters have low substrate specificities and may import a broader set of metabolites (Kube *et al.*, 2008).

Phytoplasmas share the majority of the few metabolic pathways that are still present. They can convert glucose-6-phosphate into glyceraldehyde-3-phosphate, which feeds the glycolytic pathway to make pyruvate and the phospholipids biosynthesis pathway to make phosphatidylethanolamine (Oshima *et al.*, 2004). '*Ca.* Phytoplasma mali' lacks genes of the glycolytic pathway involved in the metabolism of glyceraldehyde-3-phosphate into pyruvate, but all phytoplasmas can utilize glyceraldehyde-3-phosphate to generate phosphatidylethanolamine (Kube *et al.*, 2008), which may be converted into phosphatidylcholine, depending on the presence of a functional *pmt* gene encoding phospholipid *N*-methyltransferase (Bai *et al.*, 2006). Phosphatidylcholine is an important constituent of cell membranes, suggesting that phytoplasmas can make at least part of their own membranes. This is consistent with the phytoplasmas being phylogenetically close to acholeplasmas (Lim and Sears, 1992; Lee *et al.*, 2000), which can also synthesize phospholipids (Razin *et al.*, 1998), whereas the more distantly related mollicutes, the spiroplasmas and mycoplasmas are sterol and fatty acid auxotrophs. Nevertheless, phytoplasmas are likely to acquire exogenous lipids as well (Oshima *et al.*, 2004; Kube *et al.*, 2008).

While the four phytoplasmas seem to utilize glucose-6-phosphate, it is not clear how phytoplasmas obtain glucose-6-phosphate. None of the phytoplasmas has phosphoenolpyruvate:sugar phosphotransferase (PTS) systems, which import glucose and convert glucose into glucose-6-phosphate. The phytoplasmas do, however, have the MalEFGK ABC transporter system for the import of carbohydrates. The solute-binding protein of this transporter system (MalE) may have affinity for maltose, trehalose, sucrose and palatinose (Silva *et al.*, 2005). Affinity of MalE to trehalose is likely, as trehalose is a major sugar in the insect haemolymph. '*Ca.* Phytoplasma australiense' has a full-length *gtfA* gene that encodes sucrose phosphorylase, which can convert sucrose into glucose and fructose (Tran-Nguyen *et al.*, 2008), but enzymes that convert glucose and fructose into glucose-6-phosphate have not been identified. Furthermore, the sucrose phosphorylase gene is fragmented in '*Ca.* Phytoplasma asteris' OY-M and is completely absent from the genomes of '*Ca.* Phytoplasma asteris' AY-WB (Bai *et al.*, 2006) and '*Ca.* Phytoplasma mali' (Kube *et al.*, 2008). Thus, more investigations are needed to determine how phytoplasmas generate glucose-6-phosphate.

Phytoplasmas also share genes involved in malate import (CitS) and NAD-specific malic enzyme, which converts malate into pyruvate, thereby also generating NAD^+ and NADH. Mycoplasmas and spiroplasmas typically import lactate and are capable of converting lactate into pyruvate. Phytoplasmas lack the enzymes for conversion of lactate into pyruvate, and hence are most likely to utilize malate rather than lactate (Bai *et al.*, 2006). The use of malate is energy efficient and it can serve as the sole energy source for bacteria (Dimroth and Schink, 1998). Saving energy is apparently important, because all four phytoplasmas lack F-type ATPases (ATP synthases) that use the transmembrane potential for ATP synthesis (Kube *et al.*, 2008). In addition, '*Ca.* Phytoplasma mali' also lacks genes for the glycolytic pathway that generates NADH and 2 × ATP. The glycolytic pathway is present in the other three phytoplasmas (Oshima *et al.*, 2004; Bai *et al.*, 2006; Tran-Nguyen *et al.*, 2008).

Consistent with the phylogeny (discussed above), '*Ca.* Phytoplasma mali' is the most divergent among the four phytoplasmas. It has a linear chromosome, whereas the other three phytoplasmas have circular chromosomes (Table 2.1). It also has the smallest genome, with 536 coding sequences, compared with over 700 coding sequences for the other phytoplasmas. One striking observation is that the numbers of single ORFs are similar among '*Ca.* Phytoplasma asteris' OY-M and AY-WB and '*Ca.* Phytoplasma australiense', ranging from 482 to 486 ORFs, but is only 408 ORFs for '*Ca.* Phytoplasma mali' (Table 2.1). The majority of the single-copy ORFs encode proteins that partake in basic metabolic pathways and transporter systems (Bai *et al.*, 2006), thereby providing more evidence that '*Ca.* Phytoplasma mali' is the most reduced in its metabolic capacities among the four phytoplasmas (Kube *et al.*, 2008).

Repeats

Despite their small genomes, '*Ca.* Phytoplasma asteris' OY-M and AY-WB and '*Ca.* Phytoplasma australiense' contain a substantial number of multiple-copy

ORFs (Table 2.1) (Oshima *et al.*, 2004; Bai *et al.*, 2006; Jomantiene and Davis, 2006; Jomantiene *et al.*, 2007; Tran-Nguyen *et al.*, 2008). The majority of these multiple-copy ORFs are organized in clusters named potential mobile units (PMUs) (Table 2.1) (Bai *et al.*, 2006). The AY-WB genome contains one apparent complete PMU, named PMU1, which is 20 kb in length and contains 21 predicted ORFs (Bai *et al.*, 2006). PMU1 contains genes for a full-length transposase (*tra5*) on one end and a truncated transposase sequence at the other end, and genes involved in DNA replication (*ssb*, *dnaB* and *dnaG*) and synthesis (*tmk*), recombination (*himA*), a transcription factor (*sigF*), and several genes encoding membrane-targeted and secreted proteins. One of the membrane-targeted proteins is annotated as HflB, which is a membrane-associated, Zn-dependent protease. PMU1 is also flanked by 327-bp inverted repeats. With this genetic composition and the abundant presence of PMUs in the small phytoplasma genomes, Bai *et al.* (2006) hypothesized that PMU1, and perhaps also other PMUs, may transpose in a replicative fashion. In addition, the membrane-targeted and secreted proteins may have a function in phytoplasma pathogenesis (Bai *et al.*, 2006, 2009). How PMUs contribute to phytoplasma biology is not yet clear, but it is proposed that PMUs are part of a phase-variation mechanism to allow adaptation of phytoplasmas to insect and plant hosts (Bai *et al.*, 2006). Furthermore, PMU-like regions contain virulence factors, including the secreted AY-WB protein (SAP)11, which targets plant cell nuclei (Hogenhout *et al.*, 2008; Bai *et al.*, 2009).

Most PMUs in the AY-WB and OY-M genomes are shorter than PMU1 and contain truncated genes (predicted pseudogenes) (Bai *et al.*, 2006). Multiple copies of PMUs, many of which carry pseudogenes, were also found in the '*Ca.* Phytoplasma australiense' genome (Tran-Nguyen *et al.*, 2008). '*Ca.* Phytoplasma mali' contains one PMU that is similar to PMU1 of '*Ca.* Phytoplasma asteris' AY-WB (Kube *et al.*, 2008). This PMU lacks the inverted repeats and genes encoding DNA-binding protein HU and the full-length transposase (Kube *et al.*, 2008) and therefore is unlikely to have replicative transposase activity. The '*Ca.* Phytoplasma mali' contains a second PMU-like remnant containing four ORFs (Kube *et al.*, 2008). The low number of PMUs correlates to the fewer fragmented genes in '*Ca.* Phytoplasma mali' compared with the other phytoplasmas (Table 2.1). This translates into a higher average ORF size of 955 bp for '*Ca.* Phytoplasma mali' relative to the other phytoplasma genomes, which have average ORF sizes ranging between 776 bp and 833 bp (Table 2.1). Despite the few PMUs, the '*Ca.* Phytoplasma mali' genome contains 11 copies of *hflB* (Kube *et al.*, 2008), which, except for one *hflB* gene, are mostly part of PMU or PMU derivatives in the other phytoplasmas. The numerous *hflB* genes in all four phytoplasma genomes is striking as most bacteria harbour only one copy of this gene and suggests that *hflB* has an important role in phytoplasma biology.

An irregular GC skew, indicating high genome plasticity, is observed for '*Ca.* Phytoplasma asteris' OY-M and AY-WB genomes (Bai *et al.*, 2006; Kube *et al.*, 2008). In contrast, the '*Ca.* Phytoplasma mali' genome has a regular GC skew, indicating that this genome is stable (Kube *et al.*, 2008). As evidenced in a scatter plot genome alignment of '*Ca.* Phytoplasma asteris' OY-M and

AY-WB, the genome segments most prone to reorganization are those rich in PMUs and PMU-like sequences (Bai *et al.*, 2006). The OY-M and AY-WB genomes are clearly less congruent than the genomes of, for example, *E. coli* and *Salmonella typhimurium* (Fig. 2.2B, C), although, based on a phylogeny of 16S ribosomal DNA sequences, *E. coli* and *S. typhimurium* are more distant from each other compared with '*Ca.* Phytoplasma asteris' OY-M and AY-WB (Fig. 2.2A).

The majority of the metabolic genes are clustered together in ±250 kb of the '*Ca.* Phytoplasma asteris' genome, and the PMUs tend to congregate into tandem or multiple repeats in other genome sections (Fig. 2.2B) (Bai *et al.*, 2006). It is possible that PMUs are more likely to recombine with the highly similar sequences of other PMUs. Furthermore, insertion elements of the IS3 family, such as *tra5*, preferentially insert in their own inverted repeats (Mahillon and Chandler, 1998). An alternative explanation is that phytoplasmas with PMU insertions in the ±250 kb metabolic genes section are disadvantaged, because they are more likely to lose essential genes due to recombination among the PMUs.

'*Ca.* Phytoplasma mali' contains *recA*, *recG*, *recO*, *RecR*, *RecU*, *ruvA* and *ruvB*, which are mostly missing from '*Ca.* Phytoplasma asteris' AY-WB and OY-M and '*Ca.* Phytoplasma australiense' (Bai *et al.*, 2006; Kube *et al.*, 2008; Tran-Nguyen *et al.*, 2008). These genes are involved in DNA recombination and repair. Thus, '*Ca.* Phytoplasma mali' appears capable of RecA-mediated homologous recombination. The lack of a RecA-mediated recombination system may be an advantage for '*Ca.* Phytoplasma asteris' AY-WB and OY-M and '*Ca.* Phytoplasma australiense', because homologous recombination among the PMUs will lead to even greater genome instability (Bai *et al.*, 2006).

It was proposed that the repeated genes are derived from prophages. Some of the repeated genes of PMUs are also part of sequence-variable mosaics (SVMs), which are defined as genes repetitively clustered in non-randomly distributed segments (Jomantiene and Davis, 2006; Jomantiene *et al.*, 2007). Computational analyses of the OY-M and AY-WB genomes projected that SVMs coincide with gene content and organization of prophage remnant clusters, even though genes encoding canonical integrases, terminases or prohead proteases, considered to be 'cornerstone genes' in prophage identification, were not found in the predicted prophage clusters (Wei *et al.*, 2008). Nevertheless, the cryptic prophage remnants are thought to have originated from phages in the order *Caudovirales* (Wei *et al.*, 2008). The prophage clusters form genomic islands that are not interrupted by operons encoding bacterial housekeeping proteins or ribosomal RNAs and exhibit dinucleotide relative abundance and codon position GC values (Wei *et al.*, 2008). However, the remnants contain non-phage DNA segments (morons) that are inserted between two putative phage genes. Based on this evidence and absence of SVM-like structures in genomes of ancestral relatives including *Acholeplasma* spp., it was hypothesized that phage-mediated gene exchange via ancient phage attacks triggered events that launched evolution of the phytoplasma clade (Wei *et al.*, 2008). A particular bacterial group as the origin or major

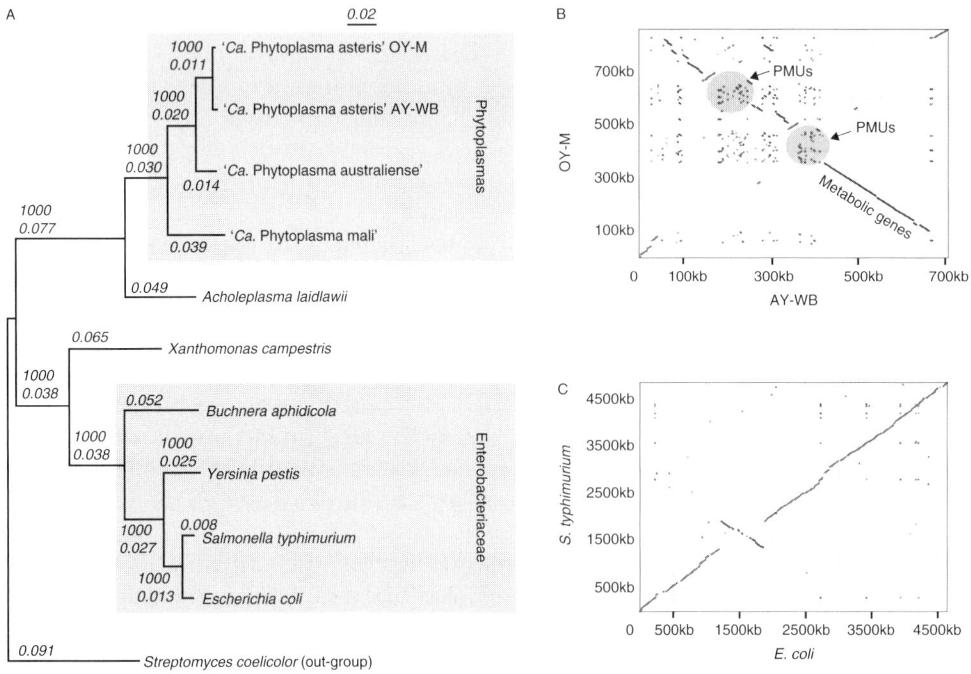

Fig. 2.2. Phytoplasmas experience more genome arrangements compared with the Enterobacteriaceae. A. Comparison of 16S ribosomal DNA phylogenies of phytoplasmas and the Enterobacteriaceae. GenBank accession numbers: '*Ca.* Phytoplasma asteris' OY-M (NC_005303, PAMr01); '*Ca.* Phytoplasma asteris' AY-WB (NC_007716, r01); '*Ca.* Phytoplasma australiense' (NC_010544, PAa_r01) '*Ca.* Phytoplasma mali' (NC_011047, ATP_R0001); *Acholeplasma laidwadii* PG-8A (NC_010163, ACL_0067); *Xanthomonas campestris* pv. *campestris* str. 8004 (NC_007086, XC_4386); *Buchnera aphidicola* str. APS (*Acyrthosiphon pisum*) (NC_002528, BU243); *Yersinia pestis* biovar Microtus str. 91001 (NC_005810, YP_r1); *Salmonella typhimurium* LT2 (NC_003197, STM_0249); *Escherichia coli* str. K-12 substr. MG1655 (NC_000913, b0201); and *Streptomyces coelicolor* A3(2) (NC_003888, SCOr03). The 16S ribosomal DNA sequences were aligned in ClustalX v. 1.83. The phylogenetic tree was generated from the aligned sequences, with all gaps in the alignments excluded. The tree was bootstrapped 1000 times and was visualized with the program NJplot (Perrière and Gouy, 1996). *S. coelicolor* was selected as the out-group. Bootstrap values were 1000 for all nodes, as indicated in the tree. The branch lengths are also indicated and are helpful to compare the distances among the phytoplasma and Enterobacteriaceae members. The bar at the top of the tree indicates branch lengths. B. Alignment of the '*Ca.* Phytoplasma asteris' OY-M and AY-WB genomes (GenBank accession numbers: NC_005303 and NC_007716). Sequences that are similar and align between the two genomes are indicated with a continuous line. The line is interrupted at locations where sequences become different. The genomes do not align at locations with a phytoplasma mobile unit (PMU), as indicated in the graph. The conserved stretch of metabolic genes is also indicated. C. Alignment of the *S. typhimurium* and *E. coli* genomes (GenBank accession numbers: NC_003197 and NC_000913). These genomes have good alignment over the entire genome lengths. Graphs in B and C were generated by the NUCmer3.0 program and mummerplot script of the MUMmer3.20 package. The postscript output of the outcluster file was imported into Adobe Photoshop CS3 for addition of labels.

source of phytoplasmal prophage genes has not yet been identified (Wei *et al.*, 2008).

Extrachromosomal DNAs

Extrachromosomal DNAs were identified in '*Ca.* Phytoplasma asteris' AY-WB and OY-M and '*Ca.* Phytoplasma australiense', but not in the '*Ca.* Phytoplasma mali' genome (Table 2.1). All of the plasmids contain genes involved in rolling-circle replication, such as *rep* and *ssb* genes. The additional two to six ORFs encode proteins with unknown functions, several of which are predicted to target the phytoplasma membrane (Bai *et al.*, 2006). In OY-M phytoplasma, it has been shown that a plasmid isolated from a non-insect-transmissible line lacks two ORFs that are present in the plasmid from the wild-type line, suggesting that these plasmid proteins may play a role in insect transmissibility and host specificity (Nishigawa *et al.*, 2002a). The *repA* genes of two of the four AY-WB plasmids (AYWB-pI and AYWB-pIII), plasmid EcOYM from OY-M and eDNA of '*Ca.* Phytoplasma australiense' show high similarity to geminivirus *repA* (Nishigawa *et al.*, 2002b; Bai *et al.*, 2006; Tran-Nguyen and Gibb, 2006). The other plasmids contain unique *rep* genes (Oshima *et al.*, 2001; Bai *et al.*, 2006). The recombination of plasmid with phytoplasma chromosome (Bai *et al.*, 2006) and intramolecular recombination among phytoplasma plasmids have been described (Nishigawa *et al.*, 2002b; Liefting *et al.*, 2004), implying that these mechanisms could also be involved in increasing biological diversity of phytoplasmas.

Summary of comparative genome analyses

From the above it is clear that '*Ca.* Phytoplasma asteris' (AY) and '*Ca.* Phytoplasma australiense' (STOL) have more in common than each of them has to '*Ca.* Phytoplasma mali' (AP). The AY and STOL phytoplasmas have circular genomes and numerous PMU-like repeats, which appear to destabilize the genome. In contrast, AP phytoplasma has a linear genome and one copy of a PMU that lacks genes and repeats essential for transposition activity. Furthermore, the AY and STOL phytoplasmas have retained more metabolic pathways compared with AP phytoplasma. Thus, AP phytoplasma has the most condensed genome among the four phytoplasmas. These findings are in agreement with the phylogeny that places AY and STOL phytoplasmas together in Cluster I, distantly from AP phytoplasmas, which are in Cluster II (Hogenhout *et al.*, 2008).

Phytoplasma Candidate Virulence Factors

During plant infection, many phytoplasmas induce various perturbations that suggest interference with plant development. Typical symptoms include phyllody (flowers that become leafy), virescence (green coloration of plant parts that are not normally green), witches'-broom (growth of a dense mass of shoots from a single point), abnormal number of fine hairy roots, inhibition of flowering, stunting and chlorosis (Bertaccini, 2007; Hogenhout *et al.*, 2008). It is unlikely that all these symptoms occur as a consequence of deficiency of plant nutrients utilized by phytoplasmas. Indeed, plant-pathogenic spiroplasmas are, similar to phytoplasmas, also phloem-limited bacteria. The stunting and yellowing typically seen in spiroplasma-infected plants is shown to be due to the imbalance of glucose and fructose in the phloem (André *et al.*, 2005). However, spiroplasmas do not induce phyllody, virescence and witches'-broom typically induced by many phytoplasmas. Therefore, it is likely that phytoplasmas produce molecules that interfere with plant development.

It was hypothesized that some phytoplasmas secrete a variety of effector (virulence) proteins that interfere with plant processes, leading to changes in development (Hogenhout *et al.*, 2008; Bai *et al.*, 2009). Production of these effectors could be an advantage to phytoplasmas, because some of the common symptoms lead to more vegetative plant tissues (phyllody, virescence, witches'-broom, hairy roots), thereby generating more phloem network for phytoplasma replication. In addition, the insect vectors, including leafhoppers, planthoppers and psyllids, feed from and lay eggs in vegetative plant parts including leaves and roots. There are several studies providing evidence that phytoplasma-infected plants are more attractive to insect vectors. These include production of volatiles that attract insect vectors (Mayer *et al.*, 2008a, b), and higher rates of insect survival (Madden and Nault, 1983; Madden *et al.*, 1984; Purcell, 1988) and reproduction (Beanland *et al.*, 2000). Thus, it appears that phytoplasmas manipulate plants to become better hosts for themselves and their insect vectors.

The complete genome sequence of '*Ca.* Phytoplasma asteris' AY-WB was mined for the presence of effector proteins. The approach for finding effector proteins was based on several assumptions. These were that: (i) effector proteins are probably secreted into the extracellular environment of phytoplasmas as they have to interact with host components; and (ii) phytoplasmas depend on the SecA-dependent system for secretion of proteins. The latter was based on the finding that the phytoplasma genome contains a complete set of genes required for the SecA-dependent secretion system (Kakizawa *et al.*, 2004). Furthermore, phytoplasmas do not encode type-III or type-IV secretion systems (Oshima *et al.*, 2004; Bai *et al.*, 2006; Kube *et al.*, 2008; Tran-Nguyen *et al.*, 2008) typically found in Gram-negative bacteria but absent from Gram-positive bacteria and the mollicutes. Proteins secreted via the SecA-dependent pathway in prokaryotes and eukaryotes typically have an N-terminal signal peptide (SP) that can be between 20 and 50 amino acids in length and consists of a consecutive stretch of positive, hydrophobic and polar amino acids. The SP is cleaved during the protein export process across the bacterial cell wall, leading to the presence of a mature protein without SP in the extracellular environment. Computer software, such as SignalP (Menne *et al.*, 2000; Bendtsen *et al.*, 2004; Schneider and Fechner, 2004), has been developed for recognition of the SP and cleavage sites of SPs in proteins. This software was successfully used to identify candidate effector proteins in the genome of '*Ca.* Phytoplasma asteris' AY-WB (Fig. 2.3A) (Bai *et al.*, 2009).

The SignalP software led to the detection of 20 secreted proteins that remain attach to the cell membrane after secretion and 56 proteins that appear to be released into the extracellular environment after secretion (Fig. 2.3A) (Bai *et al.*, 2009). The 20 proteins were characterized based on the presence of additional transmembrane (TM) regions in addition to the SP. This list contains the antigenic membrane protein (AMP) that is shown to be involved in attachment of phytoplasmas to microfilaments in the insect gut (Suzuki *et al.*, 2006). This interaction is correlated with insect vector specificity for phytoplasma transmission (Suzuki *et al.*, 2006). The AMP is under strong positive selection, providing additional evidence for involvement of this protein in phytoplasma–host interactions (Kakizawa *et al.*, 2006). Thus, the list contains at least confirmed virulence factor. The 56 proteins do not have additional TM domains besides the SP. These 56 proteins are likely to be released into the plant and insect hosts where they can interact with host cell components for manipulation of host processes. The 56 proteins were named secreted AY-WB proteins (SAP) and are considered to be candidate effector proteins (Bai *et al.*, 2009).

One of the AY-WB candidate effector proteins, named SAP11, also contains a nuclear localization signal (NLS) (Fig. 2.3A) (Bai *et al.*, 2009). Since bacteria do not have nuclei, it is likely that SAP11 targets the nuclei of plant or insect cells. *Agrobacterium*-mediated transient expression assays in *Nicotiana benthamiana* leaves showed that SAP11 tagged at the N-terminus to green fluorescent protein (GFP) or yellow fluorescent protein (YFP) accumulates in plant cell nuclei and is dependent on an intact NLS and the host cell protein α-importin for the nuclear localization (Bai *et al.*, 2009). Immunocytology studies showed that (unmodified) SAP11 produced by AY-WB in the phloem accumulates in nuclei of plant cells beyond the phloem in AY-WB-infected plants (Bai *et al.*, 2009). However, SAP11 does not accumulate in nuclei of insect cells, but accumulates at seemingly high abundance in the salivary gland cells and canalicule (cell vacuoles that lead to the main salivary duct and feeding mouthparts) of AY-WB-infected *Macrosteles quadrilineatus* (Bai *et al.*, 2007), the leafhopper vector of AY-WB (Zhang *et al.*, 2004). Thus, mining of the AY-WB genome led to the identification of phytoplasma protein that targets specific cell organelles of plant and insect hosts. However, it is not yet clear how SAP11 manipulates plants and insects to the benefit of AY-WB.

We hypothesize that candidate effector proteins evolved to improve phytoplasma–host interactions in a number of ways (Fig. 2.3B). They can interfere with plant development, leading to more phloem network for phytoplasma replication. For example, phytoplasma effectors could play a role in DNA methylation processes that downregulate the expression of developmental genes involved in flower development (Pacros *et al.*, 2007). They can also downregulate plant immune defence responses to phytoplasmas and phytoplasma insect vectors, and manipulate the production of plant volatiles and secondary metabolites to attract insect vectors. In insects, the effector

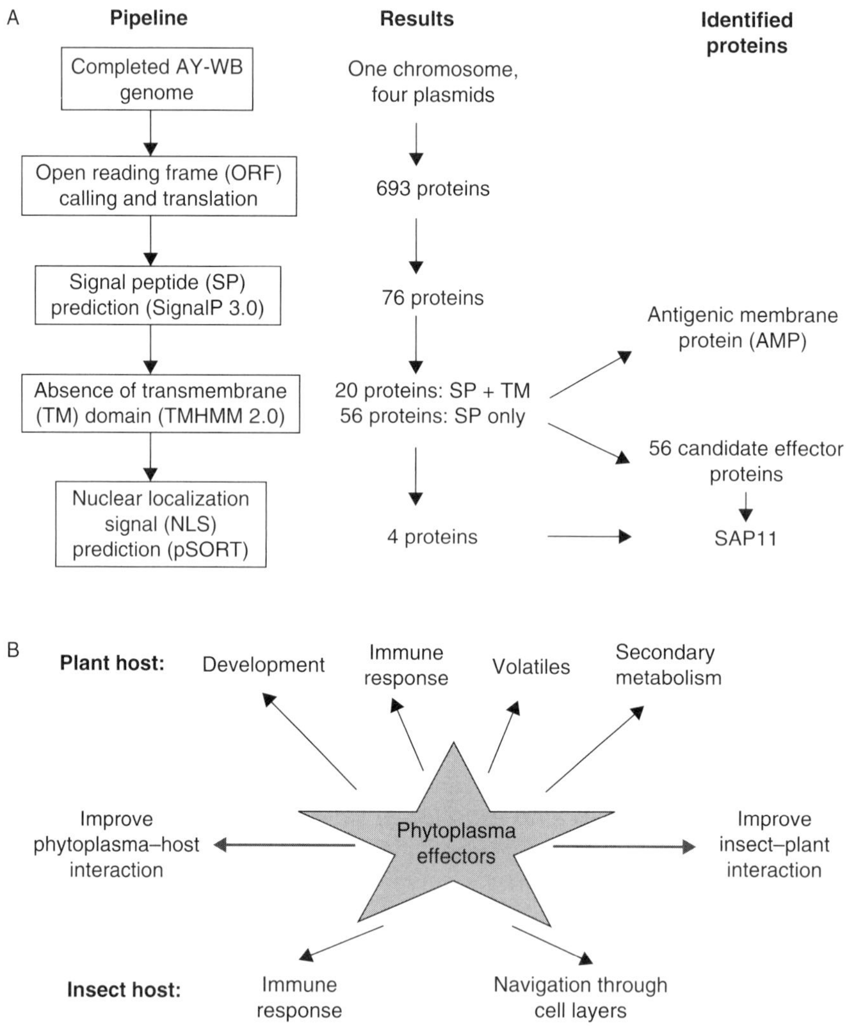

Fig. 2.3. Identification of phytoplasma virulence factors (effector proteins) using bioinformatics. A. Bioinformatics pipeline used for the identification of '*Ca.* Phytoplasma asteris' AY-WB effectors. AMP and SAP11 have confirmed interactions with host components (see text) (Suzuki *et al.*, 2006; Bai *et al.*, 2009). B. Schematic illustration of the possible functions of phytoplasma effectors. Effectors perturb the development, immune response, volatile production and secondary metabolism of the plant host and the immune response of insects. They can also aid phytoplasma navigation through the various cell layers of insect hosts. The overall effect is that phytoplasma fitness is enhanced through manipulation of plant and insect hosts and insect–plant interaction.

proteins may downregulate immune responses to phytoplasmas and help phytoplasma navigation through insect cell layers. The abundant presence of SAP11 in leafhopper salivary gland cells and canaliculi points to a possible role of effector proteins in improving plant–insect interactions.

Conclusions

Even though phytoplasmas cannot be obtained in a pure culture in cell-free media, four phytoplasma genomes have been successfully sequenced and annotated to completion. Host DNA contaminations, together with the fact that phytoplasma DNA is AT rich and often repeat rich, have led to certain challenges in sequencing and assembly of phytoplasma genomes. In the future, some of these difficulties can be overcome with next-generation sequencing technologies, such as 454 Sequencing™ (Roche Diagnostics Corporation) or Illumina® sequencing technology combined with the conventional sequencing of phytoplasma DNA.

Comparative analyses of sequenced phytoplasma genomes have revealed that, while they share some common general features, the genomes are diverse and appear to evolve fast. Among the four sequenced genomes, the one of '*Ca.* Phytoplasma mali' is most divergent, having a linear chromosome harbouring the lowest number of coding sequences and metabolic pathways. '*Ca.* Phytoplasma asteris' OY-M and AY-WB and '*Ca.* Phytoplasma australiense' have multiple repeats that are organized as clusters, named PMUs, of which some are full length and are putative active composite transposons. These genomes contain a high frequency of PMU derivatives that consist of fragmented genes. '*Ca.* Phytoplasma mali' possesses an incomplete PMU that is unlikely to have transposition activity and also has a very low number of PMU derivatives with fragmented genes. Thus, '*Ca.* Phytoplasma mali' is most condensed in terms of repeats. This translates into higher genome stability, as evidenced by a more regular GC skew and longer average ORF length in '*Ca.* Phytoplasma mali' versus the other three phytoplasma genomes.

The functional genomics research on phytoplasmas resulted in the identification of a number of membrane-targeted virulence proteins. The secreted virulence proteins are called effectors. One of the '*Ca.* Phytoplasma asteris' AY-WB effectors targets plant cell nuclei and is produced at high levels in insect salivary glands. This and other effectors are likely to manipulate processes in the plant and insect hosts to enhance phytoplasma replication and survival, and can also enhance insect performance on plant hosts. Functional analysis of phytoplasma effector proteins is an exciting avenue for future research that will help to elucidate how phytoplasmas can persist in nature.

Acknowledgements

We thank Joe Win for help with installing the MUMmer 3.20 package. Saskia A. Hogenhout is supported by The John Innes Centre and The Gatsby

Charitable Foundation. The John Innes Centre is grant aided by the Biotechnology and Biological Sciences and Research Council (BBSRC).

References

André, A., Maucourt, M., Moing, A., Rolin, D. and Renaudin, J. (2005) Sugar import and phytopathogenicity of *Spiroplasma citri*: glucose and fructose play distinct roles. *Molecular Plant–Microbe Interactions* 18, 33–42.

Bai, X., Zhang, J., Ewing, A., Miller, S.A., Radek, A.J., Shevchenko, D.V., Tsukerman, K., Walunas, T., Lapidus, A., Campbell, J.W. and Hogenhout, S.A. (2006) Living with genome instability: the adaptation of phytoplasmas to diverse environments of their insect and plant hosts. *Journal of Bacteriology* 188, 3682–3696.

Bai, X., Ammar, E.-D. and Hogenhout, S.A. (2007) A secreted effector protein of AY-WB phytoplasma accumulates in nuclei and alters gene expression of host plant cells, and is detected in various tissues of the leafhopper *Macrosteles quadrilineatus*. *Bulletin of Insectology* 60, 217–218.

Bai, X., Correa, V., Toruño, T., Ammar, E.-D., Kamoun, S. and Hogenhout, S.A. (2009) AY-WB phytoplasma secretes a protein that targets plant cell nuclei. *Molecular Plant–Microbe Interactions* 22, 18–30.

Beanland, L., Hoy, C.W., Miller, S.A. and Nault, L.R. (2000) Influence of Aster Yellows Phytoplasma on the fitness of aster leafhopper (Homoptera: Cicadellidae). *Annals of the Entomological Society of America* 93, 271–276.

Bendtsen, J.D., Nielsen, H., von Heijne, G. and Brunak, S. (2004) Improved prediction of signal peptides: SignalP 3.0. *Journal of Molecular Biology* 340, 783–795.

Bertaccini, A. (2007) Phytoplasmas: diversity, taxonomy, and epidemiology. *Frontiers in Bioscience* 12, 673–689.

Dimroth, P. and Schink, B. (1998) Energy conservation in the decarboxylation of dicarboxylic acids by fermenting bacteria. *Archives of Microbiology* 170, 69–77.

Hogenhout, S.A., Oshima, K., Ammar, E.-D., Kakizawa, S., Kingdom, H.N. and Namba, S. (2008) Phytoplasmas: bacteria that manipulate plants and insects. *Molecular Plant Patholology* 9, 403–423.

Jomantiene, R. and Davis, R.E. (2006) Clusters of diverse genes existing as multiple, sequence-variable mosaics in a phytoplasma genome. *FEMS Microbiology Letters* 255, 59–65.

Jomantiene, R., Zhao, Y. and Davis, R.E. (2007) Sequence-variable mosaics: composites of recurrent transposition characterizing the genomes of phylogenetically diverse phytoplasmas. *DNA and Cell Biology* 26, 557–564.

Kakizawa, S., Oshima, K., Nishigawa, H., Jung, H.Y., Wei, W., Suzuki, S., Tanaka, M., Miyata, S., Ugaki, M. and Namba, S. (2004) Secretion of immunodominant membrane protein from onion yellows phytoplasma through the Sec protein-translocation system in *Escherichia coli*. *Microbiology* 150, 135–142.

Kakizawa, S., Oshima, K., Jung, H.Y., Suzuki, S., Nishigawa, H., Arashida, R., Miyata, S., Ugaki, M., Kishino, H. and Namba, S. (2006) Positive selection acting on a surface membrane protein of the plant-pathogenic phytoplasmas. *Journal of Bacteriology* 188, 3424–3428.

Kube, M., Schneider, B., Kuhl, H., Dandekar, T., Heitmann, K., Migdoll, A.M., Reinhardt, R. and Seemüller, E. (2008) The linear chromosome of the plant-pathogenic mycoplasma '*Ca.* P. mali'. *BMC Genomics* 9, 306.

Lee, I.M., Davis, R.E. and Gundersen-Rindal, D.E. (2000) Phytoplasma: phytopathogenic mollicutes. *Annual Review of Microbiology* 54, 221–255.

Liefting, L.W., Shaw, M.E. and Kirkpatrick, B.C. (2004) Sequence analysis of two plasmids from the phytoplasma beet leaf hopper-transmitted virescence agent. *Microbiology* 150, 1809–1817.

Lim, P.O. and Sears, B.B. (1992) Evolutionary relationships of a plant-pathogenic mycoplasma-like organism and *Acholeplasma laidlawii* deduced from two ribosomal protein gene sequences. *Journal of Bacteriolology* 174, 2606–2611.

Madden, L. and Nault, L.R. (1983) Differential pathogenicity of corn stunting mollicutes to leafhopper vectors in *Dalbulus* and *Baldulus* species. *Phytopathology* 73, 1608–1614.

Madden, L., Nault, L.R., Heady, S.E. and Styer, W.E. (1984) Effect of maize stunting mollicutes on survival and fecundity of *Dalbulus* leafhopper vectors. *Annals of Applied Biology* 105, 431–441.

Mahillon, J. and Chandler, M. (1998) Insertion sequences. *Microbiology and Molecular Biology Reviews* 62, 725–774.

Mayer, C.J., Vilcinskas, A. and Gross, J. (2008a) Phytopathogen lures its insect vector by altering host plant odor. *Journal of Chemical Ecology* 34, 1045–1049.

Mayer, C.J., Vilcinskas, A. and Gross, J. (2008b) Pathogen-induced release of plant allomone manipulates vector insect behavior. *Journal of Chemical Ecology* DOI 10.1007/s10886-008-9564-6.

Menne, K.M., Hermjakob, H. and Apweiler, R. (2000) A comparison of signal sequence prediction methods using a test set of signal peptides. *Bioinformatics* 16, 741–742.

Nishigawa, H., Oshima, K., Kakizawa, S., Jung, H.-Y., Kuboyama, T., Miyata, S., Ugaki, M. and Namba, S. (2002a) A plasmid from a non-insect-transmissible line of a phytoplasma lacks two open reading frames that exist in the plasmid from the wild-type line. *Gene* 298, 195–201.

Nishigawa, H., Oshima, K., Kakizawa, S., Jung, H.-Y., Kuboyama, T., Miyata, S., Ugaki, M. and Namba, S. (2002b) Evidence of intermolecular recombination between extrachromosomal DNAs in phytoplasma: a trigger for the biological diversity of phytoplasma? *Microbiology* 148, 1389–1396.

Oshima, K., Kakizawa, S., Nishigawa, H., Kuboyama, T., Miyata, S., Ugaki, M. and Namba, S. (2001) A plasmid of phytoplasma encodes a unique replication protein having both plasmid- and virus-like domains: clue to viral ancestry or result of virus/plasmid recombination? *Virology* 285, 270–277.

Oshima, K., Kakizawa, S., Nishigawa, H., Jung, H.-Y., Wei, W., Suzuki, S., Arashida, R., Nakata, D., Miyata, S., Ugaki, M. and Namba, S. (2004) Reductive evolution suggested from the complete genome sequence of a plant-pathogenic phytoplasma. *Nature Genetics* 36, 27–29.

Pacros, P., Hernould, M., Teyssier, E., Eveillard, S. and Renaudin, J. (2007) Stolbur phytoplasma-infected tomato showed alteration of SIDEF methylation status and deregulation of methyltransferase gene expression. *Bulletin of Insectology* 60, 221–222.

Perrière, G. and Gouy, M. (1996) www-Query: An on-line retrieval system for biological sequence banks. *Biochimie* 78, 364–369.

Purcell, A.H. (1988) Increased survival of *Dalbulus maidis* DeLong and Wolcott, a specialist on maize on nonhost plants infected with mollicute plant pathogens. *Entomologia Experimentalis et Applicata* 46, 187–196.

Razin, S., Yogev, D. and Naot, Y. (1998) Molecular biology and pathogenicity of mycoplasmas. *Microbiology and Molecular Biology Reviews* 62, 1094–1156.

Schneider, G. and Fechner, U. (2004) Advances in the prediction of protein targeting signals. *Proteomics* 4, 1571–1580.

Silva, Z., Sampaio, M.M., Henne, A., Bohm, A., Gutzat, R., Boos, W., da Costa, M.S. and Santos, H. (2005) The high-affinity maltose/trehalose ABC transporter in the extremely thermophilic bacterium *Thermus thermophilus* HB27 also recognizes sucrose and palatinose. *Journal of Bacteriolology* 187, 1210–1218.

Suzuki, S., Oshima, K., Kakizawa, S., Arashida, R., Jung, H.Y., Yamaji, Y., Nishigawa, H., Ugaki, M. and Namba, S. (2006) Interaction between the membrane protein of a pathogen and insect microfilament complex determines insect-vector specificity. *Proceedings of the National Academy of Sciences of the USA* 103, 4252–4257.

Tran-Nguyen, L.T.T. and Gibb, K.S. (2006) Extrachromosomal DNA isolated from tomato big bud and '*Ca.* P. australiense' phytoplasma strains. *Plasmid* 56, 153–166.

Tran-Nguyen, L.T.T., Kube, M., Schneider, B., Reinhardt, R. and Gibb, K.S. (2008) Comparative genome analysis of '*Ca.* P. australiense' (subgroup *tuf*-Australia I; *rp-A*) and '*Ca.* Phytoplasma asteris' strains OY-M and AY-WB. *Journal of Bacteriology* 190, 3979–3991.

Wei, W., Davis, R.E., Jomantiene, R. and Zhao, Y. (2008) Ancient, recurrent phage attacks and recombination shaped dynamic sequence-variable mosaics at the root of phytoplasma genome evolution. *Proceedings of the National Academy of Sciences of the USA* 105, 11827–11832.

Zhang, J., Hogenhout, S.A., Nault, L.R., Hoy, C.W. and Miller, S.A. (2004) Molecular and symptom analyses of phytoplasma strains from lettuce reveal a diverse population. *Phytopathology* 94, 842–849.

3 Functional Genomics of Phytoplasmas

Shigeyuki Kakizawa, Kenro Oshima and Shigetou Namba

The University of Tokyo, Japan

Introduction

Since phytoplasmas cannot be cultured *in vitro*, it is difficult to analyse the infection system or the virulence mechanism of phytoplasmas. However, recent advances of sequencing technology have enabled us to clarify the complete genomic sequences of uncultured bacteria. Until now, four complete genomic sequences of phytoplasmas have been determined (Oshima *et al.*, 2004; Bai *et al.*, 2006; Kube *et al.*, 2008; Tran-Nguyen *et al.*, 2008). Phytoplasmas possess ca. 500–840 genes in their genomes, and approximately 40–50% of them encode hypothetical proteins whose function has not yet been established. Plant-pathogenic bacteria generally possess a number of different types of pathogenicity genes (Abramovitch *et al.*, 2006; Jones and Dangl, 2006), which are often clustered in the chromosome as a 'pathogenicity island'. A type-III secretion system is also essential for many phytopathogenic bacteria, for delivering effector proteins into host cells (Grant *et al.*, 2006). However, phytoplasma genomes possess no genes homologous to any of these known pathogenicity genes, suggesting the presence of novel mechanism(s) for the interaction between phytoplasmas and their hosts (Oshima *et al.*, 2002, 2004, 2007). Since phytoplasmas are cell-wall-less bacteria and reside intracellularly within the host cell, the membrane proteins or secreted proteins of phytoplasmas seem to function directly in the cytoplasm of the host plant and insect cells (Hogenhout *et al.*, 2008). Therefore, in order to understand the phytoplasma–host interactions, it is important to identify the functions of membrane proteins or secreted proteins encoded in phytoplasma genomes. In this section, we review recent reports regarding the secretion system and the membrane protein of phytoplasmas.

The Phytoplasma Sec System is Functional

There are at least five independent protein export systems in bacteria (Economou, 1999). In *Escherichia coli*, these systems secrete various proteins such as toxins, adhesins and hydrolytic enzymes. Of these systems, only the Sec system is essential for cell viability. In *Bacillus subtilis*, the Sec pathway is thought to be the most important of four distinct transport pathways (Tjalsma *et al.*, 2000).

The Sec protein translocation system is best characterized in *E. coli*, comprising at least 11 proteins and one RNA species (Economou, 1999). Among these proteins, SecY, SecE, SecG and SecA constitute a translocase complex that acts as export machinery at the cytoplasmic membrane.

SecA, SecY and SecE are required for protein translocation and cell viability in *E. coli* (Economou, 1999), and protein translocation activity can be reconstituted *in vitro* with only these three proteins (Akimaru *et al.*, 1991), whereas SecG is not essential. SecYEG heterotrimers form a membrane-spanning pore. Secretory proteins, which are substrates of the Sec system, have the signal peptide in their N-terminal. When a secretory protein is transcribed, its signal peptide is recognized by the signal recognition particle (SRP). In *E. coli*, the SRP is composed of a protein component, Ffh, and 4.5S RNA. SRP targets certain proteins to the membrane, where FtsY is thought to be the SRP receptor. SecB is a Sec-system-specific chaperone, which primarily recognizes some sequence motifs present in the mature part of a secretory protein and retards its folding. SecA binds a chaperone-guided secretory protein and brings it to the SecYEG pore. SecA uses its ATPase activity to propel protein secretion in a stepwise manner.

In '*Candidatus* (*Ca.*) Phytoplasma asteris' OY strain, the genes encoding the SecA, SecY and SecE proteins, the essential components of the Sec system (Economou, 1999), have been identified (Kakizawa *et al.*, 2001, 2004), and the expression of SecA protein in phytoplasma-infected plants was confirmed (Kakizawa *et al.*, 2001; Wei *et al.*, 2004). These three genes were also identified in the AY-WB phytoplasma genome (Bai *et al.*, 2006), and the *SecY* genes were cloned from several phytoplasmas (Lee *et al.*, 2006). These results strongly suggested that the Sec system commonly exists in phytoplasmas.

An antigenic membrane protein (Amp), a type of the immunodominant membrane protein found in phytoplasmas, has the signal sequence of the Sec system in its N-terminal and its signal sequence is cleaved in OY phytoplasma (Kakizawa *et al.*, 2004), suggesting that the Sec system is functional in phytoplasma.

Prediction of Phytoplasma Secretory Proteins

The membrane proteins or secreted proteins of cell-wall-less phytoplasmas are likely to function directly in the cytoplasm of the host plant and insect cells. Therefore, it is speculated that proteins such as adhesins, proteases and hydrolytic enzymes could be transported from the phytoplasma cytoplasm to the phytoplasma cell surface or host cytoplasm via the Sec pathway, and these transported proteins may influence pathogenicity. Thus, identifying secreted proteins in the phytoplasma genome is important to understanding host–phytoplasma interactions.

Generally, the secretory proteins through the Sec system have a signal peptide in their N-terminal. The consensus signal sequence consists of three parts: a positively charged N-terminal domain with at least one arginine or lysine residue; a hydrophobic core domain, which forms an α-helix and penetrates the inner membrane; and an A-X-A consensus sequence serving as the signal peptidase I (SPaseI) cleavage site, in which two alanine residues can be replaced by other, preferably small and uncharged, residues (e.g. valine, leucine, isoleucine, and so on) (Tjalsma *et al.*, 2000). The predicted export signal sequence of OY Amp matched closely with the general Sec system signal peptides, and OY Amp seems to be exported by the Sec system of *E. coli* and phytoplasma, accompanied by the processing of its signal sequence. This suggests the commonality of the signal sequence recognition mechanisms between phytoplasma and *E. coli* (Kakizawa *et al.*, 2004). This would mean that prediction programs such as SignalP (Nielsen *et al.*, 1997) or PSORT (Nakai and Kanehisa, 1991) could be used to recognize signal sequences in phytoplasma proteins and thereby to identify secretory proteins of phytoplasma. The secretory proteins of phytoplasmas, both those that remain membrane-embedded and those that are secreted into the host cytoplasm, are expected to interact directly with host plant and insect cells and to play critical roles in host–phytoplasma interactions. Therefore, to search for proteins that are exported by phytoplasmas and to investigate the functions of exported proteins is important to elucidate host–phytoplasma interactions. In fact, it has recently been reported that '*Ca.* Phytoplasma asteris' AY-WB strain secretes a protein that targets plant cell nuclei, which is thus one of the candidate virulence factors of phytoplasma (Bai *et al.*, 2009).

Other Protein Secretion Systems in Phytoplasma

Many Gram-negative pathogens of plants and animals possess type-III secretion systems (T3SS) that can inject bacterial virulence 'effector' proteins into host cells (Cornelis and van Gijsegem, 2000). The T3SS and flagella are evolutionarily related and they share a remarkably similar basal structure. However, both T3SS and flagella are restricted to Gram-negative bacteria. Phytoplasmas phylogenetically belong to Gram-positive bacteria, thus no T3SS have been found in phytoplasmas.

The bacterial type-IV secretion system (T4SS) is another important secretion system of plant and animal pathogens. T4SS comprises a large family of translocation systems as a pilus-like structure and mediates the transfer of DNA and protein substrates across the cell envelope to bacterial or eukaryotic cells, generally through a process requiring direct cell-to-cell contact (Grohmann *et al.*, 2003). Several component proteins of the conjugative transfer system of Gram-positive and -negative bacteria have sequence similarity to the component proteins of the T4SS; therefore, the conjugative transfer system and the T4SS are thought to be ancestrally related (Grohmann *et al.*, 2003). Thus, it was suggested that the T4SS is widely distributed among Gram-negative and -positive bacteria (Christie and Cascales, 2005).

However, phytoplasmas do not have either pili or T4SS. In phytoplasma genomes, there are no genes that have homology to the component proteins of pili or T4SS, and no observation was reported about the existence of pili-like structures with electron microscopy analysis. This is in contrast to spiroplasmas, which have pili-like structures (Ammar *et al.*, 2004).

YidC is involved in the membrane integration process of newly synthesized membrane proteins. YidC was found to co-purify with components of the Sec system (Scotti *et al.*, 2000); it was thought that YidC worked in conjunction with the Sec translocase to transfer the transmembrane regions of Sec-dependent substrate proteins into the hydrophobic bilayer (Urbanus *et al.*, 2001). However, it was recently demonstrated that YidC is sufficient in promoting the membrane insertion of a membrane protein (Pf3 coat protein) *in vitro*, demonstrating that YidC can function separately from the Sec system. YidC is specifically used for the insertion of membrane proteins and not for the translocation of exported proteins (Dalbey and Kuhn, 2000; Samuelson *et al.*, 2000). Possibly, the function of YidC is to recognize hydrophobic regions of a membrane protein and to catalyse the integration of these regions in a transmembrane orientation into the membrane bilayer (Serek *et al.*, 2004). In both the OY and AYWB phytoplasma genomes, there is one gene encoding YidC (Oshima *et al.*, 2004; Bai *et al.*, 2006); thus the phytoplasma would have this YidC integration system. Because YidC is an essential protein in *E. coli* (Samuelson *et al.*, 2000), it also might have an important role in phytoplasmas.

The Major Membrane Protein of Phytoplasmas

Previous studies have proposed that a subset of membrane proteins, usually referred to as immunodominant membrane proteins (IDPs), constitutes a major portion of the total cellular membrane proteins in most phytoplasmas (Shen and Lin, 1993). Immunogold-labelling electron microscopy studies have demonstrated that IDP is located on the exterior surface of the cell membrane (Milne *et al.*, 1995). Since the mollicute membrane proteins probably play important roles in the attachment of the bacteria to their host cell surface, the IDP is a candidate for being involved in host–phytoplasma interactions. Genes encoding IDP have been isolated from several phytoplasmas (Berg *et al.*, 1999; Blomquist *et al.*, 2001; Barbara *et al.*, 2002; Morton *et al.*, 2003; Kakizawa *et al.*, 2004, 2006a). These proteins show great amino acid and antigenic variation. All of the proteins have a central hydrophilic region, which may be on the outside of the phytoplasmal cell, and one or two transmembrane domains. Thus, immunodominant membrane proteins are probably secreted across the phytoplasmal cell membrane during protein localization.

These IDPs were classified into three distinct types: (i) immunodominant membrane protein (Imp); (ii) immunodominant membrane protein A (IdpA);

and (iii) antigenic membrane protein (Amp). These IDPs show no amino acid similarity to each other and are located on different regions of the genome. All IDPs possess a central hydrophilic region, possibly external to the phytoplasma cell, but the organization of the hydrophobic transmembrane anchor is variable (Kakizawa *et al.*, 2006b). Therefore, these three types of IDPs are not orthologues of each other. The first type (Imp) is anchored by only N-terminal transmembrane regions; the second type (IdpA) has N-terminal and C-terminal transmembrane regions, and neither of them is cleaved; the third type (Amp) also has two transmembrane regions, but the N-terminal one is cleaved and only the C-terminal one serves as an anchor (Barbara *et al.*, 2002).

Interestingly, the gene encoding *imp* was observed in the genomes of the Western X-disease phytoplasma (WX) (Liefting and Kirkpatrick, 2003) and OY phytoplasma (Kakizawa *et al.*, 2009), in addition to their original IDP genes. The sequence homology of *imp* was quite low between OY and WX; however, the gene organization around *imp* is well conserved. In contrast, the orthologue of IdpA, which is the IDP of WX, has not been found in the complete genomic sequences of other phytoplasmas (Oshima *et al.*, 2004; Bai *et al.*, 2006; Kube *et al.*, 2008; Tran-Nguyen *et al.*, 2008) by using homology-based search or genomic structure-based search (Kakizawa *et al.*, 2009). In addition, the orthologue of *amp* was not found in the complete genomic sequence of '*Ca.* Phytoplasma mali'. These observations imply that the phytoplasma ancestor may have possessed *imp*, and subsequently the AY group or WX group may have obtained Amp or IdpA, respectively, during their evolution.

The Variability and Positive Selection of IDPs

It has been reported that the IDPs cloned from several strains are highly variable (Barbara *et al.*, 2002; Morton *et al.*, 2003; Kakizawa *et al.*, 2004) (Fig. 3.1).

In general, coding regions rarely show lower identities than non-coding regions because of functional constraints. However, the sequence similarity of IDP genes between phytoplasmas is lower than that of their upstream genes, downstream genes or non-coding regions (Barbara *et al.*, 2002; Kakizawa *et al.*, 2004), suggesting that IDPs have been subjected to strong divergent selective pressures. In addition, the extracellular hydrophilic domains of IDPs are more divergent than the transmembrane domains or the export signal sequences, implying that the host–phytoplasma interactions promote the variability of IDPs (Barbara *et al.*, 2002; Kakizawa *et al.*, 2004). Moreover, it has been reported that the sequence identities of *imp* in several phytoplasmas were not correlated with that of 16S rDNA, which suggests that the variability of IDPs reflects some factors other than evolutionary time (Morton *et al.*, 2003).

Recently, a positive selection mechanism was observed on the Amp proteins (Kakizawa *et al.*, 2006a). Positive selection means that substitutions in a certain gene offer fitness advantages to the organism. Thus, if a positive selection was observed on a protein, one could suggest that the protein has an important role in the evolution of an organism and influences the organism's

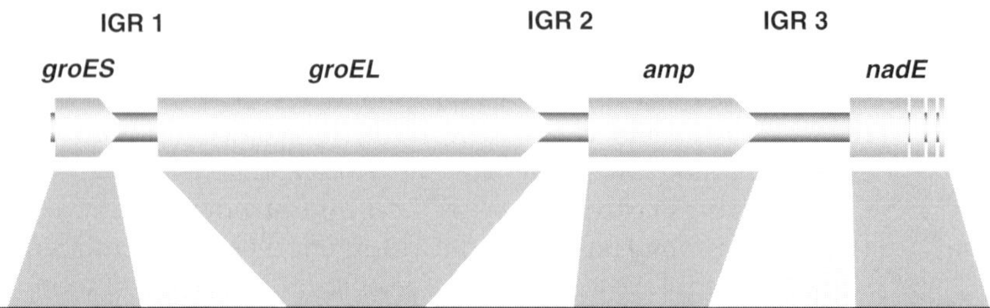

	groES	IGR 1	groEL	IGR 2	amp	IGR 3	nadE
Similarities at DNA level	99–100	100	99	90–100	89–98	92–100	98–99
Similarities at amino acid level	97–100		99		77–98		97–100

Fig. 3.1. Similarities at both the nucleotide and amino acid levels of the *amp* and other genes among six AY-group phytoplasma strains (Kakizawa *et al.*, 2006a,b). IGR, intergenic region. The *amp* gene is more variable than other proteins and the intergenic regions.

fitness directly. Many examples of positive selection at the molecular sequence level have been reported previously (Hughes and Nei, 1988; Tanaka and Nei, 1989; Nielsen and Yang, 1998; Bishop *et al.*, 2000; Jiggins *et al.*, 2002; Urwin *et al.*, 2002; Andrews and Gojobori, 2004). Most of these proteins play very important roles in their respective organisms and probably influence their fitness directly. Analysing the positively selected proteins from an organism should contribute to our understanding of the evolution of these organisms and the proteins (Ohta, 1992).

Most of the positively selected amino acids were in the central hydrophilic domain of the Amp (Kakizawa *et al.*, 2006a). This observation suggests that amino acid substitutions in the Amp offer fitness advantages to phytoplasmas, and also suggests that Amp plays an important role in host–pathogen interactions. The positive selection on Amp might be due to the interaction between the phytoplasma and its extracellular environment, the host cytoplasm. Recently, positive selection was also observed in several *imps* by molecular evolutional analysis (Kakizawa *et al.*, 2009), suggesting that *imp* has some important roles in host–phytoplasma interactions. However, it remains unclear whether the positive selection pressure was derived from the hosts, either insects or plants. The detailed analysis of the cause of positive pressure and the clarification of the meanings of the variability of the IDPs are still problems to be solved.

Amp Forms the Complex with Insect Microfilaments

It has been reported that interaction between the Amp (a type of immunodominant membrane protein) of OY phytoplasma and the insect microfilament complex determines insect vector specificity (Suzuki *et al.*, 2006) (Plate 1). OY phytoplasma was localized to the microfilaments of the visceral smooth muscle surrounding the insect's intestinal tract, and the Amp forms a complex with three insect proteins, actin, myosin heavy chain and myosin light chain, both *in vitro* and *in vivo*. Amp–microfilament complex (AM complex) formation was correlated with the phytoplasma-transmitting capability of leafhoppers, suggesting that the interaction between Amp and insect microfilament complexes plays a major role in determining the transmissibility of phytoplasmas.

The interactions between surface proteins of microbes and host microfilaments were often reported. For example, in mammal-pathogenic bacteria, such as *Listeria*, *Salmonella* and *Shigella*, several interactions between the surface membrane proteins of the bacteria and the microfilaments of host cells have been reported (Tilney and Portnoy, 1989; Gouin *et al.*, 1999; Hayward and Koronakis, 1999, 2002; van Nhieu *et al.*, 1999; Zhou *et al.*, 1999; Juris *et al.*, 2000; Pantaloni *et al.*, 2001; Delahay and Frankel, 2002; Cossart *et al.*, 2003), and the ability to form a complex with host cell microfilaments seems to be intimately involved in host cell determination. Bacterial motility within the cytoplasm of infected epithelial cells depends on the actin polymerization machinery through which the bacterium gains a propulsive force to spread within the cytoplasm and into adjacent epithelial cells (Goldberg, 2001). In *Shigella*, VirG (a surface-exposed outer-membrane protein of the bacterium) is a critical virulence factor for this actin-based motility, and the interaction between VirG and its specific host ligand neural Wiskott–Aldrich syndrome protein (N-WASP), a protein that regulates the actin cytoskeleton, determines the host cell type, allowing actin-based spreading (Suzuki *et al.*, 2002). Taken together with the complex formation of the phytoplasma Amp and the insect host microfilament, the interactions between a bacterial membrane protein and the microfilament of a host cell are a general system that is important for successful bacterial infection.

The infection of insect hosts by either phytoplasmas or spiroplasmas involves several steps (Hogenhout *et al.*, 2008). First, through the stylet, the insect ingests mollicutes from plant phloem elements and the bacteria attach to the gut epithelial cells of the insect host. Next, they enter the gut cells, multiply, cross the intestinal wall and enter the haemolymph, where they multiply and circulate to other tissues. Finally, they penetrate the salivary gland, multiply and are injected into the plant phloem when the insect feeds, resulting in transmission to a new plant host. A previous study has shown that, for phytoplasma, the ability to pass through the insect intestine (including the epithelial cells, connective tissue and visceral muscle) and salivary gland is an important factor in its vector determination (Purcell *et al.*, 1981). In the case of spiroplasmas, the salivary gland is a specific barrier that spiroplasmas must cross in order to be transmitted (Markham and Townsend, 1979). The formation of the AM complex would be necessary for passage through these host barriers.

Spiroplasma citri is thought to invade from the gut epithelium of the insect host, *Circulifer tenellus* (Baker), through a process of receptor-mediated cell endocytosis (Kwon *et al.*, 1999). Receptors on leafhopper gut epithelial cells probably recognize specific spiroplasma membrane proteins. Several candidate *S. citri* attachment protein genes have been isolated, including the immunodominant membrane protein (spiralin) (Foissac *et al.*, 1997; Duret *et al.*, 2003), P58 (Ye *et al.*, 1997), SARP1 (Berg *et al.*, 2001) and P32 of pSci6 plasmid (Berho *et al.*, 2006). It has been reported that a defective mutant of spiralin was less effective in its transmissibility (Fletcher *et al.*, 1996) and that spiralin binds to glycoproteins of its insect vector *Circulifer haematoceps* (Mulsant & Rey) (Killiny *et al.*, 2005). Although no homology was detected between Amp and spiralin, the immunodominant membrane proteins of phytoplasmas and spiroplasmas would commonly play an important role in transmission by insect vectors.

Recently, it has been reported that the plasmid pSci6 confers insect transmissibility to a non-transmissible strain of *S. citri* (Berho *et al.*, 2006). The pSci6 plasmid encodes P32 protein, which was thought to be associated with insect transmissibility. However, when only p32 gene was transformed to the non-transmissible strain of *S. citri*, its transmissibility was not restored. Thus, pSci6-encoded determinants other than P32 might be essential for the insect transmissibility, and the detailed analysis of pSci6 is expected. In phytoplasma, the AM complex is important for insect transmissibility, but it is not enough to explain the whole process of insect transmission, which includes passing through two barriers of the insect host. Thus, other factors would also be necessary. Previously, a plasmid-encoded gene, ORF3, has been suggested to be important for insect transmission (Oshima *et al.*, 2001a, b; Nishigawa *et al.*, 2002). Further detailed analysis of the ORF3 is also important to clarify the mechanism of insect transmission.

Amp–Microfilament Complex Determines Insect Vector Specificity

Insect-transmissible pathogens can cause devastating damage to humans, animals and plants, because these pathogens can be transmitted rapidly over a wide area. In nature, most insect-transmissible pathogens, including phytoplasmas, are transmitted by specific insect vectors and not by other insects, even if they are closely related (Lee *et al.*, 2000; Alavi *et al.*, 2003). Therefore, the scale of damage caused by a pathogen is determined largely by the number of insect vector species that are capable of transmitting the pathogen (Lee *et al.*, 2000). Elucidating the determination mechanisms of insect-vector specificity may enable the rapid discrimination of vector and non-vector insects and allow the monitoring or forecasting of the spread and infection route of the pathogen.

In general, each phytoplasma species is transmitted by a specific vector insect, whereas the plant-host ranges are broad. For example, phytoplasmas infect at least 700 species of plants in 98 families (Lee *et al.*, 2000; Hogenhout *et al.*, 2008), and AY can infect ca. 161 plant species in 120 genera and 39 families (McCoy *et al.*, 1989). The majority of phytoplasmas can infect the periwinkle plant (Lee *et al.*, 1998). In contrast, the specificity for the insect host is usually much stricter. For example, *Macrosteles striifrons* (Fallen) transmits OY phytoplasma but cannot propagate rice yellow dwarf (RYD) phytoplasma ('*Ca.* Phytoplasma oryzae', RYD strain). In contrast, *Nephotettix cincticeps* Uhler transmits RYD phytoplasma but cannot be infected with OY phytoplasma. Therefore, insect vector determination is an important factor influencing the entire host range of phytoplasmas in nature. It has been reported that interactions between the Amp of OY phytoplasma and the insect microfilament complex are related to insect vector specificity (Suzuki *et al.*, 2006) (Plate 1). Thus, the formation of the AM complex is important for the host range of phytoplasmas in nature and the extent of crop damage by phytoplasmas.

Although the AM complex is important for insect vector specificity, many things are still unclear. First, whether the AM complex is associated with these three possible steps or not is still unclear and should be clarified. Moreover, as mentioned above, there are two major barriers, the insect intestine and salivary gland, to pass through for successful infection by a phytoplasma. It is also still unclear which barriers are related to the AM complex. Secondly, what is the protein that directly interacts with Amp? To date, three proteins that bind Amp have been identified: actin, myosin light chain and myosin heavy chain (Suzuki *et al.*, 2006). It is also not known whether Amp can bind to one of these three components or whether additional factors involving the AM complex exist.

Thirdly, why is the Amp so variable? As mentioned above, positive selection was observed in Amp, meaning that amino acid substitutions in the Amp offer fitness advantages to phytoplasmas. Although one of the functions of Amp is to form a complex with the insect microfilament, this information cannot explain the variability of Amp. Previously, several studies have reported that host–bacterium interactions promote the variability of bacterial membrane proteins (Deitsch *et al.*, 1997). In the example of the membrane protein OspC of *Borrelia burgdorferi* (the causative agent of Lyme disease) (Wilske *et al.*, 1996), the attachment protein, which has an important role in infection, must co-evolve with the host and might also need to change itself positively to adapt to a new host species. Based on these examples, one could extrapolate to the phytoplasmas and suggest divergent selective pressures such as adaptation to avoid the insect vector immune system (similar to several membrane proteins of animal pathogens (Deitsch *et al.*, 1997; Jiggins *et al.*, 2002; Urwin *et al.*, 2002; Horino *et al.*, 2003)) or attachment to host proteins, which is an important step in establishing infection for several pathogens (Andrews and Gojobori, 2004). Further analysis will be needed to clarify the cause of positive selection on Amp and the relationships between the function and variability of Amp.

References

Abramovitch, R.B., Anderson, J.C. and Martin, G.B. (2006) Bacterial elicitation and evasion of plant innate immunity. *Nature Reviews Molecular Cell Biology* 7, 601–611.

Akimaru, J., Matsuyama, S.I., Tokuda, H. and Mizushima, S. (1991) Reconstitution of a protein translocation system containing purified SecY, SecE, and SecA from *Escherichia coli*. *Proceedings of the National Academy of Sciences of the USA* 88, 6545–6549.

Alavi, Y., Arai, M., Mendoza, J., Tufet-Bayona, M., Sinha, R., Fowler, K., Billker, O., Franke-Fayard, B., Janse, C.J., Waters, A. and Sinden, R.E. (2003) The dynamics of interactions between *Plasmodium* and the mosquito, a study of the infectivity of *Plasmodium berghei* and *Plasmodium gallinaceum*, and their transmission by *Anopheles stephensi*, *Anopheles gambiae* and *Aedes aegypti*. *International Journal of Parasitology* 33, 933–943.

Ammar, E., Fulton, D., Bai, X.D., Meulia, T. and Hogenhout, S.A. (2004) An attachment tip and pili-like structures in insect- and plant-pathogenic spiroplasmas of the class Mollicutes. *Archives of Microbiology* 181, 97–105.

Andrews, T.D. and Gojobori, T. (2004) Strong positive selection and recombination drive the antigenic variation of the PilE protein of the human pathogen *Neisseria meningitidis*. *Genetics* 166, 25–32.

Bai, X.D., Zhang, J.H., Ewing, A., Miller, S.A., Radek, A.J., Shevchenko, D.V., Tsukerman, K., Walunas, T., Lapidus, A., Campbell, J.W. and Hogenhout, S.A. (2006) Living with genome instability, the adaptation of phytoplasmas to diverse environments of their insect and plant hosts. *Journal of Bacteriology* 188, 3682–3696.

Bai, X.D., Correa, V.R., Toruno, T.Y., Ammar, E.D., Kamoun, S. and Hogenhout, S.A. (2009) AY-WB phytoplasma secretes a protein that targets plant cell nuclei. *Molecular Plant–Microbe Interactions* 22, 18–30.

Barbara, D.J., Morton, A., Clark, M.F. and Davies, D.L. (2002) Immunodominant membrane proteins from two phytoplasmas in the aster yellows clade (chlorante aster yellows and clover phyllody) are highly divergent in the major hydrophilic region. *Microbiology (UK)* 148, 157–167.

Berg, M., Davies, D.L., Clark, M.F., Vetten, H.J., Maier, G., Marcone, C. and Seemüller, E. (1999) Isolation of the gene encoding an immunodominant membrane protein of the apple proliferation phytoplasma, and expression and characterization of the gene product. *Microbiology (UK)* 145, 1937–1943.

Berg, M., Melcher, U. and Fletcher, J. (2001) Characterization of *Spiroplasma citri* adhesion related protein SARP1, which contains a domain of a novel family designated sarpin. *Gene* 275, 57–64.

Berho, N., Duret, S., Danet, J.L. and Renaudin, J. (2006) Plasmid pSci6 from *Spiroplasma citri* GII-3 confers insect transmissibility to the non-transmissible strain *S. citri* 44. *Microbiology (UK)* 152, 2703–2716.

Bishop, J.G., Dean, A.M. and Mitchell-Olds, T. (2000) Rapid evolution in plant chitinases, molecular targets of selection in plant–pathogen coevolution. *Proceedings of the National Academy of Sciences of the USA* 97, 5322–5327.

Blomquist, C.L., Barbara, D.J., Davies, D.L., Clark, M.F. and Kirkpatrick, B.C. (2001) An immunodominant membrane protein gene from the Western X-disease phytoplasma is distinct from those of other phytoplasmas. *Microbiology (UK)* 147, 571–580.

Christie, P.J. and Cascales, E. (2005) Structural and dynamic properties of bacterial Type IV secretion systems (review). *Molecular Membrane Biology* 22, 51–61.

Cornelis, G.R. and van Gijsegem, F. (2000) Assembly and function of type III secretory systems. *Annual Review of Microbiology* 54, 735–774.

Cossart, P., Pizarro-Cerda, J. and Lecuit, M. (2003) Invasion of mammalian cells by *Listeria* monocytogenes, functional mimicry to subvert cellular functions. *Trends in Cell Biology* 13, 23–31.

Dalbey, R.E. and Kuhn, A. (2000) Evolutionarily related insertion pathways of bacterial, mitochondrial, and thylakoid membrane proteins. *Annual Review of Cellular and Developmental Biology* 16, 51–87.

Deitsch, K.W., Moxon, E.R. and Wellems, T.E. (1997) Shared themes of antigenic variation and virulence in bacterial, protozoal, and fungal infections. *Microbiology and Molecular Biology Reviews* 61, 281–293.

Delahay, R.M. and Frankel, G. (2002) Coiled-coil proteins associated with type III secretion systems, a versatile domain revisited. *Molecular Microbiology* 45, 905–916.

Duret, S., Berho, N., Danet, J.L., Garnier, M. and Renaudin, J. (2003) Spiralin is not essential for helicity, motility, or pathogenicity but is required for efficient transmission of *Spiroplasma citri* by its leafhopper vector *Circulifer haematoceps*. *Applied and Environmental Microbiology* 69, 6225–6234.

Economou, A. (1999) Following the leader, bacterial protein export through the Sec pathway. *Trends in Microbiology* 7, 315–320.

Fletcher, J., Shaw, M.E., Baker, G.R., Dugan, K.J., Ye, F.C., Sha, Y.H., Zuck, P.D. and Myers, G.D. (1996) Molecular characterization of *Spiroplasma citri* BR3 lines that differ in transmissibility by the leafhopper *Circulifer tenellus*. *Canadian Journal of Microbiology* 42, 124–131.

Foissac, X., Bove, J.M. and Saillard, C. (1997) Sequence analysis of *Spiroplasma phoeniceum* and *Spiroplasma kunkelii* spiralin genes and comparison with other spiralin genes. *Current Microbiology* 35, 240–243.

Goldberg, M.B. (2001) Actin-based motility of intracellular microbial pathogens. *Microbiology and Molecular Biology Reviews* 65, 595–626.

Gouin, E., Gantelet, H., Egile, C., Lasa, I., Ohayon, H., Villiers, V., Gounon, P., Sansonetti, P.J. and Cossart, P. (1999) A comparative study of the actin-based motilities of the pathogenic bacteria *Listeria monocytogenes*, *Shigella flexneri* and *Rickettsia conorii*. *Journal of Cell Science* 112, 1697–1708.

Grant, S.R., Fisher, E.J., Chang, J.H., Mole, B.M. and Dangl, J.L. (2006) Subterfuge and manipulation, Type III effector proteins of phytopathogenic bacteria. *Annual Review of Microbiology* 60, 425–449.

Grohmann, E., Muth, G. and Espinosa, M. (2003) Conjugative plasmid transfer in Gram-positive bacteria. *Microbiology and Molecular Biology Reviews* 67, 277–301.

Hayward, R.D. and Koronakis, V. (1999) Direct nucleation and bundling of actin by the SipC protein of invasive *Salmonella*. *Embo Journal* 18, 4926–4934.

Hayward, R.D. and Koronakis, V. (2002) Direct modulation of the host cell cytoskeleton by *Salmonella* actin-binding proteins. *Trends in Cell Biology* 12, 15–20.

Hogenhout, S.A., Oshima, K., Ammar, E.D., Kakizawa, S., Kingdom, H.N. and Namba, S. (2008) Phytoplasmas, bacteria that manipulate plants and insects. *Molecular Plant Pathology* 9, 403–423.

Horino, A., Sasaki, Y., Sasaki, T. and Kenri, T. (2003) Multiple promoter inversions generate surface antigenic variation in *Mycoplasma penetrans*. *Journal of Bacteriology* 185, 231–242.

Hughes, A.L. and Nei, M. (1988) Pattern of nucleotide substitution at major histocompatibility complex class-I loci reveals overdominant selection. *Nature* 335, 167–170.

Jiggins, F.M., Hurst, G.D.D. and Yang, Z.H. (2002) Host–symbiont conflicts, positive selection on an outer membrane protein of parasitic but not mutualistic Rickettsiaceae. *Molecular Biology and Evolution* 19, 1341–1349.

Jones, J.D.G. and Dangl, J.L. (2006) The plant immune system. *Nature* 444, 323–329.

Juris, S.J., Rudolph, A.E., Huddler, D., Orth, K. and Dixon, J.E. (2000) A distinctive role for the *Yersinia* protein kinase, actin binding, kinase activation, and cytoskeleton disruption. *Proceedings of the National Academy of Sciences of the USA* 97, 9431–9436.

Kakizawa, S., Oshima, K., Kuboyama, T., Nishigawa, H., Jung, H.Y., Sawayanagi, T., Tsuchizaki, T., Miyata, S., Ugaki, M. and Namba, S. (2001) Cloning and expression analysis of phytoplasma protein translocation genes. *Molecular Plant–Microbe Interactions* 14, 1043–1050.

Kakizawa, S., Oshima, K., Nighigawa, H., Jung, H.Y., Wei, W., Suzuki, S., Tanaka, M., Miyata, S., Ugaki, M. and Namba, S. (2004) Secretion of immunodominant membrane protein from onion yellows phytoplasma through the Sec protein-translocation system in *Escherichia coli*. *Microbiology (UK)* 150, 135–142.

Kakizawa, S., Oshima, K., Jung, H.Y., Suzuki, S., Nishigawa, H., Arashida, R., Miyata, S., Ugaki, M., Kishino, H. and Namba, S. (2006a) Positive selection acting on a surface membrane protein of the plant-pathogenic phytoplasmas. *Journal of Bacteriology* 188, 3424–3428.

Kakizawa, S., Oshima, K. and Namba, S. (2006b) Diversity and functional importance of phytoplasma membrane proteins. *Trends in Microbiology* 14, 254–256.

Kakizawa, S., Oshima, K., Ishii, Y., Hoshi, A., Maejima, K., Jung, H.Y., Yamaji, Y. and Namba, S. (2009) Cloning of immunodominant membrane protein genes of phytoplasmas and their *in planta* expression. *FEMS Microbiology Letters* 293, 92–101.

Killiny, N., Castroviejo, M. and Saillard, C. (2005) *Spiroplasma citri* spiralin acts *in vitro* as a lectin binding to glycoproteins from its insect vector *Circulifer haematoceps*. *Phytopathology* 95, 541–548.

Kube, M., Schneider, B., Kuhl, H., Dandekar, T., Heitmann, K., Migdoll, A.M., Reinhardt, R. and Seemüller, E. (2008) The linear chromosome of the plant-pathogenic mycoplasma '*Candidatus* Phytoplasma mali'. *BMC Genomics* 9, 306.

Kwon, M.O., Wayadande, A.C. and Fletcher, J. (1999) *Spiroplasma citri* movement into the intestines and salivary glands of its leafhopper vector, *Circulifer tenellus*. *Phytopathology* 89, 1144–1151.

Lee, I.M., Gundersen-Rindal, D.E. and Bertaccini, A. (1998) Phytoplasma, ecology and genomic diversity. *Phytopathology* 88, 1359–1366.

Lee, I.M., Davis, R.E. and Gundersen-Rindal, D.E. (2000) Phytoplasma, phytopathogenic mollicutes. *Annual Review of Microbiology* 54, 221–255.

Lee, I.M., Zhao, Y. and Bottner, K.D. (2006) SecY gene sequence analysis for finer differentiation of diverse strains in the aster yellows phytoplasma group. *Molecular and Cellular Probes* 20, 87–91.

Liefting, L.W. and Kirkpatrick, B.C. (2003) Cosmid cloning and sample sequencing of the genome of the uncultivable mollicute, Western X-disease phytoplasma, using DNA purified by pulsed-field gel electrophoresis. *FEMS Microbiology Letters* 221, 203–211.

Markham, P.J. and Townsend, R. (1979) Experimental vectors of spiroplasmas. In: Maramorosch, K. and Harris, K.F. (eds) *Leafhopper Vectors and Plant Disease Agents*. Academic Press, New York, pp. 413–455.

McCoy, R.E., Caudwell, A., Chang, C.J., Chen, T.A., Chiykowski, L.N., Cousin, M.T., Dale, J.L., de Leeuw, G.T.N., Golino, D.A., Hackett, K.J., Kirkpatrick, B.J., Marwitz, R., Petzold, H., Sinha, R.C., Sugiura, M., Whitcomb, R.F., Yang, I.L., Zhu, B.M. and Seemüller, E. (1989) Plant diseases associated with mycoplasma-like organisms. In: Whitcomb, R.F. and Tully, J.G. (eds) *The Mycoplasmas*, Vol. V. Academic Press, New York, pp. 546–640.

Milne, R.G., Ramasso, E., Lenzi, R., Masenga, V., Sarindu, N. and Clark, M.F. (1995) Preembedding and postembedding immunogold labeling and electron-microscopy in plant host tissues of 3 antigenically unrelated MLOs – primula yellows, tomato big bud and bermudagrass white leaf. *European Journal of Plant Pathology* 101, 57–67.

Morton, A., Davies, D.L., Blomquist, C.L. and Barbara, D.J. (2003) Characterization of homologues of the apple proliferation immunodominant membrane protein gene from three related phytoplasmas. *Molecular Plant Pathology* 4, 109–114.

Nakai, K. and Kanehisa, M. (1991) Expert system for predicting protein localization sites in Gram-negative bacteria. *Proteins* 11, 95–110.

Nielsen, H., Engelbrecht, J., Brunak, S. and von Heijne, G. (1997) Identification of prokaryotic and eukaryotic signal peptides and prediction of their cleavage sites. *Protein Engineering* 10, 1–6.

Nielsen, R. and Yang, Z.H. (1998) Likelihood models for detecting positively selected amino acid sites and applications to the HIV-1 envelope gene. *Genetics* 148, 929–936.

Nishigawa, H., Oshima, K., Kakizawa, S., Jung, H.Y., Kuboyama, T., Miyata, S., Ugaki, M. and Namba, S. (2002) A plasmid from a non-insect-transmissible line of a phytoplasma lacks two open reading frames that exist in the plasmid from the wild-type line. *Gene* 298, 195–201.

Ohta, T. (1992) The nearly neutral theory of molecular evolution. *Annual Review of Ecology and Systematics* 23, 263–286.

Oshima, K., Kakizawa, S., Nishigawa, H., Kuboyama, T., Miyata, S., Ugaki, M. and Namba, S. (2001a) A plasmid of phytoplasma encodes a unique replication protein having both plasmid- and virus-like domains. Clue to viral ancestry or result of virus/plasmid recombination? *Virology* 285, 270–277.

Oshima, K., Shiomi, T., Kuboyama, T., Sawayanagi, T., Nishigawa, H., Kakizawa, S., Miyata, S., Ugaki, M. and Namba, S. (2001b) Isolation and characterization of derivative lines of the onion yellows phytoplasma that do not cause stunting or phloem hyperplasia. *Phytopathology* 91, 1024–1029.

Oshima, K., Miyata, S., Sawayanagi, T., Kakizawa, S., Nishigawa, H., Jung, H.Y., Furuki, K., Yanazaki, M., Suzuki, S., Wei, W., Kuboyama, T., Ugaki, M. and Namba, S. (2002) Minimal set of metabolic pathways suggested from the genome of onion yellows phytoplasma. *Journal of General Plant Pathology* 68, 225–236.

Oshima, K., Kakizawa, S., Nishigawa, H., Jung, H.Y., Wei, W., Suzuki, S., Arashida, R., Nakata, D., Miyata, S., Ugaki, M. and Namba, S. (2004) Reductive evolution suggested from the complete genome sequence of a plant-pathogenic phytoplasma. *Nature Genetics* 36, 27–29.

Oshima, K., Kakizawa, S., Arashida, R., Ishii, Y., Hoshi, A., Hayashi, Y., Kagiwada, S. and Namba, S. (2007) Presence of two glycolytic gene clusters in a severe pathogenic line of *Candidatus* Phytoplasma asteris. *Molecular Plant Pathology* 8, 481–489.

Pantaloni, D., Le Clainche, C. and Carlier, M.F. (2001) Cell biology – mechanism of actin-based motility. *Science* 292, 1502–1506.

Purcell, A.H., Richardson, J. and Finlay, A. (1981) Multiplication of the agent of X-disease in a non-vector leafhopper *Macrosteles fascifrons*. *Annals of Applied Biology* 99, 283–298.

Samuelson, J.C., Chen, M.Y., Jiang, F.L., Moller, I., Wiedmann, M., Kuhn, A., Phillips, G.J. and Dalbey, R.E. (2000) YidC mediates membrane protein insertion in bacteria. *Nature* 406, 637–641.

Scotti, P.A., Urbanus, M.L., Brunner, J., de Gier, J.W.L., von Heijne, G., van der Does, C., Driessen, A.J.M., Oudega, B. and Luirink, J. (2000) YidC, the *Escherichia coli* homologue of mitochondrial Oxa1p, is a component of the Sec translocase. *EMBO Journal* 19, 542–549.

Serek, J., Bauer-Manz, G., Struhalla, G., van den Berg, L., Kiefer, D., Dalbey, R. and Kuhn, A. (2004) *Escherichia coli* YidC is a membrane insertase for Sec-independent proteins. *EMBO Journal* 23, 294–301.

Shen, W.C. and Lin, C.P. (1993) Production of monoclonal-antibodies against a mycoplasmalike organism associated with sweet-potato witches-broom. *Phytopathology* 83, 671–675.

Suzuki, S., Oshima, K., Kakizawa, S., Arashida, R., Jung, H.Y., Yamaji, Y., Nishigawa, H., Ugaki, M. and Namba, S. (2006) Interaction between the membrane protein of a pathogen and insect microfilament complex determines insect-vector specificity. *Proceedings of the National Academy of Sciences of the USA* 103, 4252–4257.

Suzuki, T., Mimuro, H., Suetsugu, S., Miki, H., Takenawa, T. and Sasakawa, C. (2002) Neural Wiskott–Aldrich syndrome protein (N-WASP) is the specific ligand for Shigella VirG among the WASP family and determines the host cell type allowing actin-based spreading. *Cell Microbiology* 4, 223–233.

Tanaka, T. and Nei, M. (1989) Positive Darwinian selection observed at the variable-region genes of immunoglobulins. *Molecular Biology and Evolution* 6, 447–459.

Tilney, L.G. and Portnoy, D.A. (1989) Actin-filaments and the growth, movement, and spread of the intracellular bacterial parasite, *Listeria monocytogenes*. *Journal of Cell Biology* 109, 1597–1608.

Tjalsma, H., Bolhuis, A., Jongbloed, J.D.H., Bron, S. and van Dijl, J.M. (2000) Signal peptide-dependent protein transport in *Bacillus subtilis*, a genome-based survey of the secretome. *Microbiology and Molecular Biology Reviews* 64, 515–547.

Tran-Nguyen, L.T.T., Kube, M., Schneider, B., Reinhardt, R. and Gibb, K.S. (2008) Comparative genome analysis of '*Candidatus* Phytoplasma australiense' (subgroup tuf-Australia I; rp-A) and '*Ca*. Phytoplasma asteris' strains OY-M and AY-WB. *Journal of Bacteriology* 190, 3979–3991.

Urbanus, M.L., Scotti, P.A., Froderberg, L., Saaf, A., de Gier, J.W.L., Brunner, J., Samuelson, J.C., Dalbey, R.E., Oudega, B. and Luirink, J. (2001) Sec-dependent membrane protein insertion, sequential interaction of nascent FtsQ with SecY and YidC. *EMBO Reports* 2, 524–529.

Urwin, R., Holmes, E.C., Fox, A.J., Derrick, J.P. and Maiden, M.C.J. (2002) Phylogenetic evidence for frequent positive selection and recombination in the meningococcal surface antigen PorB. *Molecular Biology and Evolution* 19, 1686–1694.

van Nhieu, G.T., Caron, E., Hall, A. and Sansonetti, P.J. (1999) IpaC induces actin polymerization and filopodia formation during *Shigella* entry into epithelial cells. *EMBO Journal* 18, 3249–3262.

Wei, W., Kakizawa, S., Suzuki, S., Jung, H.Y., Nishigawa, H., Miyata, S., Oshima, K., Ugaki, M., Hibi, T. and Namba, S. (2004) *In planta* dynamic analysis of onion yellows phytoplasma using localized inoculation by insect transmission. *Phytopathology* 94, 244–250.

Wilske, B., Busch, U., Fingerle, V., JaurisHeipke, S., Mursic, V.P., Rossler, D. and Will, G. (1996) Immunological and molecular variability of OspA and OspC. Implications for *Borrelia* vaccine development. *Infection* 24, 208–212.

Ye, F.C., Melcher, U. and Fletcher, J. (1997) Molecular characterization of a gene encoding a membrane protein of *Spiroplasma citri*. *Gene* 189, 95–100.

Zhou, D., Mooseker, M.S. and Galan, J.E. (1999) Role of the *S. typhimurium* actin-binding protein SipA in bacterial internalization. *Science* 283, 2092–2095.

4 Prospects of Multiple Gene-based Systems for Differentiation and Classification of Phytoplasmas

ING-MING LEE, YAN ZHAO AND ROBERT E. DAVIS

Molecular Plant Pathology Laboratory, Beltsville, USA

Introduction

Phytoplasmas, formerly termed mycoplasma-like organisms (MLOs), are minute cell-wall-less prokaryotes that are associated with diseases in several hundred plant species (Lee *et al.*, 2000; Bertaccini 2007; Hogenhout *et al.*, 2008). Since the discovery of these unique plant pathogens more than four decades ago, attempts to culture them in cell-free media have failed, making it difficult to determine the taxonomic status of phytoplasmas by the traditional methods applied to cultured prokaryotes. Modern mollicute systematics has adopted a polyphasic taxonomy system, developed over the last two decades, which is based on phenotypic, genotypic and phylogenetic criteria for classification of members of the *Mollicutes* (Weisburg *et al.*, 1989; Murray *et al.*, 1990; Vandamme *et al.*, 1996; Razin *et al.*, 1998). Comprehensive phylogenetic studies based on 16S rRNA and other housekeeping genes have readily placed phytoplasmas in the class of *Mollicutes* (Lim and Sears, 1989; Namba *et al.*, 1993; Gundersen *et al.*, 1994; Seemüller *et al.*, 1998; Lee *et al.* 2000, 2006b; Zhao *et al.*, 2005; Martini *et al.*, 2007; Hodgetts *et al.*, 2008). However, because of the paucity of accessible phenotypic criteria, it was inevitable that phytoplasma taxonomy would be heavily based on molecular characteristics and phylogeny. Molecular-based analyses introduced during the last two decades have proven to be more accurate and reliable than biological criteria long used for phytoplasma identification (Lee *et al.*, 2000). PCR-based assays developed in the late 1980s and early 1990s further facilitated the detection and classification of phytoplasmas by providing a much more sensitive means for phytoplasma detection (see references in Lee *et al.*, 2000).

The highly conserved 16S rRNA gene sequence has been used as the primary molecular tool for classification of phytoplasmas. A total of 19 distinct groups, termed 16S rRNA groups (16Sr groups), based on actual RFLP analysis of PCR-amplified 16S rDNA sequences or 29 groups based on *in silico*

RFLP analysis have been identified (Lee *et al.*, 1998, 2000; Wei *et al.*, 2007). Many 16Sr subgroups were further classified by this approach. It was proposed that each group represents at least one phytoplasma species (Gundersen *et al.*, 1994). At present, species designation is primarily based on dissimilarity of 16S rDNA sequences among phytoplasmas. An arbitrary threshold of 2.5% dissimilarity was applied as a guideline for electing a new species (IRPCM, 2004). Because of the highly conserved nature of the 16S rRNA gene, this guideline may exclude many ecologically or biologically distinct phytoplasma strains, some of which may warrant designation as a new taxon. Additional unique biological properties, such as insect vectors and plant hosts, as well as other molecular criteria need to be included for speciation. Over the last decade, epidemiological studies revealed that diverse phytoplasma strains that are very closely related based on analysis of 16S rRNA gene sequences were involved in similar diseases associated with various cultivars of a given crop grown in various geographical regions. To facilitate the development of disease control measures, it is important to know the correlation between various phytoplasma strains and their unique ecological niches. Often these strains cannot be readily differentiated by analysis of 16S rRNA gene sequences. This underscores a need to seek out additional markers for finer differentiation of closely related phytoplasma strains. In the last decade, several other conserved genes or specific genomic DNA fragments have been employed as supplementary molecular markers for finer differentiation of closely related strains. This chapter will summarize the recent progress on this account and will propose additional potential genes for classification of phytoplasmas at different taxonomic ranks. Comments will include the potential use of multiple gene-based systems for defining a taxon at species or strain level.

16S rRNA Gene-based System for Classification of Phytoplasmas

Several molecular markers have been employed for differentiation and classification of phytoplasmas. The 16S rRNA gene is the most widely used marker in the phytoplasma research community and proves to be very useful in preliminary classification of phytoplasmas. Several universal or generic oligonucleiotide primer pairs based on the 16S rRNA gene, the 16S–23S intergenic spacer region and partial 23S rRNA gene sequences have been designed, which allow amplification of >1200 bp to near full-length 16S rRNA gene sequences of all phytoplasmas associated with various plants and insect vectors (Lee *et al.*, 1993; Namba *et al.*, 1993; Schneider *et al.*, 1993; Gundersen and Lee, 1996; Smart *et al.*, 1996). Restriction fragment length polymorphism (RFLP) analysis of PCR-amplified 16S rRNA gene sequences using selected restriction enzymes was employed by Schneider *et al.* (1993) and Lee *et al.* (1993) for classification of phytoplasmas. Based on RFLP analyses with 17 restriction enzymes, Lee *et al.* (1993, 1998, 2000) constructed a comprehensive classification scheme for phytoplasmas. Separation of major groups was based on similarity coefficients of collective RFLP patterns of a 1.2 kb PCR amplicon (Lee *et al.*, 1998). The similarity coefficients of RFLP patterns between two distinct groups were 90% or below. Subgroup delineation within a given group was based on restriction site analysis within this amplicon. A new subgroup was assigned if an unknown phytoplasma strain had one or more restriction sites different from those in all the existing members of the given group. The scheme has been periodically updated (Lee *et al.*, 1998, 2000, 2006a; Montano *et al.*, 2001; Arocha *et al.*, 2005; Al-Saady *et al.*, 2008; and see references in Lee *et al.*, 2004a, b; Zhao *et al.*, 2009a). Thus far, it comprises 19 major phytoplasma groups and about 50 subgroups. The grouping is near congruent with the phylogenetic tree constructed by analysis of 16S rRNA sequences. Each group was proposed to represent at least one phytoplasma species (Gundersen *et al.*, 1994). Recently, the scheme was further updated and upgraded, based on virtual RFLP patterns, by Wei *et al.* (2007, 2008) and Zhao *et al.* (2009b) through the use of computer-simulated RFLP analysis of vast collections of phytoplasma 16S rRNA gene sequences that were reported and deposited in GenBank. Currently, the scheme comprises 29 groups and 89 subgroups. Each subgroup is defined by unique collective RFLP patterns. The scheme, accompanied by illustrated RFLP patterns of all representative strains in print or online, has essentially provided the most comprehensive list of reference phytoplasma strains. By comparison with these patterns, one can identify an unknown phytoplasma strain either through actual (for preliminary identification) or computer-simulated virtual RFLP analysis of the 16S rRNA gene sequence. Sequencing of the 16S rRNA gene is required for performing virtual RFLP analysis. In practice, actual RFLP analysis may be the choice for preliminary characterization of unknown phytoplasmas associated with a given new disease if numerous samples need to be analysed or if there is no sequencing facility available. This updated scheme represents the most comprehensive classification system for phytoplasmas and has provided reliable molecular markers for rapid identification of phytoplasma strains.

The merits of the 16S rRNA gene-based system for phytoplasma classification lie in its highly conserved nature, so that the universal oligonucleotide primers are relatively easily designed, and in the wealth of sequences available in the GenBank database, which makes it plausible to conduct comprehensive phylogenetic studies. However, because of its highly conserved nature, the 16S rRNA gene is inadequate for finer differentiation of closely related but distinct phytoplasmas strains. It was evident that some subgroups contained more than one biologically significant strain.

16S–23S rRNA Intergenic Spacer Region (ISR)

The phytoplasma 16S–23S rRNA intergenic spacer region (about 232 bp) contains a portion that codes for the highly conserved $tRNA^{Ile}$. However, the flanking sequences that extend from the $tDNA^{Ile}$ to 16S rDNA and to 23S rDNA are variable among various phytoplasmas. The ISR can serve as a useful tool for differentiation of phytoplasma groups and subgroups. Overall, the ISR is comparable to the 16S rRNA gene sequence in its capacity for use in delineating distinct phytoplasma lineages (Smart *et al.*, 1996). Because of limited informative characters available in its relatively short sequence, ISR cannot be used to differentiate all the 16Sr subgroups. On the other hand, combined analysis of the entire16S rRNA gene plus ISR sequence proved to be useful in several cases for differentiating distinct strain types within a given 16Sr subgroup (Griffiths *et al.*, 1999; Marcone *et al.*, 2000; Padovan *et al.*, 2000; Andersen *et al.*, 2006).

Tuf Gene-based System

The *tuf* gene, encoding the elongation factor, EF-Tu, is another highly conserved gene that has been frequently used for differentiation and classification of phytoplasmas. In 1997, Schneider *et al.* designed primer pairs that can be used for amplifications of *tuf* gene sequences from most phytoplasma groups. It was found that the *tuf* gene, like the 16S rRNA gene, represents a potential marker for classification of phytoplasma groups. The nucleotide sequence similarities among the aster yellows (AY), peach X-disease and stolbur (STOL) phytoplasma groups ranged from 87.8 to 97.0%. Phytoplasma groups and subgroups can be differentiated based on RFLP analyses using several restriction enzymes. The resolving efficacy for separation of distinct lineages among phytoplasmas is slightly lower than that of the 16S rRNA gene (Schneider *et al.*, 1997; Marcone *et al.*, 2000). However, in some cases, the *tuf* gene was found to be useful in the differentiation of various ecological strains or strain variants within 16S rRNA subgroups (Langer and Maixner, 2004). For example, several strain variants were recognized within 16XII-A and 16XII-B, based on analysis of *tuf* gene sequences (Langer and Maixner, 2004; Streten and Gibb, 2005; Andersen *et al.*, 2006; Pacifico *et al.*, 2007; Riolo *et al.*, 2007; Iriti *et al.*, 2008).

Ribosomal Protein Gene-based System

Ribosomal protein (rp) genes are more variable than 16S rRNA genes and have more phylogenetically informative characters, which substantially enhance the resolving power in delineating distinct phytoplasma strains. Earlier studies on differentiation of phytoplasma strains in groups 16SrI and 16SrV indicated that analysis of rp gene sequences not only readily delineated subgroups that are consistent with 16Sr subgroups but also identified, within some subgroups, additional distinct strains (lineages) that could not be resolved by analysis of 16S rRNA gene sequences (Martini *et al.*, 2002; Lee *et al.*, 2004a). Most of the additional lineages identified have distinct biological properties. For example, maize bushy stunt (MBS) phytoplasma, classified as a member of 16SrI-B a subgroup, in which the narrow range of host plants and specific vectors are distinct from other members of the 16SrI group, represents a distinct rp subgroup. Likewise, subgroup 16SrV-C can be further differentiated into several rp subgroups based on RFLP analyses with several selected key restriction enzymes (Martini *et al.*, 2002; Lee *et al.*, 2004a).

Recently, Martini *et al.* (2007) constructed a comprehensive phylogenetic tree based on the analysis of two ribosomal protein genes, *rplV* (*rpl22*) and *rpsC* (*rps3*), from 46 phytoplasma strains representing 12 16Sr groups. This rp gene-based phylogenetic tree, which was congruent with that inferred from the 16S rRNA gene, yielded more clearly defined phylogenetic interrelationships amongst phytoplasma strains and delineated more distinct phytoplasma subclades and distinct lineages than those resolved by the 16S rRNA gene-based tree. For example, three '*Ca.* Phytoplasma' species ('*Ca.* Phytoplasma mali', '*Ca.* Phytoplasma pyri' and '*Ca.* Phytoplasma prunorum'), which share 98.9–99.1% 16S rDNA sequence similarity, shared 94.3–94.6% rp gene sequence similarity and were readily delineated by analysis of rp gene sequences. The average sequence similarities between two given 16S phytoplasma groups ranged from 50.4 to 83.5% based on rp genes compared with 85.0–96.9% based on the 16S rRNA gene. The greater sequence variation makes rp genes a better molecular tool for differentiation of distinct phytoplasma strains.

SecY Gene-based System

The *secY* gene, encoding for a protein translocase subunit, is another molecular marker that is useful for finer differentiation of phytoplasma strains. The *secY* gene sequence variability is similar to that of rp genes. The average *secY* gene sequence similarities between two given 16Sr phytoplasma groups ranges from 57.4 to 76.0% (Lee *et al.*, unpublished). *SecY* subgroups delineated based on RFLP analyses of *secY* gene sequences from groups 16SrI and 16SrV phytoplasmas generally coincided with those delineated with rp gene sequences (Lee *et al.*, 2004a, b, 2006b; Martini *et al.*, 2007). However, due to more informative characters, the resolving power of *secY* is slightly better than rp gene sequences. Complete characterization of the majority of phytoplasma groups and their representative strains is in progress (Lee *et al.*, unpublished). The *secY* gene, like the rp gene, could represent a good candidate marker for classification of phytoplasma strains.

SecA Gene and Other Genes

Another protein translocase subunit encoding gene, *secA*, was recently employed for classification of phytoplasmas (Hodgetts *et al.*, 2008). A portion of the gene sequence, about 480 bp, was PCR-amplified from various phytoplasma strains representing 12 16Sr groups. The sequence similarity ranged from 69.7 to 84.4% between two given 16Sr groups. The resolving power of the *secA* gene as a phylogenetic parameter for phytoplasma differentiation is similar to those of rp and *secY* genes.

Other genes, *nusA* (Shao *et al.*, 2006), folate genes (*folP* and *folk*) (Davis *et al.*, 2003) and *dnaB*, have been employed and have proved useful for differentiation among strains in groups 16SrI and 16SrXII. Several housekeeping genes, including *dnaA* (encoding chromosomal replication initiator protein), *polC* (encoding DNA polymerase III alpha subunit) and *dnaE* (DNA polymerase III alpha subunit), could represent additional candidate genes that may be useful for classification of phytoplasmas as well. The sequence variability of these genes is similar to, or slightly greater than, that of the *secY* or rp genes.

Perspective of Multiple Gene-based Systems

Phytoplasmas are insect-transmitted plant pathogens, and are capable of multiplication in both vector and plant hosts (Lee *et al.*, 2000). Because the susceptibility of different insect and plant species to phytoplasmas varies with the associated phytoplasma strains, including strains within a given 16Sr subgroup, the selection pressures imposed by insect and/or plant hosts on the associated phytoplasmas have facilitated evolution and/or isolation of unique phytoplasma populations or distinct strains over time. Moreover, due to varying geographical distributions of plant and vector species, geographical isolation may also contribute to this process. The three-way interactions between phytoplasmas, vectors and plant hosts contribute to the complexity of phytoplasmal ecology. These ecologically isolated phytoplasma strains often possess unique biological properties, such as specificity to plant and vector species and symptoms they induce in the affected plants. On the one hand, it is not uncommon that a given disease (e.g. grapevine yellows) could be attributed to diverse, yet closely related, phytoplasma populations (Angelini *et al.*, 2001; Leyva-López *et al.*, 2002; Martini *et al.*, 2002; Langer and Maixner, 2004; Lee *et al.*, 2006b; Botti and Bertaccini, 2007); some populations may be present predominantly in a particular cultivar or geographical region. On the other hand, closely related strains (e.g. '*Ca.* Phytoplasma asteris' strains) could cause different diseases and induce different symptoms (Lee *et al.*, 2004b). To facilitate epidemiological studies, it is essential to identify and characterize the diverse ecological strains that may be involved in the disease. The deficiency of the 16S rRNA gene-based system for finer differentiation of closely related strains underscores an urgent need to incorporate additional molecular markers in routine phytoplasma classification into this system. Several molecular markers, other than the 16S rRNA gene, identified thus far have shown much-improved resolving power in delineation of these ecological strains. The emerging multiple-gene systems should provide molecular criteria for better delineation of species and strains.

An arbitrary threshold value of 97.5% 16S rRNA gene sequence similarity was recommended by the International Research Program for Comparative Mycoplasmology, Phytoplasma/Spiroplasma Working Team (IRPCM, 2004) to separate two distinct '*Ca.* Phytoplasma' species. There are no definitive threshold values for identifying strains that share >97.5% similarity but warrant designation of new species, and for delineating strains that are ecologically distinct but closely related. Several molecular markers, described in previous sections, that have greater resolving power than the 16S rRNA gene could be selected as additional standard phylogenetic parameters for designation of these closely related but distinct biological or ecological strains. The combination of the 16S rRNA gene with one or more variable gene or DNA fragments, 16S rRNA plus *secY*, 16S rRNA plus rp or 16S rRNA plus *secA*, proved to be sufficient for clearly discriminating two closely related strains (Lee *et al.*, 2004a, b, 2006b; Martini *et al.*, 2007; Hodgetts *et al.*, 2008). The 16S rRNA plus ISR also improved the separation of closely related strains that could not be well resolved by use of the 16S rRNA gene alone (Langer and Maixner, 2004). Because of greater sequence variability, these suites of markers could facilitate finer delineation of closely related strains, including derivative variants from a given strain. As for the 16S rRNA gene, the arbitrary threshold values of sequence similarities for assigning strains at different taxonomic ranks could be established for these additional markers. Recently, multi-locus sequence typing using *secY*, *map* and *uvrB-degV* gene sequences was employed for differentiation of three distinct flavescence dorée phytoplasma strain clusters and group 16SrV phytoplasmas infecting grapevine and alder in Europe (Arnaud *et al.*, 2007).

Advances in genome sequencing and bioinformatic tools have resulted in the completion of four annotated phytoplasma genomes (Oshima *et al.*, 2004; Bai *et al.*, 2006; Kube *et al.*, 2008; Tran-Nguyen *et al.*, 2008). More phytoplasma genome sequencing projects are under way. Comparative genomics of these genomes should provide additional suites of genes from which markers will be chosen for phytoplasma delineations and descriptions at strain, species, genus and population levels. However, the multiple gene-based system suitable for classification of the whole spectrum of phytoplasmas will not be realized until a near complete sequence database on these potential molecular markers is available. As more useful molecular markers are identified and used for strain differentiation from various research groups, it is crucial to have an updated sequence (>1000 bp) database of these markers in GenBank, accessible to the phytoplasma research community.

Issues in Phytoplasma Taxonomy

Molecular tools such as monoclonal antibodies, DNA-based probes and PCR-based sensitive detection procedures have largely replaced traditional procedures based on biological properties, greatly advancing phytoplasma disease diagnostics and facilitating phytoplasma characterization. More than 1500 phytoplasma strains have been characterized and identified based on the 16S rRNA gene sequence in the last decade and a half. Phylogenetic analyses based on the wealth of 16S rRNA gene sequences available in GenBank further enhanced the notion that phytoplasmas represent a discreet clade divergent from a common ancestor closely related to *Acholeplasma* spp. and revealed the extent of diversity of these unique plant pathogens in nature (Wei *et al.*, 2007, 2008; Zhao *et al.*, 2009b). Due to the inability to obtain both *in vivo* pure cultures and accessible phenotypic criteria of phytoplasmas, it was inevitable that phytoplasma taxonomy would be heavily based on molecular characteristics and molecular phylogeny. Currently, phytoplasma classification and species nomenclature are primarily based on the 16S rRNA gene. A provisional taxonomic system for uncultured bacteria proposed by Murray and Schleifer (1994) was adopted. Thus far, 28 '*Ca.* Phytoplasma' species have been named, based on criteria recommended by the International Research Program on Comparative Mycoplasmology, Phytoplasma/Spiroplasma Working Team – Phytoplasma Taxonomy Group (IRPCM, 2004). However, a formal phytoplasma taxonomy system and eventual abandonment of the '*Candidatus* (*Ca.*) Phytoplasma' species convention is a goal for members of the International Phytoplasma Working Group.

The advance in phylogenetic studies and completion of numerous genome-sequencing projects in cultured bacteria have yielded much insight into genomic organizations and primary components that determine the genomic diversity and phenotypic properties of the bacterial kingdom. As a result, the perspectives for modern taxonomy in prokaryotes have changed (Woese *et al.*, 1980; Woese, 1987, 2000; Weisburg *et al.*, 1989; Murray *et al.*, 1990; Vandamme *et al.*, 1996; Razin *et al.*, 1998; Stackebrandt *et al.*, 2002; Brown *et al.*, 2007; Stackebrandt, 2007). There is a consensus that the highly conserved 16S rRNA gene could be employed as a primary phylogenetic parameter for bacteria speciation to replace cumbersome conventional procedures based on DNA–DNA homology (http://www.bergeys.org/; Brown *et al.*, 2007). Based on 16S rRNA gene sequences available at the time, Stackebrandt and Goebel (1994) noted that bacteria which share <97% sequence identity will not yield a total genomic DNA reassociation of >60%, regardless of the DNA–DNA hybridization methods used. This finding suggests that 97% identity based on 16S rRNA gene sequence could be a threshold value for defining a bacteria species, replacing the DNA–DNA hybridization conventionally used to estimate DNA homology of bacterial genomes.

As DNA sequencing technology continues to advance, a large number of bacterial genome sequences, including those of phytoplasmas, will become available in the very near future. Through comparative genomic studies, various suites of molecular markers with varying degrees of genetic variability that are related to biochemical, phenotypic and biological properties could be chosen for delineating and defining genus, species and strains. It is anticipated that molecular methods will be the primary means for detection and identification of bacteria in the foreseeable future. Thus, delineation of bacterial species based on molecular means should result in a classification scheme that is phylogenetically valid and less ambiguous than the conventional approach based on somewhat subjective phenotypic criteria. The absolute requirement of obtaining pure bacterial culture in artificial media for formal nomenclature of bacterial species becomes a moot point. Molecular delineation can be accomplished solely based on genomic sequence data without the need for a live bacterial culture.

The goal to eventually establish a formal phytoplasma taxonomy by a molecular-based system may be feasible after all. The 97% 16S rRNA gene sequence identity as the threshold value for defining cultured bacteria should also be applicable for uncultured phytoplasmas. Many of the proposed '*Ca.* Phytoplasma' species, especially those whose complete genome sequences are available, are justified as being designated as formal species. Evidence has indicated that using the 97% identity as the threshold for electing new species will potentially exclude many strains that do not meet the criterion but warrant species designation based on unique biological properties (Fox *et al.*, 1992). Bacteria speciation based on multiple phylogenetic parameters may overcome the deficiency of using the 16S rRNA gene as the sole parameter. Selected suites of multiple markers will eventually afford definitions of phytoplasma at strain, species or higher level. Although formal establishment of a molecular-based taxonomic system for bacteria may not be realized in the near future, the molecular-based system remains our major focus for differentiation and classification of phytoplasma strains because we cannot culture phytoplasmas in artificial media.

References

Al-Saady, N.A., Khan, A.J., Calari, A., Al-Subhi, A.M. and Bertaccini, A. (2008) '*Candidatus* Phytoplasma omanense', associated with witches' broom of *Cassia italica* (Mill.) Spreng in Oman. *International Journal of Systematic and Evolutionary Microbiology* 58, 461–466.

Andersen, M.T., Newcomb, R.D., Liefting, L.W. and Beever, R.E. (2006) Phylogenetic analysis of '*Candidatus* Phytoplasma australiense' reveals distinct populations in New Zealand. *Phytopathology* 96, 838–845.

Angelini, E., Clair, D., Borgo, M., Bertaccini, A. and Boudon-Padieu, E. (2001) Flavescence dorée in France and Italy. Occurrence of closely related phytoplasma isolates and their near relationships to Palatinate grapevine yellows and an alder yellows phytoplasma. *Vitis* 40, 79–84.

Arnaud, G., Malembic-Maher, S., Salar, P., Bonnet, P., Maixner, M., Marcone, C., Boudon-Padieu, E. and Foissac, X. (2007) Multilocus sequence typing confirms the close genetic interrelatedness of three distinct Flavescence dorée phytoplasma strain clusters and group 16SrV phytoplasmas infecting grapevine and alder in Europe. *Applied and Environmental Microbiology* 73, 4001–4010.

Arocha, Y., López, M., Piñol, B., Fernández, M., Picornell, B., Almeida, R., Palenzuela, I., Wilson, M. and Jones, P. (2005) '*Candidatus* Phytoplasma graminis' and '*Candidatus* Phytoplasma caricae', two novel phytoplasmas associated with disease of sugarcane, weeds and papaya in Cuba. *International Journal of Systematic and Evolutionary Microbiology* 55, 2451–2463.

Bai, X., Zhang, J., Ewing, A., Miller, S.A., Radek, A.J., Shevchenko, D.V., Tsukerman, K., Walunas, T., Lapidus, A. *et al.* (2006) Living with genome instability, the adaptation of phytoplasmas to diverse environments of their insect and plant hosts. *Journal of Bacteriology* 188, 3682–3696.

Bertaccini, A. (2007) Phytoplasmas, diversity, taxonomy, and epidemiology. *Frontiers of Bioscience* 12, 673–689.

Botti, S. and Bertaccini, A. (2007) Grapevine yellows in northern Italy, molecular identification of Flavescence dorée phytoplasma strains and of Bois Noir phytoplasmas. *Journal of Applied Microbiology* 103, 2325–2330.

Brown, D., Whitcomb, R.F. and Bradbury, J.M. (2007) Revised minimal standard for description of new species of the class Mollicutes (division Tenericutes). *International Journal of Systematic and Evolutionary Microbiology* 57, 2703–2719.

Davis, R.E., Jomantiene, R., Zhao, Y. and Dally, E.L. (2003) Folate biosynthesis pseudogenes, *ψfolP* and *ψfolK*, and an O-sialoglycoprotein endopeptidase gene homolog in the phytoplasma genome. *DNA Cell Biology* 22, 697–706.

Fox, G.E., Wisotzkey, J.D. and Jurtshuk, P. (1992) How close is close? 16S rRNA sequence identity may not be sufficient to guarantee species identity. *International Journal of Systematic Bacteriology* 42, 166–170.

Griffiths, H.M., Sinclair, W.A., Boudon-Padieu, E., Daire, X., Lee, I.-M., Sfalanga, A. and Bertaccini, A. (1999) Phytoplasmas associated with elm yellows. Molecular variability and differentiation from related organisms. *Plant Disease* 83, 1101–1104.

Gundersen, D.E. and Lee, I.-M. (1996) Ultrasensitive detection of phytoplasmas by nested-PCR assays using two universal primer pairs. *Phytopathologia Mediterranea* 35, 144–151.

Gundersen, D.E., Lee, I.-M., Rehner, S.A., Davis, R.E. and Kingsbury, D.T. (1994) Phylogeny of mycoplasmalike organisms (phytoplasmas), a basis for their classification. *Journal of Bacteriology* 176, 5244–5254.

Hodgetts, J., Boonham, N., Mumford, R., Harrison, N. and Dickinson, M. (2008) Phytoplasma phylogenetics based on analysis of secA and 23S rRNA gene sequences for improved resolution of candidate species of '*Candidatus* Phytoplasma'. *International Journal of Systematic and Evolutionary Microbiology* 58, 1826–1837.

Hogenhout, S.A., Oshima, K., Ammar, el-D., Kakizawa, S., Kingdom, H.N. and Namba, S. (2008) Phytoplasmas, bacteria that manipulate plants and insects. *Molecular Plant Pathology* 9, 403–423.

Iriti, M., Quaglino, F., Maffi, D., Casatti, P., Bianco, P.A. and Faoro, F. (2008) *Solunum malacoxylon*, a new natural host of *stolbur* phytoplasma. *Journal of Phytopathology* 156, 8–14.

IRPCM Phytoplasma/Spiroplasma Working Team – Phytoplasma Taxonomy Group (2004) '*Candidatus* Phytoplasma', a taxon for the wall-less, non-helical prokaryotes that colonize plant phloem and insects. *International Journal of Systematic and Evolutionary Microbiology* 54, 1243–1255.

Kube, M., Schneider, B., Kuhl, H., Dandekar, T., Heitmann, K., Migdoll, A.M., Reinhardt, R. and Seemüller, E. (2008) The linear chromosome of the plant-pathogenic mycoplasma '*Candidatus* Phytoplasma mali'. *BMC Genomics* 9, 306.

Langer, M. and Maixner, M. (2004) Molecular characterization of grapevine yellows associated with phytoplasmas of stolbur group based on RFLP-analysis of non-ribosomal DNA. *Vitis* 43, 185–189.

Lee, I.-M., Hammond, R.W., Davis, R.E. and Gundersen, D.E. (1993) Universal amplification and analysis of pathogen 16S rDNA for classification and identification of mycoplasmalike organisms. *Phytopathology* 83, 834–842.

Lee, I.-M., Gundersen-Rindal, D.E., Davis, R.E. and Bartoszyk, I.M. (1998) Revised classification scheme of phytoplasmas based on RFLP analyses of 16SrRNA and ribosomal protein gene sequences. *International Journal of Systematic Bacteriology* 48, 1153–1169.

Lee, I.-M., Davis, R.E. and Gundersen-Rindal, D.E. (2000) Phytoplasma, phytopathogenic mollicutes. *Annual Review of Microbiology* 54, 221–255.

Lee, I.-M., Martini, M., Marcone, C. and Zhu, S.F. (2004a) Classification of phytoplasma strains in the elm yellows group (16SrV) and proposition of '*Candidatus* Phytoplasma ulmi' for the phytoplasma associated with elm yellows. *International Journal of Systematic and Evolutionary Microbiology* 54, 337–347.

Lee, I.-M., Gundersen, D.E., Davis, R.E., Bottner, K.D., Marcone, C. and Seemüller, E. (2004b) '*Candidatus* Phytoplasma asteris', a novel phytoplasma taxon associated with aster yellows and related diseases. *International Journal of Systematic and Evolutionary Microbiology* 54, 1037–1048.

Lee, I.-M., Bottner, K.D., Secor, G. and Rivera-Varas, V. (2006a) '*Candidatus* Phytoplasma americanum', a phytoplasma associated with a purple top wilt disease complex. *International Journal of Systematic and Evolutionary Microbiology* 56, 1593–1597.

Lee, I.-M., Zhao, Y. and Bottner, K.D. (2006b) SecY gene sequence analysis for finer differentiation of diverse strains in the aster yellows phytoplasma group. *Molecular Cell Probes* 20, 87–91.

Leyva-López, N.E., Ochoa-Sánchez, J.C., Leal-Klevezas, D.S. and Martinez-Soriano, J.P. (2002) Multiple phytoplasmas associated with potato diseases in Mexico. *Canadian Journal of Microbiology* 48, 1062–1068.

Lim, P.O. and Sears, B.B. (1989) 16S rRNA sequence indicates that plant-pathogenic mycoplasmalike organisms are evolutionarily distinct from animal mycoplasmas. *Journal of Bacteriology* 171, 5901–5906.

Marcone, C., Lee, I.-M., Davis, R.E., Ragozzino, A. and Seemüller, E. (2000) Classification of aster yellows-group phytoplasmas based on combined analyses of rRNA and *tuf* gene sequences. *International Journal of Systematic and Evolutionary Microbiology* 50, 1703–1713.

Martini, M., Botti, S., Marcone, C., Marzachì, C., Casati, P., Bianco, P.A., Benedetti, R. and Bertaccini, A. (2002) Genetic variability among Flavescence dorée phytoplasmas from different origins in Italy and France. *Molecular Cell Probes* 16, 197–208.

Martini, M., Lee, I.-M., Bottner, K.D., Zhao, Y., Botti, S., Bertaccini, A., Harrison, N.A., Carraro, L., Marcone, C. and Osler, R. (2007) Ribosomal protein gene-based phylogeny for finer differentiation and classification of phytoplasmas. *International Journal of Systematic and Evolutionary Microbiology* 57, 2037–2051.

Montano, H.G., Davis, R.E., Dally, E.L., Hogenhout, S., Pimentel, P. and Brioso, P.S.T. (2001) '*Candidatus* Phytoplasma brasiliense', a new phytoplasma taxon associated with hibiscus witches'-broom disease. *International Journal of Systematic and Evloutionary Bacteriology* 51, 1109–1118.

Murray, R.G.E. and Schleifer, K.H. (1994) Taxonomic notes, a proposal for recording the properties of putative taxa of prokaryotes. *International Journal of Systematic Bacteriology* 44, 174–176.

Murray, R.G.E., Brenner, D.J., Colwell, R.R., De Vos, P., Goodfellow, M., Grimont, P.A.D., Pfennig, N., Stackebrandt, E. and Zavarzin, G.A. (1990) Report of the ad hoc committee on approaches to taxonomy within the proteobacteria. *International Journal of Systematic Bacteriology* 40, 213–215.

Namba, S., Kato, S., Iwanami, S., Shiozawa, H. *et al.* (1993) Phylogenetic diversity of phytopathogenic organisms. *International Journal of Systematic Bacteriology* 43, 461–467.

Oshima, K., Kakizawa, S., Nishigawa, H., Jung, H.-Y., Wei, W., Suzuki, S., Arashid, A.R., Nakata, D., Miyata, S. *et al.* (2004) Reductive evolution suggested from the complete genome sequence of a plant pathogenic phytoplasma. *Nature Genetics* 36, 27–29.

Pacifico, D., Foissac, X., Veratti, F. and Marzachi, C. (2007) Genetic diversity of Italian and Frech 'bois noir' phytoplasma isolates. *Bulletin of Insectology* 60(2), 345–346.

Padovan, A., Gibb, K. and Persley, D. (2000) Association of '*Candidatus* Phytoplasma australiense' with green petal and lethal yellows diseases in strawberry. *Plant Pathology* 49, 362–369.

Razin, S., Yogev, D. and Naot, Y. (1998) Molecular biology and pathology of mycoplasmas. *Microbiology and Molecular Biology Review* 62, 1094–1156.

Riolo, P., Landi, L., Nardi, S. and Isidoro, N. (2007) Relationships among *Hyalesthes obsoletus*, its herbaceous host plants and 'bois noir' phytoplasma strains in vineyard ecosystems in the Marche region (central-eastern Italy). *Bulletin of Insectology* 60(2), 353–354.

Schneider, B., Aherns, U., Kirkpatrick, B.C. and Seemüller, E. (1993) Classification of plant-pathogenic mycoplasma-like organisms using restriction-site analysis of PCR-amplified 16S rDNA. *Journal of General Microbiology* 139, 519–527.

Schneider, B., Gibb, K.S. and Seemüller, E. (1997) Sequence and RFLP analysis of the elongation factor Tu gene used in differentiation and classification of phytoplasmas. *Microbiology* 143, 3381–3389.

Seemüller, E., Marcone, C., Lauer, U., Ragozzino, A. and Gőschl, M. (1998) Current status of molecular classification of the phytoplasmas. *Journal of Plant Pathology* 80, 3–26.

Shao, J., Jomantiene, R., Dally, E.L., Zhao, Y., Lee, I.-M., Nuss, D.L. and Davis, R.E. (2006) Phylogeny and characterization of phytoplasmal NusA and use of the *nusA* gene in detection of group 16SrI strains. *Journal of Plant Pathology* 88, 193–201.

Smart, C.D., Schneider, B., Blomquist, C.L., Guerra, L.J., Harrison, N.A. *et al.* (1996) Phytoplasma-specific PCR primers based on sequences of 16S–23S rRNA spacer region. *Applied and Environmental Micorbiology* 62, 2988–2993.

Stackebrandt, E. (2007) Forces shaping bacterial systematics. *Microbe* 2, 283–288.

Stackebrandt, E. and Goebel, B.M. (1994) Taxonomic note, a place for DNA–DNA reassociation and 16S rRNA sequence analysis in the present species definition in bacteriology. *International Journal of Systematic Bacteriology* 44, 846–849.

Stackebrandt, E., Frederiksen, W., Garrity, G.M., Grimont, P.A., Kampfer, P., Maiden, M.C., Nesme, X., Rosselló-Móra, R., Swings, J., Truper, H.G., Vauterin, L., Ward, A.C. and Whitman, W.B. (2002) Report of the ad hoc committee for the re-evaluation of the species definition in bacteriology. *International Journal of Systematic and Evolutionary Microbiology* 52, 1043–1047.

Streten, C. and Gibb, K.S. (2005) Genetic variation in *Candidatus* Phytoplasma australiense. *Plant Pathology* 54, 8–14.

Tran-Nguyen, L.T., Kube, M., Schneider, B., Reinhardt, R. and Gibb, K.B. (2008) Comparative genome anlysis of '*Candidatus* Phytoplasma australiense' (subgroup tuf-Australia I; rp-A) and '*Ca.* Phytoplasma asteris' strains OY-M and AY-WB. *Journal of Bacteriology* 190, 3979–3991.

Vandamme, P., Pot, B., Gillis, M., Devos, P., Kersters, K. and Swings, J. (1996) Polyphasic taxonomy, a consensus approach to bacterial systematics. *Microbiology Review* 60, 407–438.

Wei, W., Davis, R.E., Lee, I.-M. and Zhao, Y. (2007) Computer-simulated RFLP analysis of 16S rRNA genes, identification of ten new phytoplasma groups. *International Journal of Systematic and Evolutionary Microbiology* 57, 1855–1867.

Wei, W., Lee, I.-M., Davis, R.E., Suo, X. and Zhao, Y. (2008) Automated RFLP pattern comparison and similarity coefficient calculation for rapid delineation of new and distinct phytoplasma 16Sr subgroup lineages. *International Journal of Systematic and Evolutionary Microbiology* 58, 2368–2377.

Weisburg, W.G., Tully, J.G., Rose, D.L., Petzel, J.P. and Oyaizu, H. *et al.* (1989) A phylogenetic analysis of the mycoplasmas, basis for their classification. *Journal of Bacteriology* 171, 6455–6467.

Woese, C.R. (1987) Bacterial evolution. *Microbiology Review* 51, 221–271.

Woese, C.R. (2000) Interpreting the universal phylogenetic tree. *Proceedings of the National Academy of Sciences of the USA* 97, 8392–8396.

Woese, C.R., Maniloff, J. and Zablen, L.B. (1980) Phylogenetic analysis of the mycoplasmas. *Proceedings of the National Academy of Sciences of the USA* 77, 494–498.

Zhao, Y., Davis, R.E. and Lee, I.M. (2005) Phylogenetic positions of '*Candidatus* Phytoplasma asteris' and *Spiroplasma kunkelii* as inferred from multiple sets of concatenated core housekeeping proteins. *International Journal of Systematic and Evolutionary Microbiology* 55, 2131–2141.

Zhao, Y., Sun, Q., Wei, W., Davis, R.E., Wu, W. and Liu, Q. (2009a) '*Candidatus* Phytoplasma tamaricis', a novel taxon discovered in witches'-broom diseased salt cedar (*Tamarix chinesis* Lour.). *International Journal of Systematic and Evolutionary Microbiology* 59 (in press).

Zhao, Y., Wei, W., Lee, I.-M., Shao, J., Suo, X. and Davis, R.E. (2009b) Construction of an interactive online phytoplasma classification tool, *i*PhyClassifier, and its application in analysis of the peach X-disease phytoplasma group (16SrIII). *International Journal of Systematic and Evolutionary Microbiology* 59 (in press).

5 Recent Advances in 16S rRNA Gene-based Phytoplasma Differentiation, Classification and Taxonomy

YAN ZHAO, WEI WEI, ROBERT E. DAVIS AND ING-MING LEE

Molecular Plant Pathology Laboratory, Beltsville, USA

Introduction

The means and ways by which microorganisms are characterized and classified evolve constantly, in line with conceptual innovations and technological advances. The history of bacterial systematics has witnessed gradual transformation of the discipline from superficial descriptions of limited observable phenotypic characters to a consensus polyphasic, multidisciplinary study that involves morphological, physiological, chemotaxonomic, serologic and genotypic characterizations (Stackebrandt and Goebel, 1994; Vandamme *et al.*, 1996; Rosselló-Mora, 2005; Stackebrandt, 2007). As exemplified by DNA–DNA reassociation kinetic analysis (Wayne *et al.*, 1987), 16S ribosomal RNA (rRNA) gene-based phylogenic tree reconstruction (Woese, 1987) and whole-genome average nucleotide identity (ANI) and average amino acid identity (AAI) studies (Konstantinidis and Tiedje, 2005a, b), the advent and creative applications of molecular and genomic technologies have fundamentally changed the landscape of bacterial systematics. Such advances have made it possible to determine objectively genetic relatedness among bacterial species and to draw distinctions between closely related strains. While the ultimate genotypic characterization relies on complete genome sequences (Coenye *et al.*, 2005), genes encoding 16S rRNA provide an easy and reliable access to critical information that validly represents organismal genealogy (Woese, 2000). 16S rRNA gene sequence information has become, and will remain for the foreseeable future, essential for taxonomic assignment and classification of prokaryotes (Brenner *et al.*, 2000; Stackebrandt *et al.*, 2002; Yarza *et al.*, 2008; *Bergey's Manual of Systematic Bacteriology* at http://www.bergeys.org/).

Phytoplasmas, characterized by possessing small, AT-rich genomes and living a transkingdom parasitic lifestyle, are non-helical, mycoplasma-like, cell-wall-less bacteria known to be pathogenic to more than a thousand plant species (Doi *et al.*, 1967; McCoy *et al.*, 1989; Marcone *et al.*, 1999; Lee *et al.*, 2000;

 Phytoplasmas: Genomes, Plant Hosts and Vectors (eds P. G. Weintraub and P. Jones)

Seemüller *et al.*, 2002; Hogenhout *et al.*, 2008). In infected plants, phytoplasmas inhabit sieve cells of phloem tissue and induce disease symptoms involving disrupted hormonal balance, impaired amino acid and carbohydrate translocation, inhibited photosynthesis and rapid senescence (Chang, 1998; Lepka *et al.*, 1999; Bertamini *et al.*, 2002a, b; Curković-Perica *et al.*, 2007). In their natural insect vectors, mainly leafhoppers and psyllids, phytoplasmas traverse the intestinal wall, circulate in haemolymph, and multiply in diverse tissues, including salivary glands, where phytoplasma cells become incorporated into saliva and subsequently injected into plants during feeding in phloem (Tsai, 1979; Seemüller *et al.*, 2002).

Phylogenetic studies of genes encoding 16S rRNA and a large set of conserved core housekeeping proteins suggest that extant phytoplasmas share a common evolutionary root and are descended from low G+C Gram-positive bacteria in the *Bacillus–Clostridium* group (Woese *et al.*, 1980; Weisburg *et al.*, 1989; Gundersen *et al.*, 1994; Sears and Kirkpatrick, 1994; Zhao *et al.*, 2005; Wei *et al.*, 2007b). After evolutionary divergence from an acholeplasma-like last common ancestor, phytoplasmas emerged as a discrete clade, and a large number of widely divergent phytoplasma lineages have evolved in adaptation to a broad range of bio- and geo-ecological niches (Gundersen *et al.*, 1994; Lee *et al.*, 2000; Davis *et al.*, 2005; Wei *et al.*, 2007b).

Biodiversity of phytoplasmas has long been recognized, even at the time when the aetiological agents of phytoplasmal diseases (yellows diseases) were mistakenly assumed to be viruses. Initially, differentiation of phytoplasmas or presumed 'yellows viruses' was based on the geographic origins of the diseases, on the identity of specific plant hosts and insect vectors, and on symptoms exhibited by diseased plants (Chiykowski, 1962; Freitag, 1964; Granados and Chapman, 1968; Chiykowski and Sinha, 1989; McCoy *et al.*, 1989). Given that the same phytoplasma strain may induce different symptoms in different hosts, and different phytoplasma strains may share a common vector(s) or cause diseases exhibiting similar symptoms, this 'guilty by affiliation' approach could not provide an accurate means for phytoplasma classification.

Early insights into genetic interrelationships among phytoplasma strains and current knowledge on evolution and phylogeny of the phytoplasma clade have been derived from DNA hybridization studies and molecular analyses of evolutionarily conserved sequences, especially genes encoding 16S rRNAs (Kirkpatrick and Fraser, 1989; Deng and Hiruki, 1991; Lee *et al.*, 1993, 1998, 2000; Davis and Sinclair, 1998; Jomantiene *et al.*, 1998, 2002; Wei *et al.*, 2007b, 2008b). Completion of genome sequencing of four phytoplasma strains (Oshima *et al.*, 2004; Bai *et al.*, 2006; Kube *et al.*, 2008; Tran-Nguyen *et al.*, 2008) promises new insights into phytoplasma genome organization and genetic diversity. The recent discovery of prophages and phage-derived genomic islands in phytoplasma genomes (Wei *et al.*, 2008a) and their relation to the dynamic sequence-variable mosaic structures (Jomantiene and Davis, 2006; Jomantiene *et al.*, 2007) sheds fresh light on the launch and evolution of the phytoplasma clade and unveils a major force driving phytoplasma genetic diversity. This chapter will focus on recent advances in 16S rRNA gene-based phytoplasma differentiation, classification and taxonomy. Molecular markers derived from non-16S rRNA gene sequences and emerging systems based on multilocus analyses for finer differentiation of closely related phytoplasmas are reviewed in Chapter 4, this volume.

'*Candidatus* Phytoplasma': a Provisional Genus-level Taxon and Beyond

In recent years, new phytoplasmas have been discovered at an increasingly rapid pace, in association with numerous emerging and re-emerging plant diseases worldwide. A Google Scholar search (using phytoplasma OR 'mycoplasmalike organism' as a search term) revealed that phytoplasma has been a subject of over 5600 scientific articles, of which 4590 appear in professional media with recorded publication dates. More than 2100 of the 4590 articles were published since 2004 (Fig. 5.1). At the time of writing, 1546 phytoplasma 16S rRNA gene sequences have been registered in and released by the GenBank (including sequences submitted through the European Molecular Biology Laboratory (EMBL), the DNA DataBank of Japan (DDBJ) and the National Center for Biotechnology Information (NCBI), USA), and this number has more than doubled in less than 3 years (Fig. 5.1). While phytoplasmas are a truly unique, coherent group of unambiguously identifiable plant-pathogenic mollicutes, attempts to isolate and cultivate phytoplasmas in cell-free media remain unsuccessful. Due to the inaccessibility of measurable phenotypic characters, differentiation and classification of phytoplasmas have been excluded from taking the consensus polyphasic taxonomic approach and

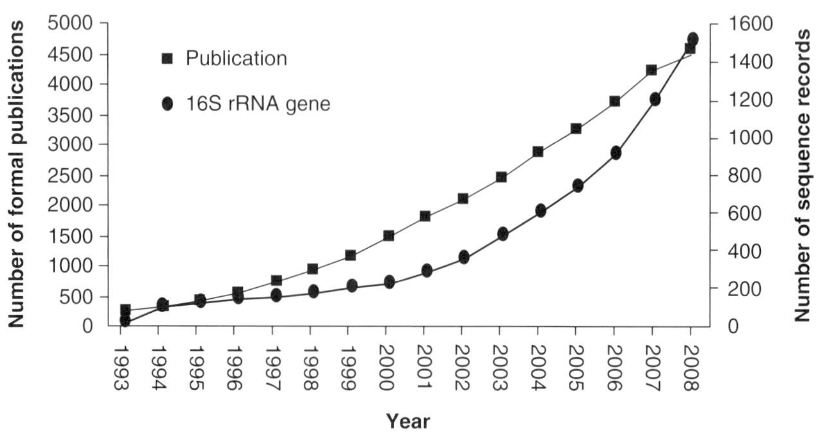

Fig. 5.1. Increase in the number of phytoplasma research publications and the number of phytoplasmal 16S rRNA gene sequence records in the GenBank from 1993 to 2008.

from using physiological- and biochemical-based phenotypic criteria that are routinely used for culturable microorganisms.

To resolve the impediments of phytoplasma taxonomy, in 2004, based on consensus among phytoplasmologists and in agreement with the International Committee of Systematic Bacteriology Subcommittee for the Taxonomy of Mollicutes, the International Research Program for Comparative Mycoplasmology, Phytoplasma/Spiroplasma Working Team – Phytoplasma Taxonomy Group adopted a taxonomic rule that had been established for recording properties of uncultured organisms (Murray and Schleifer, 1994; Murray and Stackebrandt, 1995) and proposed to erect a genus-level provisional taxon '*Candidatus* (*Ca.*) Phytoplasma' to accommodate plant-pathogenic, non-helical mollicutes (IRPCM, 2004). In order to prevent nomenclatural confusion that may arise from description of poorly differentiated taxa, the Working Team also established guidelines for naming new taxa within the genus '*Ca.* Phytoplasma': a novel '*Ca.* Phytoplasma' species description should refer to a single, unique 16S rRNA gene sequence of greater than 1200 bp and share less than 97.5% sequence similarity to that of any previously described '*Ca.* Phytoplasma' species unless the phytoplasma under consideration clearly represents an ecologically separated population (IRPCM, 2004; Firrao *et al.*, 2005). The 16S rRNA gene of a novel '*Ca.* Phytoplasma' species should possess at least one unique sequence region in addition to the signature sequence that is characteristic of phytoplasmas: 5´-CAA GAYBATKATGTKTAGCYGGDCT-3´. This phytoplasma signature sequence corresponds to the annealing site of phytoplasmal universal primer R16F2n, which, together with primer R16R2, has served as a basic tool for phytoplasma-specific 16S rDNA amplification (Gundersen and Lee, 1996). The phytoplasma strain whose 16S rRNA gene is used to describe a '*Ca.* Phytoplasma' species is called the 'reference strain'; strains whose 16S rRNA genes have even minimal difference from the corresponding reference strain are referred as 'related strains' (IRPCM, 2004). Unique identifiers other than the 16S rRNA gene and specific host information should also be included in a new species description. A novel '*Ca.* Phytoplasma' species can be named after the place or the plant host where it was discovered. Examples of '*Ca.* Phytoplasma' species' descriptions are given below:

'*Candidatus* Phytoplasma americanum' (a.mer.i.can'um. N.L. neut. adj. *americanum* pertaining to America):

APPTW12-NE is the reference strain. Related phytoplasma strains include APPTW1-TX, APPTW2-TX, APPTW9-NE, APPTW10-NE, APPTW13-NE and APPTW 1833 #6-TX, which are associated with potatoes exhibiting purple top syndrome in Texas and Nebraska, USA.

((*Mollicutes*) NC; NA; O, wall-less; NAS (GenBank accession number DQ-174122), oligonucleotide sequences of unique regions of the 16S rRNA gene are 5´-GTTTCTTCGGAAA-3´ (68–80), 5´-GTTAGAAATGACT-3´ (142–153) and 5´-GCTGGTGGCTT-3´ (1438–1448); P (*Solanum tuberosum*, phloem); M) (Lee *et al.*, 2006).

'*Candidatus* Phytoplasma tamaricis' (N.L. gen. n. *tamaricis* of *tamarix* the scientific name of salt cedar; epithet referring to the plant host).

Reference strain is SCWB1[R].

((*Mollicutes*) NC; NA; O, wall-less; NAS (GenBank accession number FJ432664), oligonucleotide sequences of unique regions of the 16S rRNA gene are: 5´-ATTAGGCATCTAGTAACTTTG-3´, 5´-TGCTCAACATTGTTGC-3´, 5´-AGCTTTGCAAAGTTG-3´, and 5´-TAACAGAGGTTATCAGAGTT-3´; P (*Tamarix chinensis*, phloem); M) (Zhao *et al.*, 2009a).

Thus far, 28 '*Ca.* Phytoplasma' species have been formally described (Table 5.1 and references therein). While most of the described '*Ca.* Phytoplasma' species possess a 16S rRNA gene that shares less than 97.5% sequence identity with that of other species, there are a few exceptions. For example, pairwise 16S rRNA gene sequence identity scores among '*Ca.* Phytoplasma mali', '*Ca.* Phytoplasma pyri' and '*Ca.* Phytoplasma prunorum' are between 98.6% and 99.0%. These three species were named on the basis of their distinctive biological properties. As causative agents of economically devastating fruit tree diseases apple proliferation, pear decline and European stone fruit yellows, respectively, '*Ca.* Phytoplasma mali', '*Ca.* Phytoplasma pyri' and '*Ca.* Phytoplasma prunorum' have different natural plant host ranges and each induces a different set of characteristic symptoms in infected plants. They are transmitted preferentially by different vectors and differ in serological properties as well as in several key molecular markers, including the 16S–23S rDNA spacer region and randomly cloned protein-encoding genes (Seemüller and Schneider, 2004). Therefore, the three phytoplasmas clearly represent mutually distinct and ecologically separated populations. Similarly, '*Ca.* Phytoplasma ulmi' and '*Ca.* Phytoplasma ziziphi' share 99.0% 16S rRNA gene sequence similarity but have been recognized as two distinct species because the two phytoplasmas occupy different ecological niches and exhibit strikingly different geographical distributions (Jung *et al.*, 2003a; Lee *et al.*, 2004b). To claim that a phytoplasma whose 16S rRNA gene shares greater than 97.5% sequence similarity with a previously named '*Ca.* Phytoplasma' species represents an ecologically separated population (thus a new species), it is required to pass all three critical tests: the two phytoplasma populations have to differ in host specificity, in vectorship and in at least one molecular or serological marker (IRPCM, 2004).

In addition to the 28 formally described '*Ca.* Phytoplasma' species, 15 other phytoplasma strains have been found either to possess 16S rRNA genes that contain the signature sequence characteristic of phytoplasmas and share less than 97.5% sequence similarity with each other and with that of any previously described species or to exhibit unique biological properties that differ substantially from those previously described species (Table 5.1). Suggestions that these phytoplasma strains fulfil the minimal requirement for a novel species description (IRPCM, 2004; Wei *et al.*, 2007b) stimulate detailed studies of these phytoplasmas and open opportunities for describing new '*Ca.* Phytoplasma' species, which will, in turn, extend our knowledge on the breadth of the provisional genus and the genetic diversity of phytoplasmas.

An alignment of 16S rRNA gene sequences from the reference strains of 28 formally described and 15 potentially new '*Ca.* Phytoplasma' species

Table 5.1. '*Candidatus* Phytoplasma' species.

Species name	Reference strain	GenBank no.	Reference
Formally described '*Ca.* Phytoplasma' species			
'*Ca.* Phytoplasma asteris'	OAY	M30790	Lee *et al.*, 2004a
'*Ca.* Phytoplasma aurantifolia'	WBDL	U15442	Zreik *et al.*, 1995
'*Ca.* Phytoplasma australasiae'	PpYC	Y10097	White *et al.*, 1998
'*Ca.* Phytoplasma.ulmi'	EY1	AY197655	Lee *et al.*, 2004b
'*Ca.* Phytoplasma ziziphi'	JWB-G1	AB052876	Jung *et al.*, 2003a
'*Ca.* Phytoplasma trifolii'	CP	AY390261	Hiruki and Wang, 2004
'*Ca.* Phytoplasma fraxini'	AshY1	AF092209	Griffiths *et al.*, 1999
'*Ca.* Phytoplasma phoenicium'	A4	AF515636	Verdin *et al.*, 2003
'*Ca.* Phytoplasma mali'	AP15	AJ542541	Seemüller and Schneider, 2004
'*Ca.* Phytoplasma pyri'	PD1	AJ542543	Seemüller and Schneider, 2004
'*Ca.* Phytoplasma spartii'	SpaWB	X92869	Marcone *et al.*, 2004a
'*Ca.* Phytoplasma prunorum'	ESFY-G1	AJ542544	Seemüller and Schneider, 2004
'*Ca.* Phytoplasma oryzae'	RYD-J	AB052873	Jung *et al.*, 2003b
'*Ca.* Phytoplasma australiense'	AUSGY	L76865	Davis *et al.*, 1997
'*Ca.* Phytoplasma japonicum'	JHP	AB010425	Sawayanagi *et al.*, 1999
'*Ca.* Phytoplasma fragariae'	StrawY	DQ086423	Valiunas *et al.*, 2006
'*Ca.* Phytoplasma cynodontis'	BGWL-C1	AJ550984	Marcone *et al.*, 2004b
'*Ca.* Phytoplasma brasiliense'	HibWB26	AF147708	Montano *et al.*, 2001
'*Ca.* Phytoplasma graminis'	SCYLP	AY725228	Arocha *et al.*, 2005
'*Ca.* Phytoplasma caricae'	PAY	AY725234	Arocha *et al.*, 2005
'*Ca.* Phytoplasma americanum'	APPTW12-NE	DQ174122	Lee *et al.*, 2006
'*Ca.* Phytoplasma castaneae'	CnWB	AB054986	Jung *et al.*, 2002
'*Ca.* Phytoplasma rhamni'	BWB	X76431	Marcone *et al.*, 2004a
'*Ca.* Phytoplasma pini'	Pin127S	AJ632155	Schneider *et al.*, 2005
'*Ca.* Phytoplasma allocasuarinae'	AlloY	AY135523	Marcone *et al.*, 2004a
'*Ca.* Phytoplasma lycopersici'	THP	EF199549	Arocha *et al.*, 2007
'*Ca.* Phytoplasma omanense'	IM-1	EF666051	Al-Saady *et al.*, 2008
'*Ca.* Phytoplasma tamaricis'	SCWB1	FJ432664	Zhao *et al.*, 2009b
Proposed potentially new or incidentally cited species			
'*Ca.* Phytoplasma pruni'†	WX	L04682	IRPCM, 2004
'*Ca.* Phytoplasma palmae'†	LY3	U18747	IRPCM, 2004
'*Ca.* Phytoplasma cocostanzaniae'†	LD	X80117	Tymon *et al.*, 1998; IRPCM, 2004
'*Ca.* Phytoplasma vitis'†	FD	AF176319	IRPCM, 2004
'*Ca.* Phytoplasma luffae'†	LfWB	AF086621	IRPCM, 2004
'*Ca.* Phytoplasma solani'†	STOL	AF248959	IRPCM, 2004
'*Ca.* Phytoplasma cocosnigeriae'†	LDN	Y14175	Tymon *et al.*, 1998; IRPCM, 2004
Mexican periwinkle virescence phytoplasma§	MPV	AF248960	IRPCM, 2004
Chinaberry yellows phytoplasma§	CbY1	AF495882	IRPCM, 2004

(*continued*)

Table 5.1. *continued*

Species name	Reference strain	GenBank no.	Reference
Buckland valley grapevine yellows phytoplasma§	BVGY	AY083605	Constable *et al.*, 2002; Wei *et al.*, 2007b
Sorghum bunchy shoot phytoplasma§	SBS	AF509322	Blanche *et al.*, 2003; Wei *et al.*, 2007b
Weeping tea witches'-broom phytoplasma§	WTWB	AF521672	Wei *et al.*, 2007b
Sugarcane phytoplasma D3T1§	Mauritius D3T1	AJ539179	Wei *et al.*, 2007b
Sugarcane phytoplasma D3T2§	Mauritius D3T2	AJ539180	Wei *et al.*, 2007b
Derbid phytoplasma§	YLS	AY744945	Wei *et al.*, 2007b

†According to Rule 28b of the Bacteriological Code, this is an incidental citation and does not constitute prior citation; §no name has been suggested for this potentially new '*Ca.* Phytoplasma' species.

revealed that pairwise sequence similarity scores ranged from 67.6 to 99.0%. At the high end, the scores above 97.5% were contributed by genetically related but ecologically distinct '*Ca.* Phytoplasma' species, like those mentioned earlier. At the low end, the scores below 80.1% arose from unusual alignment caused by the peculiar 16S rRNA gene sequence of '*Ca.* Phytoplasma lycopersici' (GenBank accession no. EF199549). Excluding these extreme scores at both ends, 16S rRNA genes of the overwhelming majority of the '*Ca.* Phytoplasma' species share 80.1–97.5% sequence similarity with each other. On the other hand, the same set of 16S rRNA genes share 84.5–91.0% sequence similarity with that of *Acholeplasma palmae* (L33734), the closest known relative of phytoplasmas. These scores indicate that phytoplasma lineages have evolved divergently from a common ancestor after the emergence of the phytoplasma clade and leave phytoplasma researchers/taxonomists with a dilemma: either all extant phytoplasma species do not belong to a single genus or *Acholeplasma palmae* is a member of the phytoplasma genus. Resolving this dilemma is an important subject of phytoplasma taxonomy in the near future. Since the 16S rRNA gene-derived phylogenetic tree is a valid representation of organismal genealogy (Woese, 2000), the topology of the phytoplasma subtree seems to hold a clue to this dilemma.

As shown in Fig. 5.2, the phytoplasma clade is divided into three distinct subclades. Within each subclade, phytoplasmas are grouped coherently with a much narrower 16S rRNA gene sequence similarity score range and with higher sequence similarity scores with each other than with *Acholeplasma palmae*. Preliminary studies on dinucleotide relative abundance (DRA) of 16S rDNA sequences also indicate that DNA physical signatures of the phytoplasmas in the three subclades are noticeably different. These results point to a need for further investigation as to whether the three phytoplasma subclades should be ranked at the genus level and, consequently, '*Candidatus* Phytoplasma' should be promoted to a family-level taxon. Availability of more phytoplasma genome sequences will certainly facilitate this investigation

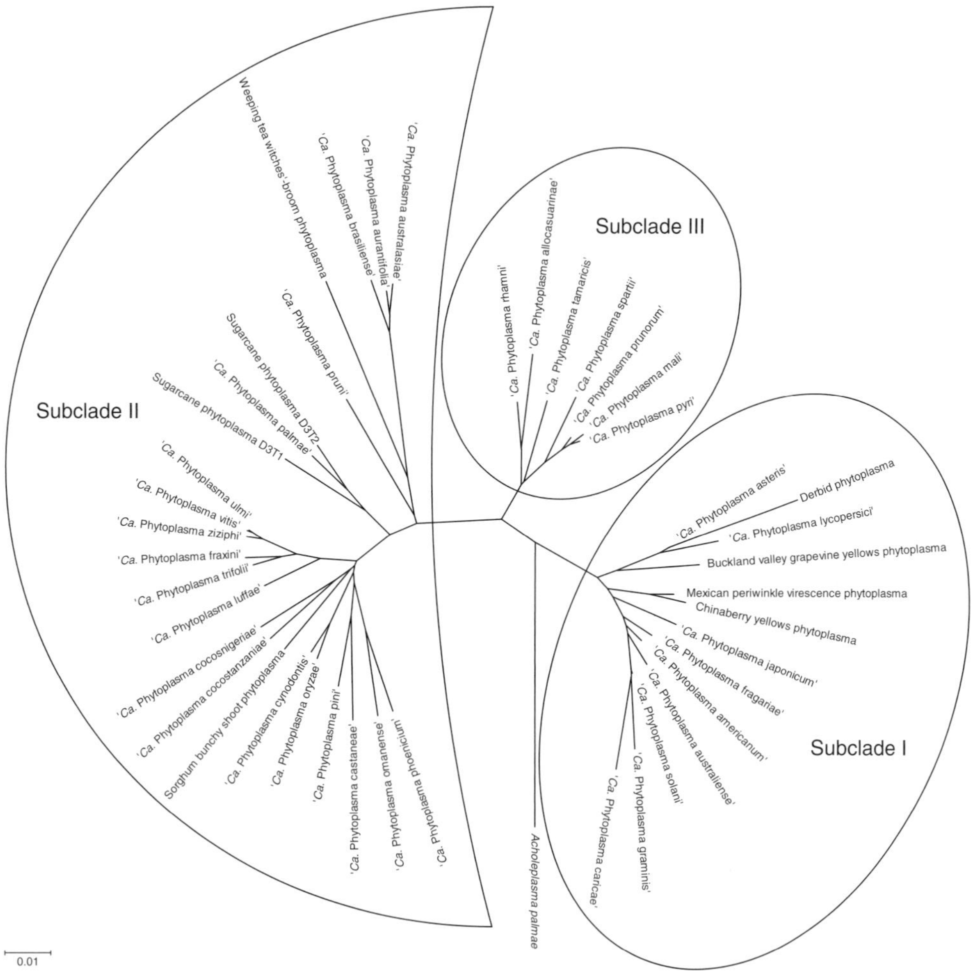

Fig. 5.2. Phylogenetic tree inferred from analysis of 16S rRNA gene sequences. Minimum evolution analysis was conducted using the close neighbour interchange (CNI) algorithm implemented in the Molecular Evolutionary Genetics Analysis program (MEGA4, Tamura *et al.*, 2007). The initial tree for the CNI search was obtained by the neighbour-joining method. The reliability of the analysis was subjected to a bootstrap test with 1000 replicates. The taxa used in the phylogenetic tree reconstruction are reference strains of each '*Candidatus* Phytoplasma' species (see Table 5.1). *Acholeplasma palmae* served as an outgroup during the phylogenetic tree reconstruction. The scale bar represents the number of nucleotide substitutions per site.

by allowing structural comparisons of core housekeeping genes (Zhao *et al.*, 2005), analyses of whole-genome ANI and AAI (Konstantinidis and Tiedje, 2005a, b), identification of phylogenetic footprints of regulatory elements (GuhaThakurta, 2006; Janky and van Helden, 2008) and comparisons of lineage-specific gene content, including decayed and laterally transferred genes (Davis *et al.*, 2003b, 2005; Lee *et al.*, 2005; Wei *et al.*, 2008a).

Group and Subgroup Classification: from Conventional to *In Silico*

In parallel to '*Candidatus*' species assignment, phytoplasmas are classified into groups and subgroups, based on their genetic relatedness. Group and subgroup classification is necessary for proper study of and easy reference to diverse phytoplasma strains, even after the provisional phytoplasma taxonomy is already in place. It is quite obvious that the number of known phytoplasmas and the ones being discovered in fields on a daily basis far exceeds the number of phytoplasmas that can be described as '*Candidatus*' species, and the term 'related strains' cannot sufficiently distinguish or describe multiple strains that represent distinct lineages.

Two major phytoplasma group-classification schemes have been developed and extensively cited in literature; both schemes are based on genetic information coded in 16S rRNA genes. One classification scheme is based on phylogenetic analysis of 16S rRNA gene sequences (Kuske and Kirkpatrick, 1992; Namba *et al.*, 1993; Gundersen *et al.*, 1994; Schneider *et al.*, 1995; Seemüller *et al.*, 1998), and 20 distinct phylogenetic groups have been delineated. The way phytoplasmas are grouped under this classification scheme is in excellent alignment with '*Ca.* Phytoplasma' species delineations (Jung *et al.*, 2002; Firrao *et al.*, 2005). However, this phylogeny-based scheme lacks a mechanism of dealing with subgroup-level classification; therefore, it does not address the needs of differentiating distinct phytoplasma lineages within a phylogenetic group or strain cluster.

A more pragmatic classification scheme is based on restriction fragment length polymorphism (RFLP) analysis of polymerase chain reaction (PCR)-amplified 16S rRNA gene fragments. The idea of using illustrative RFLP patterns to differentiate and classify phytoplasmas was conceived in the early 1990s (Lee *et al.*, 1993; Schneider *et al.*, 1993). With the RFLP profiling approach, Lee and colleagues constructed a comprehensive classification scheme (Lee *et al.*, 1993, 1998, 2000) that made possible the accurate and reliable identification and classification of a wide range of phytoplasmas. Since this PCR-RFLP analysis-based classification scheme exploits an adequate subset of characters present in the 16S rRNA genes, namely recognition sites of a defined set of 17 restriction enzymes, phytoplasma groups delineated using this scheme (16Sr groups) are consistent with 16S rRNA gene phylogeny. Favourably, this RFLP analysis-based scheme offers a mechanism, by distinguishing subtle pattern differences, to identify and differentiate distinct subgroup lineages among phytoplasmas within individual groups. Over the last 15 years, this PCR-RFLP analysis-based classification system has gradually gained acceptance among phytoplasmologists and stimulated phytoplasma research. Since the inception of the scheme, 19 groups (16SrI through 16SrXVIII, 16SrXXIX) and more than 40 subgroups have been delineated through conventional RFLP analysis (Gundersen *et al.*, 1994; Lee *et al.*, 1998, 2000, 2004a, b; Al-Saady *et al.*, 2008). Recently, a significant expansion of the scheme has been achieved through the use of computer-simulated RFLP analysis (Wei *et al.*, 2007b, c, 2008b; Cai *et al.*, 2008; Quaglino *et al.*, 2009; Zhao *et al.*, 2009a, b).

Computer-simulated RFLP analysis or virtual RFLP analysis emerged as an alternative approach for phytoplasma differentiation and classification at a time when the conventional RFLP analysis met with new challenges. In the last few years, emerging phytoplasmal diseases worldwide brought numerous new phytoplasmas into light and raised expectations that the number of 16S rRNA gene RFLP groups (16Sr groups) and subgroups could rise considerably, warranting expansion of the existing classification scheme. However, attempts to update the classification scheme using conventional RFLP analysis were hindered by lack of a complete or near-complete collection of phytoplasma strains as sources of DNA and by increasing difficulty of visual comparisons of multiple RFLP patterns, emphasizing the need for a method to circumvent the obstacles. On the other hand, recent technological advancements paved the way for the virtual RFLP approach to update the phytoplasma classification scheme: the cost of DNA sequencing reduced dramatically while the accuracy of the sequencing data improved significantly, and software-engineering tools became available to develop novel bioinformatic programs for handling nucleotide sequence data.

By mimicking laboratory restriction enzyme digestion and subsequent gel electrophoresis, computer-simulated 16S rDNA analysis produces virtual RFLP patterns, allowing high-throughput differentiation and identification of phytoplasma strains. The virtual RFLP analysis approach was a success from the beginning. In the first trial, through analysis of the then-available 800 phytoplasmal 16S rRNA gene sequences, Wei *et al.* (2007b) were able to delineate ten new phytoplasma groups (16SrXIX through 16SrXXVIII) and to classify hundreds of previously unclassified strains in a single study. Expansion of the classification scheme to include the ten new groups was justified by their distinct 16S rDNA RFLP patterns and their lower-than-threshold coefficients of pattern similarity with other groups. Recognition of the new groups was further strengthened by their distinct cladistic positions in the phylogenetic tree. Significantly, each of the ten new 16Sr groups was represented by at least one either formally described or potential '*Ca.* Phytoplasma' species (Wei *et al.*, 2007b), being consistent with the earlier proposal that each 16Sr group represents at least one species (Lee *et al.*, 1998). Representative strains of the ten new groups shared less than 97.5% 16S rDNA sequence similarity with each other and with any previously described '*Candidatus* Phytoplasma' species; therefore, each may be recognized as a new '*Candidatus* Phytoplasma' species, in accordance with the IRCPM recommendation and guidelines (IRPCM, 2004).

To streamline virtual RFLP analysis, a suite of computer programs were developed (Wei *et al.*, 2008b; Zhao *et al.*, 2009b) to perform sequential functions from recognition of correct input (phytoplasma-specific 16S rDNA F2nR2 region) for analysis to generation of numerical and visual outputs. These programs not only simplify the entire virtual RFLP analysis procedure but also ensure accurate operations. For example, in analysing RFLP patterns (conventional gels and virtual images alike) the most time-consuming and error-prone process is to visually compare multiple patterns, to count similar and dissimilar bands, and to calculate pairwise pattern similarity coefficients.

This is especially cumbersome when dealing with a large number of RFLP patterns. The *RFLP_pattern_comparison* program (Wei *et al.*, 2008b) completely eliminates the need for visual pattern comparison and manual similarity coefficient calculation. It accepts multiple nucleotide sequences in FASTA format, scans through the sequences for the recognition sites of a defined list of restriction enzymes, records the length of each restriction fragment, and performs pairwise comparisons of the recorded fragment lengths resulting from virtual digestions by each enzyme. Based on summarized numbers of similar and dissimilar fragments, the program calculates a similarity coefficient (*F*) for each pair of phytoplasma strains according to the formula of Nei and Li (1979). Since similarity coefficient values are influenced by both the number and the particular set of restriction enzymes selected for RFLP analysis, to be consistent with the existing phytoplasma classification scheme, the default enzyme list for the program included the same 17 restriction enzymes that are used for actual gel-based conventional RFLP analysis: *Alu*I, *Bam*HI, *Bfa*I, *Bst*UI (*Tha*I), *Dra*I, *Eco*RI, *Hae*III, *Hha*I, *Hinf*I, *Hpa*I, *Hpa*II, *Kpn*I, *Sau*3AI (*Mbo*I), *Mse*I, *Rsa*I, *Ssp*I and *Taq*I (Lee *et al.*, 1998; Wei *et al.*, 2008b).

The streamlined virtual RFLP analysis has been used to delineate new phytoplasma subgroup lineages and to update classification of phytoplasmas affiliated with the clover proliferation phytoplasma group (16SrVI) (Wei *et al.*, 2008b), the peanut witches'-broom phytoplasma group (16SrII) (Cai *et al.*, 2008), the stolbur (STOL) phytoplasma group (16Sr XII) (Quaglino *et al.*, 2009) and the X-disease phytoplasma group (16SrIII) (Zhao *et al.*, 2009b). As a result, dozens of new subgroup lineages have been delineated, expanding the phytoplasma classification system to 30 groups and more than 100 subgroups (Table 5.2). Virtual RFLP analysis of strains in these phytoplasma groups has revealed complex phytoplasma population structures and previously unexpected levels of phytoplasma inter- and intraspecies genetic diversity. For instance, it has been found that cactus witches'-broom (CaWB) disease can be associated with infections by diverse phytoplasmas affiliated with the aster yellows (AY) phytoplasma group (16SrI) and with the peanut witches'-broom group (16SrII) (Wei *et al.*, 2007a; Cai *et al.*, 2008). Those group 16SrII CaWB phytoplasmas belong to nine distinct subgroups, of which eight coexist in the south-western China province of Yunnan (Cai *et al.*, 2008).

Virtual RFLP analysis has also revealed a relationship between the extent of CaWB phytoplasma genetic diversity and ecosystem characteristics: the greatest breadth of CaWB phytoplasma genetic diversity was found in unmanaged ecosystems in Wenshan and Kunming regions, which are in a subtropical, mid-altitude highland (1260–1895 m) zone, having a monsoon climate with warm temperatures (average annual temperature of 15 °C) (Cai *et al.*, 2008). One implication of the result is that an epicentre of group 16SrII phytoplasma evolution may exist in or near the regions. Alternatively, the observed strain diversity originated elsewhere. However, in either case, ecological niches in unmanaged areas of the Wenshan/Kunming regions may favour strain diversity. In the unmanaged ecosystems, the cacti were apparently propagated through true seed, suggesting that, assuming lack of seed transmission, insect vector(s) may carry diverse phytoplasma strains to or

Table 5.2. Phytoplasma 16S ribosomal RNA RFLP groups.[a]

Group	Number of subgroups	Number of '*Ca.* Phytoplasma' species
16SrI: Aster yellows group	11	1 ('*Ca* Phytoplasma asteris')
16SrII: Peanut witches'-broom group	12	2 ('*Ca* Phytoplasma aurantifolia' and '*Ca.* Phytoplasma australasiae')
16SrIII: X-disease group	19	1 ('*Ca* Phytoplasma pruni'†)
16SrIV: Coconut lethal yellows group	6	1 ('*Ca* Phytoplasma palmae'†)
16SrV: Elm yellows group	6	3 ('*Ca* Phytoplasma ulmi', '*Ca* Phytoplasma ziziphi' and '*Ca* Phytoplasma vitis'†)
16SrVI: Clover proliferation group	8	1 ('*Ca* Phytoplasma trifolii')
16SrVII: Ash yellows group	3	1 ('*Ca* Phytoplasma fraxini')
16SrVIII: Loofah witches'-broom group	1	1 ('*Ca* Phytoplasma luffae'†)
16SrIX: Pigeon pea witches'-broom group	4	1 ('*Ca* Phytoplasma phoenicium')
16SrX: Apple proliferation group	5	4 ('*Ca* Phytoplasma mali', '*Ca* Phytoplasma pyri', '*Ca* Phytoplasma prunorum' and '*Ca* Phytoplasma spartii')
16SrXI: Rice yellow dwarf group	3	1 ('*Ca* Phytoplasma oryzae')
16SrXII: Stolbur group	7	4 ('*Ca* Phytoplasma australiense', '*Ca* Phytoplasma japonicum', '*Ca* Phytoplasma fragariae' and '*Ca* Phytoplasma solani'†)
16SrXIII: Mexican periwinkle virescence group	2	1§
16SrXIV: Bermudagrass white leaf group	1	1 ('*Ca* Phytoplasma cynodontis')
16SrXV: Hibiscus witches'-broom group	1	1 ('*Ca* Phytoplasma brasiliense')
16SrXVI: Sugarcane yellow leaf syndrome group	1	1 ('*Ca* Phytoplasma graminis')
16SrXVII: Papaya bunchy top group	1	1 ('*Ca* Phytoplasma caricae')
16SrXVIII: American potato purple top wilt group	1	1 ('*Ca* Phytoplasma americanum')
16SrXIX: Japanese chestnut witches'-broom group[b]	1	1 ('*Ca* Phytoplasma castaneae')
16SrXX: Buckthorn witches'-broom group[c]	1	1 ('*Ca.* Phytoplasma rhamni')
16SrXXI: Pine shoot proliferation group	1	1 ('*Ca.* Phytoplasma pini')
16SrXXII: Nigerian coconut lethal decline (LDN) group	1	1 ('*Ca.* Phytoplasma cocosnigeriae'†)
16SrXXIII: Buckland Valley grapevine yellows group	1	1§
16SrXXIV: Sorghum bunchy shoot group	1	1§

(*continued*)

Table 5.2. *continued*

Group	Number of subgroups	Number of '*Ca.* Phytoplasma' species
16SrXXV: Weeping tea tree witches'-broom group	1	1[§]
16SrXXVI: Mauritius sugarcane yellows D3T1 group	1	1[§]
16SrXXVII: Mauritius sugarcane yellows D3T2 group	1	1[§]
16SrXXVIII: Havana derbid phytoplasma group	1	1[§]
16SrXXIX: Cassia witches'-broom group[d]	1	1 ('*Ca.* Phytoplasma omanense')
16SrXXX: Salt cedar witches'-broom group	1	1 ('*Ca.* Phytoplasma tamaricis')

[a]This table is compiled based on the following references: Lee *et al.*, 1998; Bertaccini, 2007; Wei *et al.*, 2007b, 2008b; Cai *et al.*, 2008; Harrison *et al.*, 2008; Meneguzzi *et al.*, 2008; Lee and Bottner, 2009; Quaglino *et al.*, 2009; Zhao *et al.*, 2009a, b. [b]In the report by Jung *et al.* (2002), Japanese chestnut witches'-broom phytoplasma was assigned to group VI according to DNA sequence homology, rather than results from RFLP analysis. In accordance with the more widely accepted RFLP-based classification system, this phytoplasma was reassigned to group 16SrXIX by Wei *et al.* (2007b). [c]Buckthorn witches'-broom phytoplasma is most closely related to phytoplasmas in the apple proliferation group and was previously classified in group 16SrX (Lee *et al.*, 1998). Recently, this taxon was assigned to a new 16Sr group (16SrXX) on the basis of its lower-than-threshold RFLP pattern similarity coefficient values with all known phytoplasmas in the group 16SrX and other groups (Wei *et al.*, 2007b), in accordance with the principle that governs the 16S rDNA RFLP-based classification scheme. [d]The original reference (Al-Saady *et al.*, 2008) reported Cassia witches'-broom phytoplasma as the reference member of a new group, designated as group 16SrXIX. However, the group number 16SrXIX had been previously published (Wei *et al.*, 2007b) to accommodate a different phytoplasma, Japanese chestnut witches'-broom phytoplasma. Therefore, Cassia witches'-broom phytoplasma was assigned to a new group, 16SrXXIX (Zhao *et al.*, 2009a). [†]According to Rule 28b of the Bacteriological Code, this is an incidental citation and does not constitute prior citation. [§]No name has been suggested for this potentially new '*Ca.* Phytoplasma' species.

from neighbouring plant species. If the extent of CaWB strain diversity observed in Yunnan, China were the result of human activities (introduction), one should expect that the new subgroups found in Yunnan also exist in other region(s) of the world, but these subgroups have not been reported elsewhere. Since the New World is apparently the epicentre of cactus evolution (Nyffeler, 2002), it would be interesting to learn whether a broad diversity of group 16SrII phytoplasma strains occurs in the Americas.

In addition, virtual RFLP analysis has led to identification of new 16SrIII subgroup lineages (Lee and Bottner, 2009; Zhao *et al.*, 2009b) that infect potato and cause potato purple top syndrome, a disease complex that is also attributed to infections by other phytoplasmas belonging to at least five different phytoplasma 16Sr groups (16SrI, 16SrII, 16SrVI, 16SrXII and 16SrXVIII)

(Lee and Bottner, 2009). Virtual RFLP analysis has also allowed delineation of numerous new 16SrXII subgroup lineages (Quaglino *et al.*, 2009) that are associated with grapevine yellows syndrome, a disease complex that has been linked to infections by no less than five distinct species affiliated with four phytoplasma 16Sr groups (16SrI, 16SrIII, 16SrV and 16SrXII). Extensive genetic diversity of phytoplasma strains and coexistence of diverse phytoplasma strains in the same host species and/or in the same geographic location can either be interpreted as ongoing evolution of phytoplasmas in adaptation to their geo- and bio-ecological niches or be explained by differences in vector species involvement. In either case, coexistence of diverse phytoplasma strains in the same bio- and geo-ecological niche may favour phytoplasma–phytoplasma, phytoplasma–insect vector and phytoplasma–plant host interactions that provide increased opportunities for genetic recombination and the emergence of new phytoplasmal plant diseases.

An important issue that requires attention in conducting virtual RFLP analysis for phytoplasma classification is *rrn* interoperon sequence heterogeneity. The genomes of all four completely sequenced phytoplasma strains and numerous reference strains of '*Ca.* Phytoplasma' species harbour two ribosomal RNA operons, *rrn*A and *rrn*B (IRPCM, 2004; Oshima *et al.*, 2004; Bai *et al.*, 2006; Kube *et al.*, 2008; Tran-Nguyen *et al.*, 2008). In many strains, the sequences of the two *rrn* operons differ from each other (Lee *et al.*, 1993, 1998; Schneider and Seemüller, 1994; Firrao *et al.*, 1996; Liefting *et al.*, 1996; Davis and Sinclair, 1998; Harrison *et al.*, 2002; Jomantiene *et al.*, 2002; Davis *et al.*, 2003a). For those phytoplasma strains with two heterogeneous *rrn* operons, if the sequence variations between the two operons fall into restriction sites within the 16S rRNA gene F2nR2 region, two different virtual 16Sr RFLP pattern types will result from virtual RFLP analysis, which could cause erroneous assignment of the same phytoplasma to two different 16S rRNA subgroups. The same issue has been raised in conventional RFLP analysis also (Davis *et al.*, 2003a). To help resolve this issue, a three-letter subgroup designation was proposed (Wei *et al.*, 2008b). For example, paulownia witches'-broom (PaWB) phytoplasma, a member of the previously delineated subgroup 16SrI-D, possesses two sequence-heterogeneous rRNA operons, displaying two different 16Sr RFLP patterns, 16SrI-B and 16SrI-D. According to the three-letter designation proposal, the subgroup status of PaWB was suggested to be 16SrI-(B/D)D. In this subgroup designation, the first and second letters (in parentheses) denote the RFLP pattern types of *rrnA* and *rrnB*, respectively, and the third letter designates the 16Sr subgroup. In conventional RFLP analysis, a composite banding pattern may arise from two sequence-heterogeneous 16S rRNA genes. A composite pattern is suspected when the sum of the sizes of DNA fragments is greater than the expected size of the F2nR2 region (1.25 kb). In virtual RFLP analysis, such composite patterns can be reconstructed by superimposing two individual virtual gel images that are derived by analysis of two individual sequence-heterogeneous *rrn* operons, as illustrated for Canadian '*Fragaria multicipita*' phytoplasma strain MC (16SrVI-(B/G)B) (Wei *et al.*, 2008b), for dandelion virescence phytoplasma strain DanVir (16SrIII-(P/O)P) and for cirsium white leaf phytoplasma strain CirWL (16SrIII-(R/B)R) (Zhao *et al.*, 2009b).

In contrast to conventional RFLP analysis, which has typically been done in the absence of prior nucleotide sequence information, virtual RFLP analysis is a nucleotide-sequence-based operation. A legitimate question one may ask here is: with the availability of sequence information, whether RFLP analysis still remains useful for phytoplasma differentiation and classification. The answer to this question is positive. First, the already established phytoplasma 16S rDNA RFLP patterns have become authoritative expositions for scientists in the phytoplasma research community and have served as standard keys for phytoplasma strain identification and classification. Secondly, although other sequence-based analyses such as pairwise sequence comparisons and phylogenetic analyses can be used to assess genetic relationships among phytoplasma strains, neither does percentage sequence similarity score from pairwise comparisons nor do tree topologies from phylogenetic analyses directly reveal informative sites along the sequences or the 'visible' genetic markers provided by RFLP analysis. While RFLP analysis remains a valuable tool for studying microbial diversity and classification, the method by which RFLP analysis is carried out has evolved (Moyer *et al.*, 1996; Edwards and Turco, 2005; Ricke *et al.*, 2005; Abdo *et al.*, 2006). As demonstrated by the examples above, the value of virtual RFLP analysis is evident in delineation of new phytoplasma group and subgroup lineages and in unveiling complex phytoplasma population structures and genetic diversity.

Since virtual RFLP analysis is a nucleotide-sequence-based analysis, any error in an input sequence which misrepresents the phytoplasma strain under study could result in erroneous group/subgroup classification. While sequence errors may arise at various stages during PCR amplification, plasmid multiplication and DNA sequencing, they usually occur randomly and can be rectified by sample replications. To ensure reliable results from virtual RFLP analysis, it is highly recommended that consistent sequence data from at least two independent samples, i.e. from two or more infected plants or insect individuals, be obtained. If only one infected plant or insect sample is available for study, consistent sequence data from at least two independently cloned DNA segments derived from two separate PCRs must be obtained. Each clone (plasmid) should be sequenced in both directions and a minimum of 3X coverage per base position achieved (Zhao *et al.*, 2009b). With credible sequence data, virtual gel patterns generated by computer-simulated RFLP analysis can faithfully replicate the classical, authoritative patterns that had been established by conventional RFLP analysis (Wei *et al.*, 2007b). New pattern types derived from virtual RFLP analysis have also been confirmed by actual enzymatic digestions followed by gel electrophoresis (Cai *et al.*, 2008; Lee and Bottner, 2009; Quaglino *et al.*, 2009; Zhao *et al.*, 2009a). These previously established and newly recognized RFLP patterns will serve as standard keys for future identification and classification of rapidly growing numbers of phytoplasmas by either computer-simulated or conventional RFLP analyses.

*i*PhyClassifier: an Online Tool for Phytoplasma Taxonomic Assignment and Classification

Recently, an interactive phytoplasma research tool, *i*PhyClassifier, has been launched on the internet (http://www.plantpathology.ba.ars.usda.gov/cgi-bin/resource/iphyclassifier.cgi), transforming phytoplasma classification from individual laboratory procedures to a real-time World Wide Web operation. The *i*PhyClassifier server is equipped with a suite of bioinformatic programs and 16S rDNA databases. The overall operational process of *i*PhyClassifier is outlined in Fig. 5.3. The aim of *i*PhyClassifier operation is to provide meaningful suggestions on tentative '*Ca.* Phytoplasma' species' (or related strain) assignment and 16Sr group/subgroup classification status for any phytoplasma strain under study (Zhao *et al.*, 2009b).

Upon receiving query sequence(s) from users, the first step of *i*PhyClassifier operation is to invoke internal bioinformatic programs and search databases. Based on the criteria given, *i*PhyClassifier will determine whether or not a query 16S rDNA sequence is from a phytoplasma. If not, the operation will abort; otherwise, *i*PhyClassifier will either assign the query strain tentatively to an existing '*Ca.* Phytoplasma' species as a related strain or suggest that the query represents a potentially new '*Ca.* Phytoplasma' species, depending on the sequence similarity scores. In accordance with the convention on 16S rRNA gene-sequence-based prokaryotic species delineation (Murray and Schleifer, 1994; Stackebrandt and Goebel, 1994), *i*PhyClassifier implements the recommendation of the IRPCM Phytoplasma Taxonomy Group (2004) and presets 97.5% 16S rRNA gene sequence similarity as the cut-off value for new '*Ca.* species' recognition. Since the generally conserved 16S rRNA gene sequences contain pockets of hypervariable regions, the sequence similarity score calculation should be based upon comparison of full- or near-full-length 16S rRNA genes. It requires that each query sequence covers at least 1200 positions within a 16S rRNA gene.

The second step of the *i*PhyClassifier operation is to automatically trim each query sequence to the full-length F2nR2 region, using regular expressions that match primer pair R16F2n/R16R2. This step is critical because, in the 16S rRNA gene-based phytoplasma classification scheme, strains are classified into groups and subgroups strictly based on RFLP patterns derived from 16S rRNA gene F2nR2 fragments (Lee *et al.*, 1998, 2000; Wei *et al.*, 2007b, 2008b).

The third step of *i*PhyClassifier operation is to simulate restriction digestions on trimmed F2nR2 fragments, compare the RFLP pattern types derived from each query strain to those derived from representative strains of established phytoplasma 16Sr groups and subgroups, and calculate pairwise RFLP pattern similarity coefficients. In this step, *i*PhyClassifier presets 0.97 as the threshold similarity coefficient for delineation of a new subgroup RFLP pattern type within a given group (Wei *et al.*, 2008b; Zhao *et al.*, 2009b). Thus, if the virtual F2nR2 RFLP pattern derived from a 16S rRNA gene of a phytoplasma strain under study has a 0.97 or lower similarity coefficient with 16S rRNA genes of all existing representative or reference strains of the given group, a new subgroup pattern type is recognized. Adoption of 0.97 as the

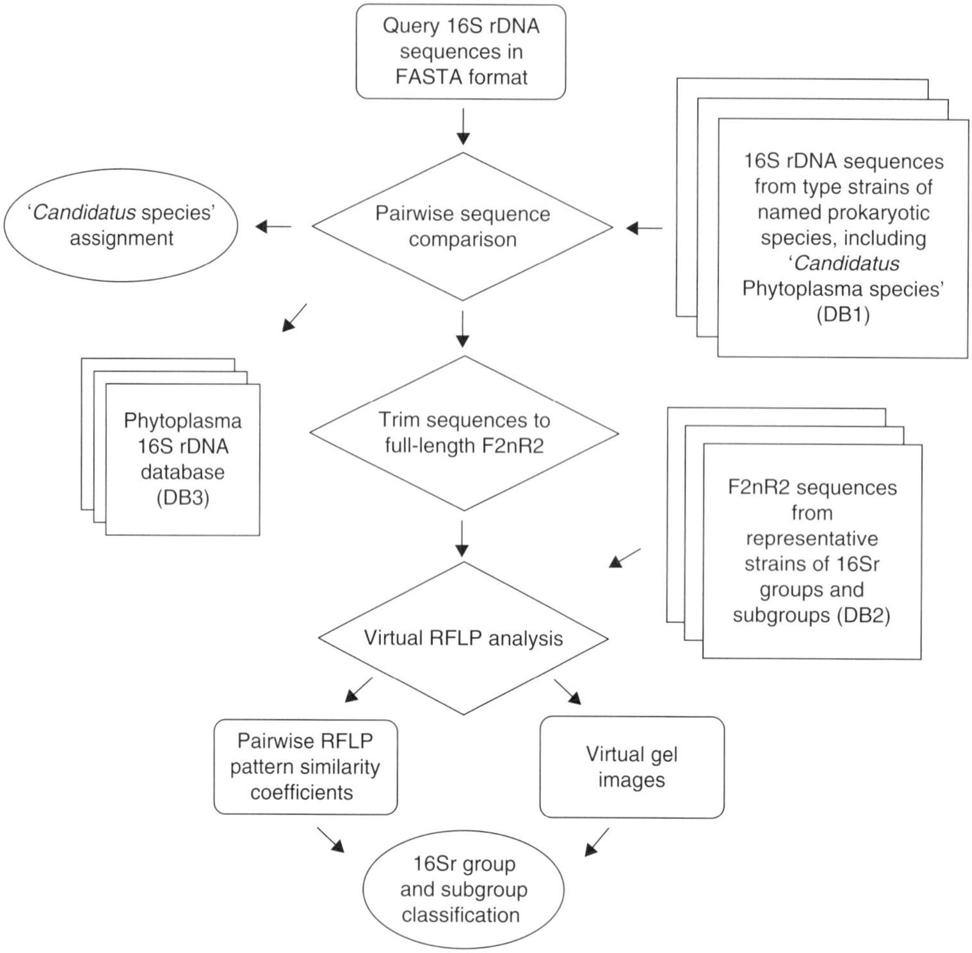

Fig. 5.3. Diagrammatic representation of the operational process of *i*PhyClassifier. Rectangles represent input and output files; squares represent databases; diamonds represent computational operations; and ovals represent recommendations on tentative 16Sr group/subgroup classification status and '*Candidatus* Phytoplasma' species' assignment. DB1, a set of full- or near-full-length 16S rRNA gene sequences from reference strains of all formally described '*Candidatus* Phytoplasma' species, reference strains of IRPCM Phytoplasma Taxonomy Group proposed (2004) but yet to be formally described '*Candidatus* Phytoplasma' species, reference strains of potentially new '*Candidatus* Phytoplasma' species identified in our previous study (Wei *et al.*, 2007b) and all type strains of other named prokaryotic species; DB2, a set of F2nR2 sequences from representative strains of established phytoplasma 16Sr groups and subgroups; and DB3, a set of F2nR2 sequences compiled from all phytoplasma 16S rRNA sequences currently deposited in the GenBank.

threshold similarity coefficient for new subgroup delineation is warranted because it reflects precisely the existing subgroup classification scheme, in which as little as one restriction site difference can distinguish a new subgroup. A similarity coefficient of 0.85 or less with all previously recognized subgroups signals that the strain under study may represent a new 16Sr group, in agreement with all previously designated groups. RFLP patterns that have a similarity coefficient of 0.99 or 0.98 to the standard pattern type of the designated representative or reference member in a given subgroup are considered as variants of the standard pattern type. These variants or minor pattern types are denoted with one or two stars (* or **) following their corresponding subgroup letters, for example 16SrI-A* ($F = 0.99$) and 16SrI-A** ($F = 0.98$), as suggested previously (Wei *et al.*, 2008b). The threshold similarity coefficients for new subgroup and group pattern type delineations are strictly based on the use of a specific set of 17 restriction enzymes originally established for classification of phytoplasmas using actual gel electrophoresis-based RFLP analysis (Lee *et al.*, 1998). The output of this operational step is assignment of the strain under study into an existing subgroup or erection of a new subgroup or a new group. Subgroup designation of strains with heterogeneous *rrn* operons requires F2nR2 sequence information from both *rrn* operons.

Concomitant with similarity coefficient calculation, which generates numerical output of the RFLP pattern analysis, *i*PhyClassifier also provides visual output, i.e. virtual gel images resulting from the RFLP pattern analysis. The gel images reveal informative sites or molecular markers along the 16S rRNA gene, converting sequence information into accessible 'virtual phenotypic characters' for phytoplasma strain differentiation and classification. Virtual RFLP patterns of reference strains of all phytoplasma 16Sr groups and subgroups are available online at the *i*PhyClassifier web site.

The framework of *i*PhyClassifier can be easily expanded to accommodate virtual RFLP analysis of full-length 16S rRNA genes (Duduk *et al.*, 2004) and other genes or multiple DNA loci (see Chapter 4, this volume), appending additional RFLP markers to 16Sr group/subgroup classification results and permitting finer distinction among closely related phytoplasmas.

Tech-on-the-move: New Approaches Applied to Phytoplasma Fingerprinting

Phytoplasma population structures are complex and dynamic, and a single geo-biological niche can contain multiple phytoplasma species and distinct lineages. Therefore, it is highly desirable to have sensitive and reliable means for simultaneous detection of a wide range of phytoplasma species as well as for precise differentiation of closely related strains. Recent applications of an array of molecular techniques have yielded new 16S rRNA gene-sequence-based diagnostic tools, the power and precision of which not only allowed quick identification and genotyping of phytoplasma strains but also advanced our knowledge of distribution and physiology of phytoplasmas in host plants.

Oligonucleotide microarray is a powerful analytical tool for biodiversity and gene expression studies. An oligonucleotide microarray chip typically consists of a series of immobilized 21- to 60-nucleotide-long probes that can hybridize to target DNA or RNA sequences. In recent years, 16S rRNA gene-sequence-based oligonucleotide microarrays have been successfully used in analysis of microbial communities and detection of pathogenic bacteria (Maynard *et al.*, 2005; Huyghe *et al.*, 2008). The five-probe array (zips 1, 3, 5, 11 and 13) devised by Frosini *et al.* (2002) was among the first attempts to apply oligo-array technology to phytoplasma detection and differentiation. In combination with ligase detection reactions, the five-probe array was proven to be effective in detecting and distinguishing grapevine-infecting phytoplasmas belonging to two different 16S ribosomal groups (16SrV and 16SrXII). Later on, Nicolaisen and Bertaccini (2007) devised an oligonucleotide microarray that can simultaneously detect almost all known phytoplasmas and differentiate phytoplasmas from nine major 16Sr groups: 16SrI, 16SrII, 16SrIII, 16SrV, 16SrVI, 16SrVII, 16SrIX, 16SrX and 16SrXII. The array probes were designed based on mutually distinct sequences that represent each of the nine 16Sr groups and on phytoplasma-universal 16S rDNA sequences. The lengths and positions of the oligonucleotides were adjusted so that the probes had relatively uniform hybridization characteristics (a T_m of 59 ± 2 °C) and had low probabilities of forming dimmers or hairpin structures. These pioneer researches provide prototypes for future design of sophisticated microarray chips for multi-locus phytoplasma genotyping and genetic diversity studies.

Single-strand conformation polymorphism (SSCP) analysis is electrophoretic separation of single-stranded DNA (ssDNA) molecules based on subtle difference(s) in sequence (Orita *et al.*, 1989). As little as one base substitution in ssDNA can result in different three-dimensional structures due to alteration of intra-strand base-pairing, thus visible mobility shifts in a non-denaturing gel. Coupled with PCR amplification, SSCP analysis can be used to rapidly identify sequence variations in short stretches of DNA (usually 175 to 250 nt) without prior knowledge of actual sequences. While having long been used in detection of point mutations and single nucleotide polymorphisms in other biological systems, it was only very recently that this technique was applied to phytoplasma research – analysis of genetic variability among grapevine-infecting phytoplasma isolates (Šeruga-Musić *et al.*, 2008). The study detected a total of nine SSCP profiles among 70 grapevine-infecting phytoplasma isolates. Two SSCP profiles were identified within a short 16S rRNA gene fragment, and the other two and five profiles were found in other genes. All 70 phytoplasma isolates were previously classified into the same subgroup, i.e. subgroup A of the STOL group (16SrXII-A). None of the polymorphisms identified by SSCP analysis can be detected by RFLP analysis.

Quantitative real-time PCR (qPCR) is an evolving technology built upon the traditional end-point PCR methods (Whitman and Dunbar, 2008). Through continuous fluorometric monitoring and instantaneous quantitation of amplification products, qPCRs allow quantitative analysis of targeted nucleic acid molecules in biological samples. The technology has led to development of numerous routine diagnostic tools for sensitive detection and accurate quantification of various pathogens in clinical, agricultural and environmental specimens (Hughes *et al.*, 2006; Wei *et al.*, 2006; Fountaine *et al.*, 2007; Bhagwat *et al.*, 2008; Smith and Osborn, 2009). In the last few years, qPCR technology has been exploited for detection and differentiation of phytoplasmas in tissues of infected plants and insect vectors. Christensen *et al.* (2004) designed primers and a fluorogenic probe based on conserved 16S rRNA gene sequences and devised a phytoplasma-universal qPCR for successful detection and quantification of a wide range of phytoplasma strains, covering ten 16Sr groups and 18 subgroups (16SrI-B, 16SrI-C, 16SrII-A, 16SrIII-A, 16SrIII-B, 16SrIII-H, 16SrV-A, 16SrV-B, 16SrV-C, 16SrV-D, 16SrV-E, 16SrVI, 16SrVII, 16SrIX, 16SrX-A, 16SrX-B, 16SrXI, 16SrXII-A). Real-time PCR assays were also developed, using primers and probes targeted to 16S rDNA sequences that are unique to specific phytoplasma groups, species or lineages. Two such assays, developed by Torres *et al.* (2005) and Galetto *et al.* (2005), respectively, were able to specifically detect quarantine phytoplasmas belonging to the apple proliferation group (16SrX) and flavescence dorée phytoplasma (16Sr-V), and another such assay, developed by Hren *et al.* (2007), was able to differentiate flavescence dorée (16SrV) and bois noir (BN) (16SrXII-A) phytoplasmas effectively, both of which are aetiological agents of the devastating grapevine yellows disease complex. In addition, non-ribosomal RNA gene-targeted qPCR assays were also developed for strain-specific detection and quantification (Chapter 4, this volume). We expect that, as more molecular markers become available, multiplex qPCR assays with multi-locus targets will be devised for phytoplasma genotyping, population structure and phytoplasma–host interaction studies in the near future.

Conclusion

Due to the inability to cultivate phytoplasmas in cell-free media and the consequent inaccessibility of measurable phenotypic characters suitable for polyphasic characterizations, molecular analyses of conserved gene sequences have become rational means for phytoplasma taxonomy and classification. Genes encoding 16S ribosomal RNAs are highly conserved across the phytoplasma clade yet contain ample information for differentiation of diverse phytoplasma strains and therefore have served as a primary molecular tool for phytoplasma identification, genotyping, taxonomic assignment and group/subgroup classification. Supplementary molecular markers have been identified for finer differentiation of closely related strains that cannot be distinguished by 16S rRNA gene sequence alone (see Chapter 10, this volume). As more phytoplasma genome information becomes available, multi-locus sequence analyses will certainly provide additional information to the 16S rRNA gene-based phylogenetic backbone and enhance the resolving power in delineating distinct phytoplasma lineages and closely related strains.

References

Abdo, Z., Schuette, U.M., Bent, S.J., Williams, C.J., Forney, L.J. and Joyce, P. (2006) Statistical methods for characterizing diversity of microbial communities by analysis of terminal restriction fragment length polymorphisms of 16S rRNA genes. *Environmental Microbiology* 8, 929–938.

Al-Saady, N.A., Khan, A.J., Calari, A., Al-Subhi, A.M. and Bertaccini, A. (2008) '*Candidatus* Phytoplasma omanense', associated with witches'-broom of *Cassia italica* (Mill.) Spreng. in Oman. *International Journal of Systematic and Evolutionary Microbiology* 58, 461–466.

Arocha, Y., López, M., Piñol, B., Fernández, M., Picornell, B., Almeida, R., Palenzuela, I., Wilson, M.R. and Jones, P. (2005) '*Candidatus* Phytoplasma graminis' and '*Candidatus* Phytoplasma caricae', two novel phytoplasmas associated with diseases of sugarcane, weeds and papaya in Cuba. *International Journal of Systematic and Evolutionary Microbiology* 55, 2451–2463.

Arocha, Y., Antesana, O., Montellano, E., Franco, P., Plata, G. and Jones, P. (2007) '*Candidatus* Phytoplasma lycopersici', a phytoplasma associated with 'hoja de perejil' disease in Bolivia. *International Journal of Systematic and Evolutionary Microbiology* 57, 1704–1710.

Bai, X.D., Zhang, J.H., Ewing, A., Miller, S.A. Radek, A.J., Shevchenko, D.V., Tsukerman, K., Walunas, T., Lapidus, A., Campbell, J.W. and Hogenhout, S.A. (2006) Living with genome instability: the adaptation of phytoplasmas to diverse environments of their insect and plant hosts. *Journal of Bacteriology* 188, 3682–3696.

Bertaccini, A. (2007) Phytoplasmas: diversity, taxonomy, and epidemiology. *Frontiers of Bioscience* 12, 673–689.

Bertamini, M., Grando, M.S., Muthuchelian, K. and Nedunchezhian, N. (2002a) Effect of phytoplasmal infection on photosystem II efficiency and thylakoid membrane protein changes in field grown apple (*Malus pumila*) leaves. *Physiological and Molecular Plant Pathology* 61, 349–356.

Bertamini, M., Nedunchezhian, N., Tomasi, F. and Grando, M.S. (2002b) Phytoplasma (Stolbur-subgroup (Bois Noir-BN)) infection inhibits photosynthetic pigments, ribulose-1,5-bisphosphate carboxylase and photosynthetic activities in field grown grapevine (*Vitis vinifera* L. cv. Chardonnay) leaves. *Physiological and Molecular Plant Pathology* 61, 357–366.

Bhagwat, A.A., Patel, J., Chua, T., Chan, A., Cruz, S.R. and Aguilar, G.A. (2008) Detection of *Salmonella* species in foodstuffs. *Methods in Molecular Biology* 429, 33–43.

Blanche, K.R., Tran-Nguyen, L.T. and Gibb, K.S. (2003) Detection, identification and significance of phytoplasmas in grasses in northern Australia. *Plant Pathology* 52, 505–512.

Brenner, D., Staley, J. and Krieg, N. (2000) Classification of prokaryotic organisms and the concept of bacterial speciation. In: Boone, D.R., Castenholz, R.W. and Garrity, G.M. (eds) *Bergey's Manual of Systematic Bacteriology*, 2nd edn, vol. 1. Springer-Verlag, New York, pp. 27–31.

Cai, H., Wei, W., Davis, R.E., Chen, H. and Zhao, Y. (2008) Genetic diversity among phytoplasmas infecting *Opuntia* species: virtual RFLP analysis identifies new subgroups in the peanut witches'-broom phytoplasma group. *International Journal of Systematic and Evolutionary Microbiology* 58, 1448–1457.

Chang, C.-J. (1998) Pathogenicity of aster yellows phytoplasma and *Spiroplasma citri* on periwinkle. *Phytopathology* 88, 1347–1350.

Chiykowski, L.N. (1962) Clover phyllody virus in Canada and its transmission. *Canadian Journal of Botany* 40, 397–404.

Chiykowski, L.N. and Sinha, R.C. (1989) Differentiation of MLO disease by means of symptomatology and vector transmission. *Zentralblatt fur Bakteriologie* 20 (suppl.), 280–287.

Christensen, N.M., Nicolaisen, M., Hansen, M. and Schulz, A. (2004) Distribution of phytoplasmas in infected plants as revealed by real-time PCR and bioimaging. *Molecular Plant–Microbe Interactions* 17, 1175–1184.

Coenye, T., Gevers, D., Van de Peer, Y., Vandamme, P. and Swings, J. (2005) Towards a prokaryotic genomic taxonomy. *FEMS Microbiological Review* 29, 147–167.

Constable, F.E., Whiting, J.R., Gibb, K.S. and Symons, R.H. (2002) A new grapevine yellows phytoplasma from the Buckland Valley of Victoria, Australia. *Vitis* 41, 147–153.

Curković-Perica, M., Lepedus, H. and Seruga-Musić, M. (2007) Effect of indole-3-butyric acid on phytoplasmas in infected *Catharanthus roseus* shoots grown *in vitro*. *FEMS Microbiology Letters* 268, 171–177.

Davis, R.E. and Sinclair, W.A. (1998) Phytoplasma identity and disease etiology. *Phytopathology* 88, 1372–1376.

Davis, R.E., Dally, E.L., Gundersen, D.E., Lee, I.-M. and Habili, N. (1997) '*Candidatus* Phytoplasma australiense', a new phytoplasma taxon associated with Australian grapevine yellows. *International Journal of Systematic Bacteriology* 47, 262–269.

Davis, R.E., Jomantiene, R., Kalvelyte, A. and Dally, E.L. (2003a) Differential amplification of sequence heterogeneous ribosomal RNA genes and classification of the '*Fragaria multicipita*' phytoplasma. *Microbiology Research* 158, 229–236.

Davis, R.E., Jomantiene, R., Zhao, Y. and Dally, E.L. (2003b) Folate biosynthesis pseudogenes, *ψfolP* and *ψfolK*, and an O-sialoglycoprotein endopeptidase gene homolog in the phytoplasma genome. *DNA Cell Biology* 22, 697–706.

Davis, R.E., Jomantiene, R. and Zhao, Y. (2005) Lineage-specific decay of folate biosynthesis genes suggests ongoing host adaptation in phytoplasmas. *DNA Cell Biology* 24, 832–840.

Deng, S. and Hiruki, C. (1991) Amplification of 16S rRNA genes from culturable and nonculturable *Mollicutes*. *Journal of Microbiological Methods* 14, 53–61.

Doi, Y., Teranaka, M., Yora, K. and Asuyama, H. (1967) *Mycoplasma* or PLT-group-like microorganisms found in the phloem elements of plants infected with mulberry dwarf, potato witches'-broom, aster yellows, or paulownia witches'-broom. *Annals of the Phytopathological Society of Japan* 33, 259–266.

Duduk, B., Botti, S., Ivanovic, M., Krstic, B., Dukic, N. and Bertaccini, A. (2004) Identification of phytoplasmas associated with grapevine yellows in Serbia. *Journal of Phytopathology* 152, 575–579.

Edwards, I.P. and Turco, R.F. (2005) Inter- and intraspecific resolution of nrDNA TRFLP assessed by computer-simulated restriction analysis of a diverse collection of ectomycorrhizal fungi. *Mycological Research* 109, 212–226.

Firrao, G., Smart, C.D. and Kirkpatrick, B.C. (1996) Physical map of the Western X disease phytoplasma chromosome. *Journal of Bacteriology* 178, 3985–3988.

Firrao, G., Gibb, K. and Streten, C. (2005) Short taxonomic guide to the genus '*Candidatus* phytoplasma'. *Journal of Plant Pathology* 87, 249–263.

Fountaine, J.M., Shaw, M.W., Napier, B., Ward, E. and Fraaije, B.A. (2007) Application of real-time and multiplex polymerase chain reaction assays to study leaf blotch epidemics in barley. *Phytopathology* 97, 297–303.

Freitag, J.H. (1964) Interaction and mutual suppression among three strains of aster yellows virus. *Virology* 24, 401–413.

Frosini, A., Casati, P., Bianco, P.A., Bordoni, R., Consolandi, C., Castiglioni, B., Mezzelani, A., Rizzi, E., Battaglia, C., Belli, G., Rossi Bernardi, L. and De Bellis, G. (2002) Ligase detection reaction and universal array as a tool to detect grapevine infecting phytoplasmas. *Minerva Biotechnology* 14, 265–267.

Galetto, L., Bosco, D. and Marzachì, C. (2005) Universal and group-specific real-time PCR diagnosis of flavescence dorée (16Sr-V), bois noir (16Sr-XII) and apple proliferation (16Sr-X) phytoplasmas from field-collected plant hosts and insect vectors. *Annals of Applied Biology* 147, 191–201.

Granados, R.R. and Chapman, R.K. (1968) Identification of some new aster yellows virus strains and their transmission by the aster leafhopper *Macrosteles fascifrons*. *Phytopathology* 58, 1685–1692.

Griffiths, H.M., Sinclair, W.A., Smart, C.D. and Davis, R.E. (1999) The phytoplasma associated with ash yellows and lilac witches'-broom: '*Candidatus* Phytoplasma fraxini'. *International Journal of Systematic Bacteriology* 49, 1605–1614.

GuhaThakurta, D. (2006) Computational identification of transcriptional regulatory elements in DNA sequence. *Nucleic Acids Research* 34, 3585–3598.

Gundersen, D.E. and Lee, I.-M. (1996) Ultrasensitive detection of phytoplasmas by nested-PCR assays using two universal primer pairs. *Phytopathologia Mediterranea* 35, 144–151.

Gundersen, D.E., Lee, I.-M., Rehner, S.A., Davis, R.E. and Kingsbury, D.T. (1994) Phylogeny of mycoplasmalike organisms (phytoplasmas): a basis for their classification. *Journal of Bacteriology* 176, 5244–5254.

Harrison, N.A., Myrie, W., Jones, P., Carpio, M.L., Castillo, M., Doyle, M.M. and Oropeza, C. (2002) 16S rRNA interoperon sequence heterogeneity distinguishes strain populations of palm lethal yellowing phytoplasma in the Caribbean region. *Annals of Applied Biology* 141, 183–193.

Harrison, N.A., Helmick, E.E. and Elliott, M.L. (2008) Lethal yellowing-type diseases of palms associated with phytoplasmas newly identified in Florida, USA. *Annals of Applied Biology* 153, 85–94.

Hiruki, C. and Wang, K.R. (2004) Clover proliferation phytoplasma: '*Candidatus* Phytoplasma trifolii'. *International Journal of Systematic and Evolutionary Microbiology* 54, 1349–1353.

Hogenhout, S.A., Oshima, K., Ammar, el-D., Kakizawa, S., Kingdom, H.N. and Namba, S. (2008) Phytoplasmas: bacteria that manipulate plants and insects. *Molecular Plant Pathology* 9, 403–423.

Hren, M., Boben, J., Rotter, A., Kralj, P., Gruden, K. and Ravnikar, M. (2007) Real-time PCR detection systems for Flavescence dorée and Bois noir phytoplasmas in grapevine: comparison with conventional PCR detection and application in diagnostics. *Plant Pathology* 56, 785–796.

Hughes, K.J., Tomlinson, J.A., Griffin, R.L., Boonham, N., Inman, A.J. and Lane, C.R. (2006) Development of a one-step real-time polymerase chain reaction assay for diagnosis of *Phytophthora ramorum*. *Phytopathology* 96, 975–981.

Huyghe, A., Francois, P., Charbonnier, Y., Tangomo-Bento, M., Bonetti, E.J., Paster, B.J., Bolivar, I., Baratti-Mayer, D., Pittet, D., Schrenzel, J. and Geneva Study Group on Noma (GESNOMA) (2008) Novel microarray design strategy to study complex bacterial communities. *Applied and Environmental Microbiology* 74, 1876–1885.

IRPCM Phytoplasma/Spiroplasma Working Team – Phytoplasma Taxonomy Group (2004) '*Candidatus* Phytoplasma', a taxon for the wall-less, non-helical prokaryotes that colonize plant phloem and insects. *International Journal of Systematic and Evolutionary Microbiology* 54, 1243–1255.

Janky, R. and van Helden, J. (2008) Evaluation of phylogenetic footprint discovery for predicting bacterial cis-regulatory elements and revealing their evolution. *BMC Bioinformatics* 9, 37.

Jomantiene, R. and Davis, R.E. (2006) Clusters of diverse genes existing as multiple, sequence-variable mosaics in a phytoplasma genome. *FEMS Microbiology Letters* 255, 59–65.

Jomantiene, R., Davis, R.E., Maas, J. and Dally, E.L. (1998). Classification of new phytoplasmas associated with diseases of strawberry in Florida, based on analysis of 16S rRNA and ribosomal protein gene operon sequences. *International Journal of Systematic Bacteriology* 48, 269–277.

Jomantiene, R., Davis, R.E., Valiunas, D. and Alminaite, A. (2002) New group 16SrIII phytoplasma lineages in Lithuania exhibit interoperon sequence heterogeneity. *European Journal of Plant Pathology* 108, 507–517.

Jomantiene, R., Zhao, Y. and Davis, R.E. (2007) Sequence-variable mosaics: composites of recurrent transposition characterizing the genomes of phylogenetically diverse phytoplasmas. *DNA Cell Biology* 26, 557–564.

Jung, H.Y., Sawayanagi, T., Kakizawa, S., Nishigawa, H., Miyata, S., Oshima, K., Ugaki, M., Lee, J.T., Hibi, T. and Namba, S. (2002) '*Candidatus* Phytoplasma castaneae', a novel phytoplasma taxon associated with chestnut witches' broom disease. *International Journal of Systematic and Evolutionary Microbiology* 52, 1543–1549.

Jung, H.Y., Sawayanagi, T., Kakizawa, S., Nishigawa, H., Wei, W., Oshima, K., Miyata, S., Ugaki, M., Hibi, T. and Namba, S. (2003a) '*Candidatus* Phytoplasma ziziphi', a novel phytoplasma taxon associated with jujube witches'-broom disease. *International Journal of Systematic and Evolutionary Microbiology* 53, 1037–1041.

Jung, H.Y., Sawayanagi, T., Wongkaew, P., Kakizawa, S., Nishigawa, H., Wei, W., Oshima, K., Miyata, S., Ugaki, M., Hibi, T. and Namba, S. (2003b) '*Candidatus* Phytoplasma oryzae', a novel phytoplasma taxon associated with rice yellow dwarf disease. *International Journal of Systematic and Evolutionary Microbiology* 53, 1925–1929.

Kirkpatrick, B.C. and Fraser, J.D. (1989) Phylogenetic relationships of the Western X-disease mycoplasma-like organism (X-MLO) as established by 16S rRNA sequence. *Phytopathology* 79, 1138.

Konstantinidis, K., and Tiedje, J.M. (2005a) Genomic insights that advance the species definition for prokaryotes. *Proceedings of the National Academy of Sciences of the USA* 102, 2567–2572.

Konstantinidis, K. and Tiedje, J.M. (2005b) Towards a genome-based taxonomy for prokaryotes. *Journal of Bacteriology* 187, 6258–6264.

Kube, M., Schneider, B., Kuhl, H., Dandekar, T., Heitmann, K., Migdoll, A.M., Reinhardt, R. and Seemüller, E. (2008) The linear chromosome of the plant-pathogenic mycoplasma '*Candidatus* Phytoplasma mali'. *BMC Genomics* 9, 306.

Kuske, C.R. and Kirkpatrick, B.C. (1992) Phylogenetic relationships between the western aster yellows mycoplasmalike organism and other prokaryotes established by 16S rRNA gene sequence. *International Journal of Systematic Bacteriology* 42, 226–233.

Lee, I.-M. and Bottner, K.D. (2009) An emerging potato purple top disease caused by a new 16SrIII group phytoplasma in Montana. *Plant Disease* (in press).

Lee, I.-M., Hammond, R.W., Davis, R.E. and Gundersen, D.E. (1993) Universal amplification and analysis of pathogen 16S rDNA for classification and identification of mycoplasmalike organisms. *Phytopathology* 83, 834–842.

Lee, I.-M., Gundersen-Rindal, D.E., Davis, R.E. and Bartoszyk, I.-M. (1998) Revised classification scheme of phytoplasmas based on RFLP analysis of 16S rRNA and ribosomal protein gene sequences. *International Journal of Systematic Bacteriology* 48, 1153–1169.

Lee, I.-M., Davis, R.E. and Gundersen-Rindal, D.E. (2000) Phytoplasma: phytopathogenic mollicutes. *Annual Review of Microbiology* 54, 221–255.

Lee, I.-M., Gundersen-Rindal, D.E., Davis, R.E., Bottner, K.D., Marcone, C. and Seemüller, E. (2004a) '*Candidatus* Phytoplasma asteris', a novel phytoplasma taxon associated with aster yellows and related diseases. *International Journal of Systematic and Evolutionary Microbiology* 54, 1037–1048.

Lee, I.-M., Martini, M., Marcone, C. and Zhu, S.F. (2004b) Classification of phytoplasma strains in the elm yellows group (16SrV) and proposal of '*Candidatus* phytoplasma ulmi' for the phytoplasma associated with elm yellows. *International Journal of Systematic and Evolutionary Microbiology* 54, 337–347.

Lee, I.-M., Zhao, Y. and Bottner, K.D. (2005) Novel insertion sequence-like elements in phytoplasma strains of the aster yellows group are putative new members of the IS3 family. *FEMS Microbiology Letters* 242, 353–360.

Lee, I.-M., Bottner, K.D., Secor, G. and Rivera-Varas, V. (2006) '*Candidatus* Phytoplasma americanum', a phytoplasma associated with a potato purple top wilt disease complex. *International Journal of Systematic and Evolutionary Microbiology* 56, 1593–1597.

Lepka, P., Stitt, M., Moll, E. and Seemüller, E. (1999) Effect of phytoplasmal infection on concentration and translocation of carbohydrates and amino acids in periwinkle and tobacco. *Physiological and Molecular Plant Pathology* 55, 59–68.

Liefting, L.W., Andersen, M.T., Beever, R.E., Gardner, R.C. and Foster, L.S. (1996) Sequence heterogeneity in the two 16S rRNA genes of *Phormium* yellow leaf phytoplasma. *Applied and Environmental Microbiology* 62, 3133–3139.

Marcone, C., Neimark, H., Ragozzino, A., Lauer, U. and Seemüller, E. (1999) Chromosome sizes of phytoplasmas composing major phylogenetic groups and subgroups. *Phytopathology* 89, 805–810.

Marcone, C., Gibb, K.S., Streten, C. and Schneider, B. (2004a) '*Candidatus* Phytoplasma spartii', '*Candidatus* Phytoplasma rhamni' and '*Candidatus* Phytoplasma allocasuarinae', respectively associated with spartium witches'-broom, buckthorn witches'-broom and allocasuarina yellows diseases. *International Journal of Systematic and Evolutionary Microbiology* 54, 1025–1029.

Marcone, C., Schneider, B. and Seemüller, E. (2004b) '*Candidatus* Phytoplasma cynodontis', the phytoplasma associated with Bermuda grass white leaf disease. *International Journal of Systematic and Evolutionary Microbiology* 54, 1077–1082.

Maynard, C., Berthiaume, F., Lemarchand, K., Harel, J., Payment, P., Bayardelle, P., Masson, L. and Brousseau, R. (2005) Waterborne pathogen detection by use of oligonucleotide-based microarrays. *Applied and Environmental Microbiology* 71, 8548–8557.

McCoy, R.E., Caudwell, A., Chang, C.J., Chen, T.A., Chiykowski, L.N., Cousin, M.T., Dale, J.L., de Leeuw, G.T.N., Golino, D.A. *et al.* (1989) Plant diseases associated with mycoplasmalike organisms. In: Whitcomb, R.F. and Tully, J.G. (eds) *The Mycoplasmas*, Vol. 5. Academic Press, New York, pp. 545–560.

Meneguzzi, N.G., Torres, L.E., Galdeano, E., Guzmán, F.A., Nome, S.F. and Conci, L.R. (2008) Molecular characterization of a phytoplasma of the ash yellows group (16Sr VII-B) occurring in *Artemisia annua* and *Conyza bonariensis* weeds. *Agriscientia* 25, 7–15.

Montano, H.G., Davis, R.E., Dally, E.L., Hogenhout, S., Pimentel, J.P. and Brioso, P.S.T. (2001) '*Candidatus* Phytoplasma brasiliense', a new phytoplasma taxon associated with hibiscus witches' broom disease. *International Journal of Systematic and Evolutionary Microbiology* 51, 1109–1118.

Moyer, C.L., Tiedje, J.M., Dobbs, F.C. and Karl, D.M. (1996) A computer-simulated restriction fragment length polymorphism analysis of bacterial small-subunit rRNA genes: efficacy of selected tetrameric restriction enzymes for studies of microbial diversity in nature. *Applied Environmental Microbiology* 63, 2631–2636.

Murray, R.G. and Stackebrandt, E. (1995) Taxonomic note: implementation of the provisional status *Candidatus* for incompletely described procaryotes. *International Journal of Systematic Bacteriology* 45, 186–187.

Murray, R.G.E. and Schleifer, K.H. (1994) Taxonomic notes: a proposal for recording the properties of putative taxa of procaryotes. *International Journal of Systematic Bacteriology* 44, 174–176.

Namba, S., Oyaizu, H., Kato, S., Iwanami, S. and Tsuchizaki, T. (1993) Phylogenetic diversity of phytopathogenic mycoplasmalike organisms. *International Journal of Systematic Bacteriology* 43, 461–467.

Nei, M. and Li, W.-H. (1979) Mathematical model for studying genetic variation in terms of restriction endonucleases. *Proceedings of the National Academy of Sciences of the USA* 76, 5269–5273.

Nicolaisen, M. and Bertaccini, A. (2007) An oligonucleotide microarray-based assay for identification of phytoplasma 16S ribosomal groups. *Plant Pathology* 56, 332–336.

Nyffeler, R. (2002) Phylogenetic relationships in the cactus family (Cactaceae) based on evidence from *trnK/matK* and *trnL-trnF* sequences. *American Journal of Botany* 89, 312–326.

Orita, M., Iwahana, H., Kanazawa, H., Hayashi, K. and Sekiya, T. (1989) Detection of polymorphisms of human DNA by gel electrophoresis as single-strand conformation polymorphisms. *Proceedings of the National Academy of Sciences of the USA* 86, 2766–2770.

Oshima, K., Kakizawa, S., Nishigawa, H., Jung, H.-Y., Wei, W., Suzuki, S., Arashida, R., Nakata, D., Miyata, S., Ugaki, M. and Namba, S. (2004) Reductive evolution suggested from the complete genome sequence of a plant-pathogenic phytoplasma. *Nature Genetics* 36, 27–29.

Quaglino, F., Zhao, Y., Bianco, P., Wei, W., Casati, P., Durante, G. and Davis, R.E. (2009) New 16Sr subgroups and distinct SNP lineages among grapevine Bois noir phytoplasma populations. *Annals of Applied Biology* (in press).

Ricke, P., Kolb, S. and Braker, G. (2005) Application of a newly developed ARB software-integrated tool for *in silico* terminal restriction fragment length polymorphism analysis reveals the dominance of a novel *pmoA* cluster in a forest soil. *Applied and Environmental Microbiology* 71, 1671–1673.

Rosselló-Mora R. (2005) Updating prokaryotic taxonomy. *Journal of Bacteriology* 187, 6255–6257.

Sawayanagi, T., Horikoshi, N., Kanehira, T., Shinohara, M., Bertaccini, A., Cousin, M.T., Hiruki, C. and Namba, S. (1999) '*Candidatus* Phytoplasma japonicum', a new phytoplasma taxon associated with Japanese Hydrangea phyllody. *International Journal of Systematic Bacteriology* 49, 1275–1285.

Schneider, B. and Seemüller, E. (1994) Presence of two sets of ribosomal genes in phytopathogenic mollicutes. *Applied and Environmental Microbiology* 60, 3409–3412.

Schneider, B., Ahrens, U., Kirkpatrick, B.C. and Seemüller, E. (1993) Classification of plant-pathogenic mycoplasma-like organisms using restriction-site analysis of PCR-amplified 16S rDNA. *Journal of General Microbiology* 139, 519–527.

Schneider, B., Seemüller, E., Smart, C.D. and Kirkpatrick, B.C. (1995) Phylogenetic classification of plant pathogenic mycoplasmalike organisms or phytoplasmas. In: Razin, R and Tulley, J.G. (eds) *Molecular and Diagnostic Procedures in Mycoplasmology*. Academic Press, San Diego, California, pp. 369–380.

Schneider, B., Torres, E., Martín, M.P., Schroder, M., Behnke, H.D. and Seemüller, E. (2005) '*Candidatus* Phytoplasma pini', a novel taxon from *Pinus silvestris* and *Pinus halepensis*. *International Journal of Systematic and Evolutionary Microbiology* 55, 303–307.

Sears, B.B. and Kirkpatrick, B.C. (1994) Unveiling the evolutionary relationships of plant pathogenic mycoplasmalike organisms. *ASM News* 60, 307–312.

Seemüller, E. and Schneider, B. (2004) '*Candidatus* Phytoplasma mali', '*Candidatus* Phytoplasma pyri' and '*Candidatus* Phytoplasma prunorum', the causal agents of apple proliferation, pear decline and European stone fruit yellows, respectively. *International Journal of Systematic and Evolutionary Microbiology* 54, 1217–1226.

Seemüller, E., Marcone, C., Lauer, U., Ragozzino, A. and Göschl, M. (1998) Current status of molecular classification of the phytoplasmas. *Journal of Plant Pathology* 80, 3–26.

Seemüller, E., Garnier, M. and Schneider, B. (2002) Mycoplasmas of plants and insects. In: Razin, S. and Herrmann, R. (eds) *Molecular Biology and Pathology of Mycoplasmas*. Academic/Plenum Kluwer, London, pp. 91–116.

Šeruga-Musić, M., Krajačić, M. and Škorić, D. (2008) The use of SSCP analysis in the assessment of phytoplasma gene variability. *Journal of Microbiological Methods* 73, 69–72.

Smith, C.J. and Osborn, A.M. (2009) Advantages and limitations of quantitative PCR (q-PCR)-based approaches in microbial ecology. *FEMS Microbiology and Evolution* 67, 6–20.

Stackebrandt, E. (2007) Forces shaping bacterial systematics. *Microbe* 2, 283–288.

Stackebrandt, E. and Goebel, B.M. (1994) Taxonomic note: a place for DNA–DNA reassociation and 16S rRNA sequence analysis in the present species definition in bacteriology. *International Journal of Systematic Bacteriology* 44, 846–849.

Stackebrandt, E., Frederiksen, W., Garrity, G.M., Grimont, P.A., Kampfer, P., Maiden, M.C., Nesme, X., Rosselló-Móra, R., Swings, J., Truper, H.G., Vauterin, L., Ward, A.C. and Whitman, W.B. (2002) Report of the ad hoc committee for the re-evaluation of the species definition in bacteriology. *International Journal of Systematic and Evolutionary Microbiology* 52, 1043–1047.

Tamura, K., Dudley, J., Nei, M. and Kumar, S. (2007) MEGA4: molecular evolutionary genetics analysis (MEGA) software version 4.0. *Molecular Biology and Evolution* 24, 1596–1599.

Torres, E., Bertolini, E., Cambra, M., Montón, C. and Martín, M.P. (2005) Real-time PCR for simultaneous and quantitative detection of quarantine phytoplasmas from apple proliferation (16SrX) group. *Molecular and Cellular Probes* 19, 334–340.

Tran-Nguyen, L.T., Kube, M., Schneider, B., Reinhardt, R. and Gibb, K.S. (2008) Comparative genome analysis of '*Candidatus* Phytoplasma australiense' (subgroup tuf-Australia I; rp-A) and '*Ca.* Phytoplasma asteris' strains OY-M and AY-WB. *Journal of Bacteriology* 190, 3979–3991.

Tsai, J.H. (1979) Vector transmission of mycoplasmal agents of plant diseases. In: Whitcomb, R.F. and Tully, J.G. (eds) *The Mycoplasmas III, Plant and Insect Mycoplasmas*. Academic Press, New York, pp. 265–307.

Tymon, A.M., Jones, P. and Harrison, N.A. (1998) Phylogenetic relationships of coconut phytoplasmas and the development of specific oligonucleotide PCR primers. *Annals of Applied Biology* 132, 437–452.

Valiunas, D., Staniulis, J. and Davis, R.E. (2006) '*Candidatus* Phytoplasma fragariae', a novel phytoplasma taxon discovered in yellows diseased strawberry, *Fragaria* × *ananassa*. *International Journal of Systematic and Evolutionary Microbiology* 56, 277–281.

Vandamme, P., Pot, B., Gillis, M., Devos, P., Kersters, K. and Swings, J. (1996) Polyphasic taxonomy, a consensus approach to bacterial systematics. *Microbiological Reviews* 60, 407–438.

Verdin, E., Salar, P., Danet, J.L., Choueiri, E., Jreijiri, F., El Zammar, S., Gélie, B., Bové, J.M. and Garnier, M. (2003) '*Candidatus* Phytoplasma phoenicium' sp. nov., a novel phytoplasma associated with an emerging lethal disease of almond trees in Lebanon and Iran. *International Journal of Systematic and Evolutionary Microbiology* 53, 833–838.

Wayne, L.G., Brenner, D.J., Colwell, R.R., Grimont, P.A.D., Kandler, O., Krichevsky, M.I., Moore, L.H., Moore, W.E.C., Murray, R.G.E., Stackebrandt, E., Starr, M.P.

and Trüper, H.G. (1987) Report of the Ad Hoc Committee on Reconciliation of Approaches to Bacterial Systematics. *International Journal of Systematic Bacteriology* 37, 463–464.

Wei, W., Opgenorth, D.C., Davis, R.E., Chang, C.J., Summers, C.G. and Zhao, Y. (2006) Characterization of a novel adhesin gene and design of a real-time polymerase chain reaction for rapid, sensitive and specific detection of *Spiroplasma kunkelii*. *Plant Disease* 90, 1233–1238.

Wei, W., Cai, H., Chen, H.R., Davis, R.E. and Zhao, Y. (2007a) First report of a natural infection of *Opuntia* sp. by a '*Candidatus* Phytoplasma asteris'-related phytoplasma in China. *Plant Disease* 91, 461.

Wei, W., Davis, R.E., Lee, I.-M. and Zhao, Y. (2007b) Computer-simulated RFLP analysis of 16S rRNA genes: identification of ten new phytoplasma groups. *International Journal of Systematic and Evolutionary Microbiology* 57, 1855–1867.

Wei, W., Lee, I.-M., Davis, R.E., Suo, X. and Zhao, Y. (2007c) Virtual RFLP analysis of 16S rDNA sequences identifies new subgroups in the clover proliferation phytoplasma group. *Bulletin of Insectology* 60(2), 349–350.

Wei, W., Davis, R.E., Jomantiene, R. and Zhao, Y. (2008a) Ancient, recurrent phage attacks and recombination events shaped dynamic sequence-variable mosaic structures at the root of phytoplasma genome evolution. *Proceedings of the National Academy of Sciences of the USA* 105, 11827–11832.

Wei, W., Lee, I.-M., Davis, R.E., Suo, X. and Zhao, Y. (2008b) Automated RFLP pattern comparison and similarity coefficient calculation for rapid delineation of new and distinct phytoplasma 16Sr subgroup lineages. *International Journal of Systematic and Evolutionary Microbiology* 58, 2368–2377.

Weisburg, W.G., Tully, J.G., Rose, D.L., Petzel, J.P., Oyaizu, H., Yang, D., Mandelco, L., Sechrest, J., Lawrence, T.G. *et al.* (1989) A phylogenetic analysis of the mycoplasmas: basis for their classification. *Journal of Bacteriology* 171, 6455–6467.

White, D.T., Blackall, L.L., Scott, P.T. and Walsh, K.B. (1998) Phylogenetic positions of phytoplasmas associated with dieback, yellow crinkle and mosaic diseases of papaya, and their proposed inclusion in '*Candidatus* Phytoplasma australiense' and a new taxon, '*Candidatus* Phytoplasma australasia'. *International Journal of Systematic Bacteriology* 48, 941–951.

Whitman, D.F. and Dunbar, S.A. (2008) Real-time polymerase chain reaction detection methods. *Recent Patents on DNA and Gene Sequences* 2, 20–26.

Woese, C.R. (1987) Bacterial evolution. *Microbiological Reviews* 51, 221–271.

Woese, C.R. (2000) Interpreting the universal phylogenetic tree. *Proceedings of the National Academy of Sciences of the USA* 97, 8392–8396.

Woese, C.R., Maniloff, J. and Zablen, L.B. (1980) Phylogenetic analysis of the mycoplasmas. *Proceedings of the National Academy of Sciences of the USA* 77, 494–498.

Yarza, P., Richter, M., Peplies, J., Euzeby, J., Amann, R., Schleifer, K.H., Ludwig, W., Glöckner, F.O. and Rosselló-Móra, R. (2008) The All-Species Living Tree project: a 16S rRNA-based phylogenetic tree of all sequenced type strains. *Systematic and Applied Microbiology* 31, 241–250.

Zhao, Y., Davis, R.E. and Lee, I.-M. (2005) Phylogenetic positions of '*Candidatus* Phytoplasma asteris' and *Spiroplasma kunkelii* as inferred from multiple sets of concatenated core housekeeping proteins. *International Journal of Systematic and Evolutionary Microbiology* 55, 2131–2141.

Zhao, Y., Sun, Q., Wei, W., Davis, R.E., Wu, W. and Liu, Q. (2009a) '*Candidatus* Phytoplasma tamaricis', a novel taxon discovered in witches'-broom diseased salt cedar (*Tamarix chinensis* Lour.). *International Journal of Systematic and Evolutionary Microbiology* 59 (in press).

Zhao, Y., Wei, W., Lee, I.-M., Shao, J., Suo, X. and Davis, R.E. (2009b) Construction of an interactive online phytoplasma classification tool, *i*PhyClassifier, and its application in analysis of the peach X-disease phytoplasma group (16SrIII). *International Journal of Systematic and Evolutionary Microbiology* 59 (in press).

Zreik, L., Carle, P., Bové, J.M. and Garnier, M. (1995) Characterization of the mycoplasmalike organism associated with witches'-broom disease of lime and proposition of a *Candidatus* taxon for the organism, '*Candidatus* Phytoplasma aurantifolia'. *International Journal of Systematic Bacteriology* 45, 449–453.

6 Phytoplasma Phylogeny and Detection Based on Genes other than 16S rRNA

JENNIFER HODGETTS AND MATTHEW DICKINSON

University of Nottingham, UK

Introduction

Phytoplasma diagnostics and phylogenetics have historically been based on the 16S rRNA gene and the 16S–23S rRNA spacer region because of the availability of universal primers for this region. However, it is well known that phylogenetics based on a single, highly conserved gene has limitations, particularly when it comes to defining subgroups and strain differences within subgroups, and the approach that is now being used for most other organisms is to rebuild trees by combining sequence data from a range of different genes. Finding universal primers that can be used for additional genes in phytoplasmas has proven difficult because of the sequence divergence between strains. Primers for genes such as *tuf* and the *rp* operon have been designed for use within specific 16Sr groups and have been useful for subgroup discrimination, and, more recently, semi-universal primers have been designed for the *rp* operon and universal primers for the *secA* gene. Results from these genes confirm the phylogenetic groupings based on the 16S rRNA gene but provide better resolution between isolates and also confirm the extensive amount of variation in both DNA and amino acid sequences between different phytoplasmas. These new sequences are also being developed into alternative phytoplasma diagnostic systems which are aimed at providing rapid methods for both identifying the presence of phytoplasmas and assigning them to specific phylogenetic groups.

Phylogenetics and Diagnostics Based on the rRNA Operon

16S rRNA gene

The rRNA operon in bacteria consists of the 16S rRNA gene followed by an internal transcribed spacer (ITS) and the 23S rRNA gene, and, in the case of

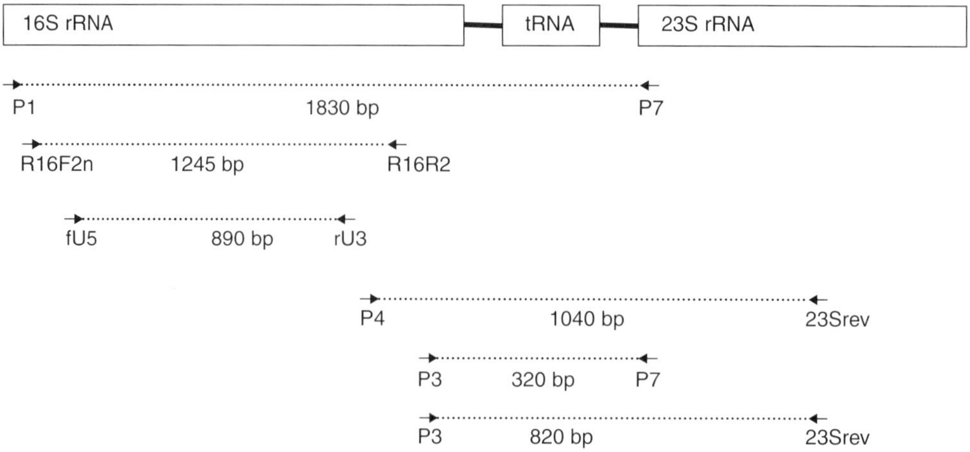

Fig. 6.1. Diagrammatic representation of the 16S–23S rRNA operon, showing the position of some of the various universal primers that have been developed for PCR amplification of this region from phytoplasmas. Primer names are given under the arrows and the sizes of the expected amplicons are shown between the dotted lines. Not drawn to scale.

phytoplasmas, there are two copies of this operon and the ITS normally contains a single tRNA within it. Since the late 1980s, numerous PCR primer combinations have been devised to amplify parts of this operon for diagnostics and phylogenetics (Fig. 6.1).

Some of these, such as P1 and P7, are universal primers that work on DNA from all phytoplasma phylogenetic groups, whilst others are group-specific (Smart *et al.*, 1996; Firrao *et al.*, 2005). However, diagnostics based on these primers can be problematic, with occasional false positives, particularly, for example, through amplification of any *Bacillus* spp. that might be present in a plant sample (Harrison *et al.*, 2002). In addition, it is important to guard against false negatives, since there is no internal control built into the diagnostic test to confirm that a negative result is due to a lack of phytoplasma and not PCR inhibition. The presence of false negatives due to PCR inhibitors can be tested through spiking controls, in which DNA preparations are spiked with DNA that is known to amplify, but this approach does not rule out the possibility that the DNA extraction might have been unsuccessful and/or the DNA degraded, and for this separate assays using primers such as those for the plant cytochrome C oxidase (COX assay) should be used. However, both of these approaches are expensive and labour intensive, especially if many samples are to be tested, for example in screening assays to identify insect vectors. More recently, real-time PCR assays have been developed for both generic and specific phytoplasma detection based on the 16S rRNA gene, and these assays have the advantage of being better automated and less labour intensive than conventional PCR, such that appropriate controls can be conducted more easily (see Galetto and Marzachì, Chapter 1, this volume). In addition, a 16S rRNA gene-based oligonucleotide array system consisting of 21- to 33-nt-long oligonucleotides has been developed to identify samples from different subgroups (Nicolaisen and Bertaccini, 2007).

Analysis of the 16S rRNA gene has resulted in the comprehensive and widely used 16Sr group classification system for phytoplasmas, which is based on restriction enzyme digest profiles of the 16S rRNA PCR products. In this system, universal primers are used to amplify a specified region of the 16S rRNA gene, and the PCR product is digested with specific restriction enzymes and the profiles analysed, normally using polyacrylamide gel electrophoresis to give good resolution of the digest products. The RFLP profiles are then compared, and specific patterns are used to classify phytoplasmas into groups and subgroups (Lee *et al.*, 1998). Prior to 2006, there were 18 defined 16Sr groups and more than 40 subgroups (Wei *et al.*, 2007). However, problems with maintaining reference strains have made it difficult to undertake conventional RFLP analysis and compare patterns by gel electrophoresis for all the different groups and subgroups and have also made it difficult to incorporate new strains and subgroups into the system. As a result, Wei *et al.* (2007, 2008) have devised a system based on sequences deposited at the National Center for Biotechnology Information's (NCBI) nucleotide sequence database, in which the restriction digestion profiles are simulated for a broad range of enzymes using *in silico* methods such as the AcaClone pDRAW32 system (http://acaclone.com). Similarity coefficients for the restriction fragments are then calculated by defined formulae (Nei and Li, 1979; Lee *et al.*, 1998; Wei *et al.*, 2008) to reflect the number of shared and distinct fragments between any given strains, and, based on a similarity coefficient of 0.85, ten new 16Sr groups have now been assigned, with groups classified as 16SrI to 16SrXXVIII. This revised phylogenetic grouping is shown in Fig. 6.2, and this figure can be used as a reference point for the group descriptions used throughout the rest of this chapter.

The *in silico* system has the potential to work well for sequences of new phytoplasmas in that they can be readily assigned to appropriate groups or subgroups. However, it does require the complete and accurate sequencing of a 1.25 kb region of the 16S rRNA gene for every test to ensure that the RFLP analysis is correct, and, whilst this is possible in some phytoplasma diagnostics laboratories, it is not appropriate as a routine diagnostic technique in laboratories in many countries where equipment and facilities are limited. There is also a real danger that slight misreads during sequencing will result in sequences being used for the *in silico* analysis that are not accurate, resulting, in turn, in inaccurate restriction profiles and phylogenetic classifications. Unfortunately, there are already a number of phytoplasma 16S rRNA gene sequences in the NCBI databases that clearly contain errors and/or are incorrectly annotated, so care needs to be taken in using this approach, and there is a case to be made that phytoplasma sequences should be independently confirmed before they are submitted to databases, particularly if there is an indication that they may result in a new group or subgroup. A further complication with basing a system on RFLP analysis can come from the presence of heterogeneity between the two copies of the 16S

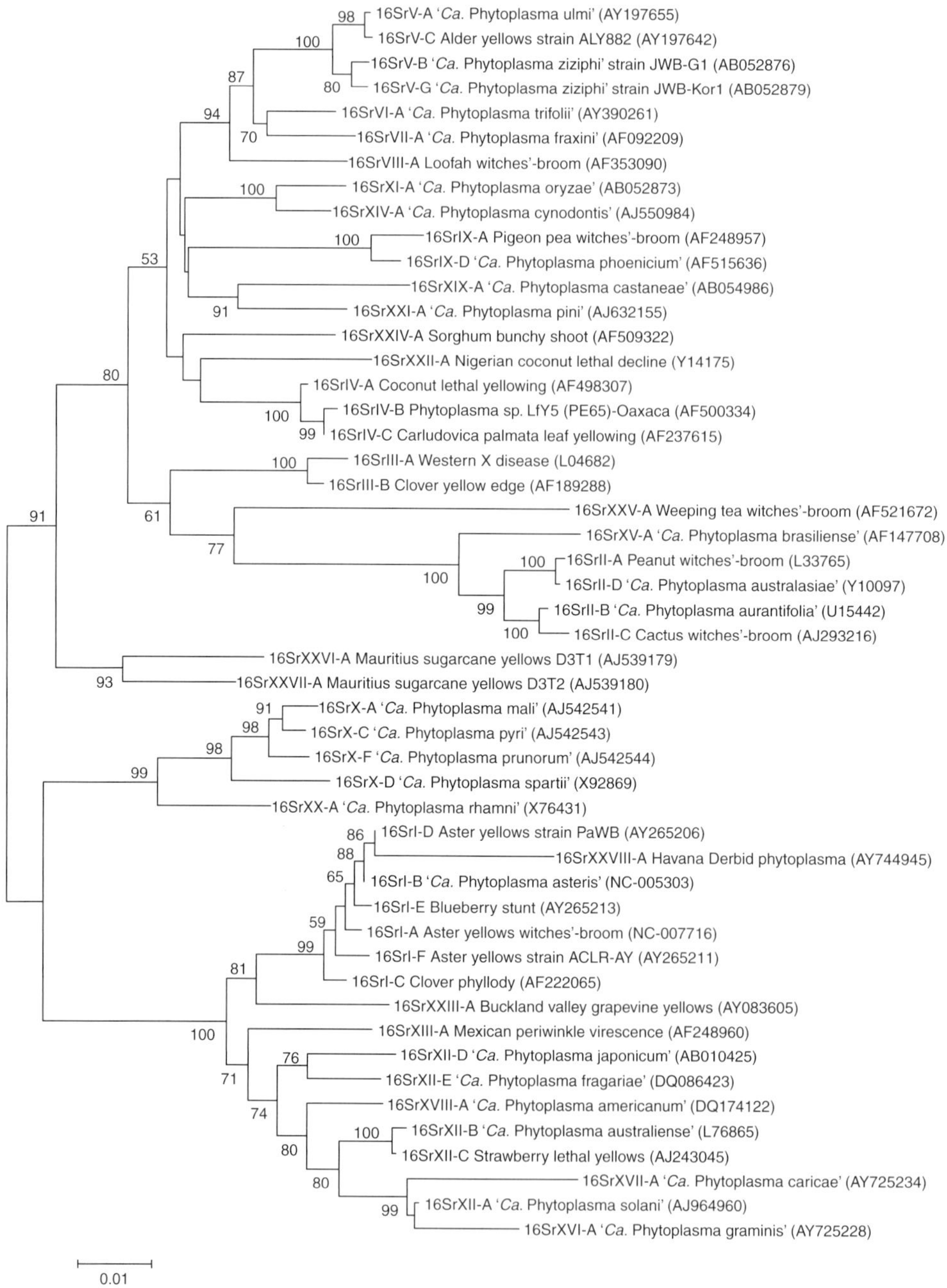

Fig. 6.2. Dendrogram, constructed by the neighbor-joining method, showing the phylogenetic relationships amongst all the 28 phytoplasma groups identified by *Wei et al.* (2007) based on the 16S rRNA gene (between primers R16F2n and R16R2). GenBank accession numbers for sequences are shown in parentheses alongside the names of the phytoplasmas. Bootstrap values greater than 50% (expressed as percentages of 1000 replications) are shown, and branch lengths are proportional to the number of inferred character state transformations. Bar, substitutions per base.

rRNA gene, as was shown for *Phormium* yellow leaf phytoplasma (Liefting *et al.*, 1996).

In parallel with the 16Sr classification system is the '*Candidatus* (*Ca.*) Phytoplasma' taxon system, in which there are currently 25 major phylogenetic groups. This approach is also based primarily on the 16S rRNA gene sequence, in that strains within a species should share at least 97.5% sequence identity within the 16S rRNA gene (IRPCM, 2004; Firrao *et al.*, 2005). However, this system also recognizes that 16S rRNA sequences alone are not sufficient for defining species, and other criteria such as host ranges and/or vectors are being used to separate species such as '*Ca.* Phytoplasma mali' and '*Ca.* Phytoplasma pyri', which would be within a single species based on the 16S rRNA sequences alone. Clearly classification systems in the future cannot be based on the 16S rRNA gene alone, and we will now detail some of the other genes that are being used to help provide a more robust approach to phytoplasma phylogenetics and classifications.

Other sequences derived from the rRNA operon

While the 16S rRNA gene-based systems have provided a solid basis for phytoplasma diagnostics and phylogenetics, there have been ongoing attempts to devise and develop alternative primers based on other phytoplasma genes, which might be better for distinguishing between different isolates and provide further definition of the groups and subgroups. Initially these studies were based on sequences adjacent to the 16S rRNA gene, in particular the 16S–23S ITS region, with a universal primer P7 developed that could be used as a reverse primer from the 23S rRNA (Smart *et al.*, 1996) (Fig. 6.1). Sequencing of the spacer region clearly showed that there was much more sequence variation in this region than in the 16S rRNA gene, presumably because there are fewer evolutionary constraints on the spacer. However, certain anomalies and inconsistencies also became apparent. Phylogenetic trees based on the ITS region alone are poorly supported by bootstrap values and place certain isolates in wrong groups (Hodgetts *et al.*, 2008), and there is also evidence that variations occur in this region between the two copies of the rRNA operon within particular phytoplasmas. For example, one of the two rRNA operons in stylosanthes little leaf has been shown to lack a $tRNA^{Ile}$ in the 16S–23S ITS region (De La Rue *et al.*, 2001).

Recently, universal primers have been developed to extend the sequencing of the phytoplasma rRNA operon further into the 23S rRNA gene (Hodgetts *et al.*, 2007). The 23Srev primer, which was developed for universal amplification of bacterial 23S rRNA genes (Anthony *et al.*, 2000), has been shown to work well on phytoplasmas in combination with phytoplasma-specific forward primers and has provided an additional 500 bp of 23S rRNA gene-sequence information for a number of different phytoplasmas beyond the P7 primer (Fig. 6.1). The phylogenetic trees based on these partial 23S rRNA gene sequences are similar to those obtained for 16S rRNA gene sequences, although there are a few minor anomalies (Hodgetts *et al.*, 2008), probably due to the relatively short nature of this 23S rRNA sequence (500 bp) when compared with the 16S rRNA gene (approx 1500 bp). The entire 23S rRNA gene is approx 2850 bp, and it may ultimately be possible to design universal primers that amplify this entire sequence for all phytoplasmas, but as yet such primers are not available.

Alternative Genes for Group and Subgroup Identification

Primers based on the 16S rRNA gene in combination with RFLP analysis have proved very useful for identification of groups and subgroups of phytoplasmas. However, sequencing and, more recently, the *in silico* methods of Wei *et al.* (2007) have shown that there are differences in the sequences between isolates within subgroups that cannot be identified by digestion with the conventional restriction enzymes and gel systems commonly used for RFLP analysis.

There are also, of course, issues when it comes to defining subgroups, particularly if these are based on a single, well-conserved gene, and by setting the similarity coefficient at different levels it is possible to change group and subgroup structures. The '*Ca.* Phytoplasma' definition system tries to take account of these issues by basing species on more than just gene sequences, using criteria such as host range and vector as well. However, the approach that is now being used in other organisms, for example, to define the 'tree of eukaryotes', is to rebuild trees that were originally based on single genes by combining sequence data for a number of different genes (Keeling *et al.*, 2005). This can be achieved by concatenating the sequence data from a number of genes into a single sequence and building the phylogeny based on this or, alternatively, the trees derived for the different genes can be combined using appropriate software into larger 'supertrees'. For example, James *et al.* (2006) have recently used a data set in which six genes are combined (nu-SSU, nu-LSU, 5.8S rRNA, *rpb1, rpb2* and *tef1*) to construct a comprehensive phylogenetic analysis of the *Fungi*.

In phytoplasmas, initial attempts to add alternative genes into phylogenetic studies were performed using genes such as the ribosomal protein genes *rpl22* and *rps3* (Lim and Sears, 1992; Gundersen *et al.*, 1994; Toth *et al.*, 1994) and the *tuf* gene (Schneider *et al.*, 1997). The primers used for these studies were generally designed based on the sequences that were available at the time for culturable mollicutes such as *Acholeplasma laidlawii*, *Spiroplasma citri* and *Mycoplasma mycoides* and were found to work well for a number of phytoplasma taxonomic groups but not for all, as is discussed in more detail below.

Subsequently, as complete genome sequences have become available for different phytoplasmas, it has been possible to try and design primers for other genes, based on sequence alignments. However, the publicly available phytoplasma genome sequences to date are from two 16SrI aster yellows (AY) phytoplasmas (onion yellows (OY) (Oshima *et al.*, 2004) and AY witches'-broom (Bai *et al.*, 2006)), 16SrXII '*Ca.* Phytoplasma australiense' (Tran-Nguyen *et al.*, 2008) and 16SrX apple proliferation (AP) (Kube *et al.*, 2008). The first three of these are in the AS branch of phytoplasmas (Wang *et al.*, 2003), which generally have genomes that are larger than the other phytoplasmas, on average by 230 kb. Sequencing and comparative genomics have shown that these three genomes are circular and they contain large numbers of potential mobile units (PMUs), which comprise up to 23% of these phytoplasma genomes and which exhibit features of bacterial pathogenicity islands. Furthermore, whilst roughly 30% of the genome shows good synteny between the 'closely related' 16SrI OY and AY witches'-broom genomes, less than 10% of the genome shows good synteny between these isolates and that of the 16SrXII '*Ca.* Phytoplasma australiense' genome. The AP phytoplasma, however, is in the WB group of phytoplasmas, which generally have smaller genomes, with less repetitive DNA, and are believed to have evolved independently. This genome is linear, has only one partial PMU sequence and shows little synteny with the AS phytoplasmas. In phylogenetic trees, AP and the other 16SrX phytoplasmas also form a quite distinct phylogenetic grouping, closer to the AS group than to the other WB phytoplasmas (Fig. 6.2). As yet there is only partial sequence information available for these other more diverse phytoplasma groups, but already the evidence suggests that these other phytoplasmas have gene sequences that are very poorly conserved in comparison to the published whole-genome sequences. Some clear evidence for this is provided by the *secA* gene sequence alignments discussed later and shown in Fig. 6.3.

The consequence of this is that, in attempts that we and others have made to align sequences for known genes from published phytoplasma sequences and to design primers based on these that are universally applicable to all phytoplasmas, it has proven very difficult to find any well-conserved sequences to design such universal primers and even more so to design the two sets of universal primers required for the nested PCR assays that are generally required for detecting phytoplasmas at low titre. We will now discuss in more detail the primers that have been designed and used to date for genes other than the rRNA operon.

rp operon genes

The initial cloning of *rp* genes from a phytoplasma was achieved by using a heterologous probe derived from the *Mycoplasma capricolum* ribosomal protein gene operon to isolate a DNA fragment encoding the proteins rpL2 and rpS19 from a phytoplasma infecting *Oenothera* (Lim and Sears, 1991). Subsequently, the same approach was used to clone a second DNA fragment, encoding proteins rpL22 and rpS3 (Lim and Sears, 1992). The *rp* genes are part of a single, large operon, the *rp* operon, which contains at least 21 genes in all the phytoplasmas sequenced to date. In the literature, these genes and the ribosomal proteins they encode are sometimes given alternative names, but the general order that they are found in the phytoplasma genome is: *rpsJ* (encoding ribosomal protein S10), *rplC* (L3), *rplD* (L4), *rplW* (L23), *rplB* (L2), *rpsS* (S19), *rplV* (L22), *rpsC* (S3), *rplP* (L16), *rpmC* (L29), *rpsQ* (S17), *rplN* (L14), *rplX* (L24), *rplE* (L5), *rpsN* (S14), *rpsH* (S8), *rplF* (L6), *rplR* (L18), *rpsE* (S5), *rpmD* (L30) and *rplO* (L15).

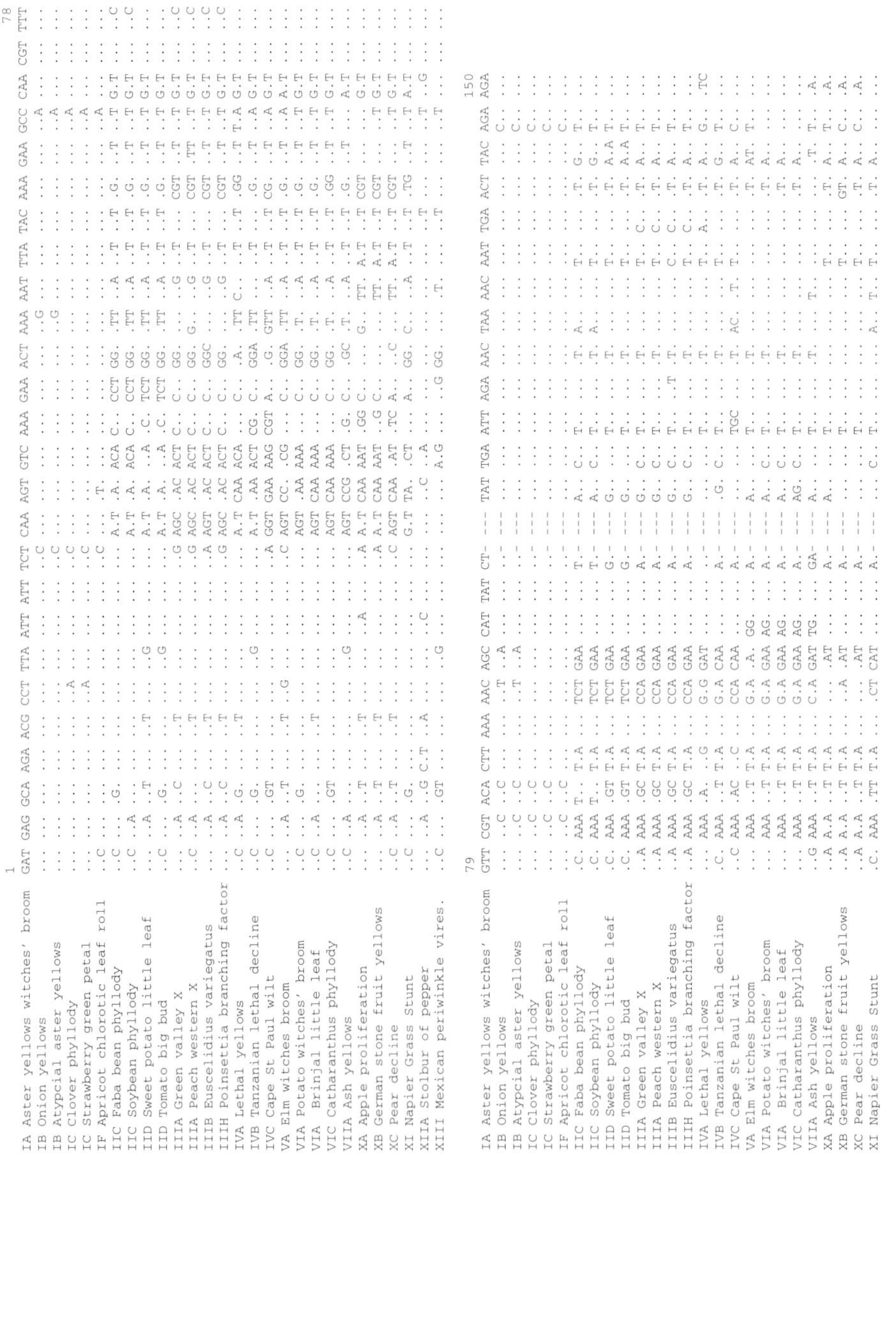

Fig. 6.3. Alignment of the first 150 bp of the *secA* gene, starting from primer secAfor2, for phytoplasmas from a range of phylogenetic groups. Sequences were aligned using CLUSTALW and dots represent amino acids identical to the aster yellows witches'-broom consensus sequence.

Based on this initial *Oenothera* phytoplasma sequencing, primers rpF1 (5′ GGA CAT AAG TTA GGT GAA TTT 3′) and rpR1 (5′ ACG ATA TTT AGT TCT TTT TGG 3′) were designed, which were initially shown to amplify the *rplV* (L22), *rpsC* (S3) and part of the *rplP* (L16) gene from *Acholeplasma laidlawii*. Gundersen *et al.* (1994) extended these studies and used these primers along with a second set of primers, rpF2 (5′ TCT CGT ACT TTT CGT GG 3′) and rpR2 (5′ ACC TTT AGC TCT TGG AA 3′), to amplify a 1245–1389 bp region encompassing the same genes from 16SrI, III, V, VII, VIII, IX and X group phytoplasmas. However, there were some phylogenetic groups, notably the 16SrII group, for which these primers failed to amplify PCR products. Additional primers have subsequently been designed such that by using the right combinations of primers it is now possible to amplify these same genes from phytoplasmas belonging to all the phylogenetic groups, and these primers are referred to as semi-universal (Martini *et al.*, 2007). In particular, the degenerate primer pair rpF1C (5′ ATG GTD GGD CAY AAR TTA GG 3′) and rp(I)R1A (5′ GTT CTT TTT GGC ATT AAC AT 3′) have been shown to amplify a 1212–1386 bp product for these genes from groups 16SrI, II, III, IV, V, VI, VII, IX, X, XII, XIII and XVIII. Primers for these genes have also been designed that are group specific, and, in addition, Martini *et al.* (2007) have designed a forward primer rpL2F3 (5′ WCC TTG GGG YAA AAA AGC TC 3′) from the *rplB* (L2) gene that was originally identified by Lim and Sears (1991). This primer works in combination with primer rp(I)R1A to amplify a 1600 bp product encompassing the 3′ end of the *rplB* (L2) gene, the *rpsS* (S19), *rplV* (L22), *rpsC* (S3), *rplP* (L16) genes and the 5′ end of the *rpmC* (L29) gene. These primers have been shown to work on groups 16SrI, III, IV, V, VI, VII, IX, X, XII, XIII and XVIII phytoplasmas but not on isolates from groups 16SrII or XI.

The PCR products amplified using these *rp* operon primers have been used in RFLP analysis in combination with 16S rRNA gene RFLP analysis to assign phytoplasmas to 16Sr-rp subgroups (Lee *et al.*, 1998). More recently, they have been used to produce phylogenetic trees, which are nearly congruent with those derived from 16S rRNA sequencing (Martini *et al.*, 2007). Furthermore, the added resolution provided by use of genes encoding proteins has enabled the subdividing of the groups into more distinct subclades than the 16S rRNA gene. For example, the 16SrV elm yellows (EY) group has been resolved into two distinct subclades, as have the 16SrIV lethal yellows group and the 16SrIX pigeon pea witches'-broom group. The 16SrII group has been resolved into four subclades, and 16SrXII and 16SrXVIII have been resolved into three subclades. It is possible that other *rp* gene universal primers could be designed in the future, based on the phytoplasma genome sequences, but as yet no such primer sequences have been published.

tuf genes

The *tuf* gene, encoding the elongation factor Tu (EF-Tu), is a well-conserved gene with a central role in translation (Schneider *et al.*, 1997), and there is a single copy of this gene in the phytoplasma genome. This gene has often been used in phylogenetic studies for other bacteria. Two primers for the amplification of the *tuf* gene, fTuf1 (5′ CAC ATT GAC CAC GGT AAA AC 3′) and rTuf1 (5′ CCA CCT TCA CGA ATA GAG AAC 3′), were initially designed for phytoplasmas based on alignment of the *Mycoplasma pneumoniae, M. genitalium, M. gallisepticum, M. hominis* and *Ureaplasma urealyticum* sequences. These primers amplified products of the expected size (1000 bp) for 16SrI AY, 16SrIII green valley X and vaccinium witches'-broom and 16SrXII stolbur (STOL) groups but failed to amplify from 16SrII faba bean phyllody and 16SrX AP (Schneider *et al.*, 1997). We have subsequently found that these primers also fail to amplify from the 16SrIV coconut lethal yellowing (LY) and 16SrXXII coconut lethal decline phytoplasmas.

Additional primers based on these early phytoplasma sequences have now been designed but none have been found that can act as universal primers. The main use of *tuf* gene primers to date has therefore been to establish subgroups within the 16Sr groups, particularly within the 16SrI AY group (Marcone *et al.*, 2000) and the 16SrXII '*Ca.* Phytoplasma australiense' group (Streten and Gibb, 2005). In a study on the AY group, the AY-specific primers fTufAy (5′ GCT AAA AGT AGA GCT TAT GA 3′) and rTufAy (5′ CGT TGT CAC CTG GCA TTA CC 3′) (Schneider *et al.*, 1997), which amplify a 940 bp product, were used on 70 phytoplasma isolates in conjunction with the 16S rRNA gene primers (Marcone *et al.*, 2000). RFLP analyses of the 16S rRNA gene PCR products divided the isolates into ten 16Sr subgroups (-A, -B, -C, -D, -E, -F, -K, -L, -M and -N), whilst the *tuf* gene RFLP profiles only resulted in seven subgroups (-A, -B, -C, -D, -E, -F and -G). In particular, the 16Sr subgroups -B, -D, -L and -M all gave the same *tuf* gene profiles. On the basis of these profiles, and in conjunction with pathological aspects, the authors concluded that the 16SrI-A, -B, -C, -D, -E, -F, -K and -N subgroups were substantially supported as distinct subgroups.

In a similar study on the '*Ca.* Phytoplasma australiense' phytoplasmas, Streten and Gibb (2005) examined 11 strains which had previously been indistinguishable using 16S rRNA gene analysis, using the same AY *tuf* gene primers (fTufAY and rTufAY) in conjunction with primers rpF1 and rpR1 for the *rplV* and *rpsC* genes. RFLP and sequence analysis were able to separate the isolates into three *tuf* gene subgroups (*tuf* Australia I, *tuf* New Zealand I and *tuf* New Zealand II) and three *rp* groups (*rp*A, *rp*B and *rp*C). By combining the data, four distinct subgroups were established.

A second elongation factor gene, *fus*, encoding EF-G, is adjacent to *tuf* in the phytoplasma genome and has also been used for the design of PCR primers (Berg and Seemüller, 1999). However, these primers, which were designed based on an AP sequence, were only able to amplify from 16SrI, III, X and XII group phytoplasmas and not from 16SrII, V or VII group phytoplasmas. There are currently no records of the use of these or other *fus* gene primers on any other phytoplasmas.

secY genes

The *secY* gene, which encodes a protein involved in the protein secretion mechanism from bacteria (in conjunction with SecA and other gene products), was initially identified in phytoplasmas in a randomly cloned 16SrV flavescence dorée DNA fragment that had originally been used as a hybridization probe (Daire *et al.*, 1997). Based on the sequencing of this fragment, a set of primers FD9f3/FD9r2 were designed to amplify a 1150 bp fragment that was found to be partly homologous to the *secY* translocase gene from mycoplasmas (Angelini *et al.*, 2003). This randomly cloned DNA fragment was subsequently used in a suppressive subtractive hybridization and genome walking approach to identify a larger clone which encoded the 3′ end of the *rplO* gene (encoding ribosomal protein L15), the entire *secY* gene along with the adjacent *map* (methionine aminopeptidase) gene, and the 5′ end of the *infA* gene, encoding initiation factor If-1 (Arnaud *et al.*, 2007). This genome walking strategy was also used in the same study on a second randomly cloned PCR product to identify a clone containing the second half of the *uvrB* gene, encoding the subunit B of exonuclease ABC, and the nearly complete *degV* gene, encoding a protein of the DegV family.

Based on these sequences, primers were designed that could be used in an analysis of the 16SrV phytoplasmas infecting grapevine and alders in Europe (Arnaud *et al.*, 2007). Nested primer sets were used to amplify a 1174 bp fragment of the *secY* gene, an 803 bp fragment of the *map* gene and a 1126 bp fragment of the *uvrB–degV* region from over 40 16SrV isolates, and phylogenetic trees were constructed from the sequences obtained. The resultant analysis consistently supported the existence of three distinct flavescence dorée strain clusters. Strain cluster FD1 had little variability and was present mainly in south-western France. Cluster FD2 showed no variability and was found in France and Italy, whilst cluster FD3 was highly variable and detected only in Italy. These results suggested that the FD2 strains may have been spread through distribution of infected propagation material between France and Italy. In addition, the analyses showed that the alder yellows phytoplasmas and German Palatinate grapevine yellows phytoplasmas belong to the same phylogenetic subclade.

The *secY* gene has also been used for differentiation of the AY group phytoplasmas (Lee *et al.*, 2006). In this study, primers were designed based on the published AY and OY sequences to amplify a 1.4 kb near-full-length *secY* gene. As far as we are aware, these primers have only been tested on AY group phytoplasmas so are probably specific for this group. Twenty representative 16SrI isolates were used in this study, representing the ten 16SrI subgroups (-A, -B, -C, -D, -E, -F, -K, -L, -M and -N). Sequencing and RFLP analysis of the resultant PCR products delineated the isolates into ten SecY groups. Two of these SecY subgroups corresponded to the 16SrI-A subgroup. Members of 16SrI-B subgroup were also split into two SecY subgroups, and the isolates from 16SrI-L and 16SrI-M also fell into one of these two SecY groupings, consistent with the findings from the earlier *tuf* gene analysis of

Marcone *et al.* (2000). The remaining 16SrI subgroups and SecY subgroups coincided with each other.

secA genes

The *secA* gene, which encodes SecA, the ATP-dependent force generator in the bacterial precursor protein translocation cascade system, was originally sequenced from OY as part of the genome sequencing project (Kakizawa *et al.*, 2001). In these studies, the 2505 bp gene, which encodes an 835 amino acid peptide, was cloned into an expression vector to produce tagged proteins that were used to raise antibodies for immunohistochemical studies. In subsequent work, Hodgetts *et al.* (2008) aligned the OY *secA* gene sequence with that of AY witches'-broom and the equivalent gene from coconut lethal yellowing, which is in the 16SrIV group. Based on these alignments, three degenerate primers were designed, SecAfor1 (5´ GAR ATG AAA ACT GGR GAA GG 3´), SecAfor2 (5´ GAY GAR GSW AGA ACK CCT 3´) and SecArev3 (5´ GTT TTR GCA GTT CCT GTC ATN CC 3´), for use in a semi-nested PCR assay in an attempt to amplify a part of the *secA* gene from a broad range of phytoplasma phylogenetic groups.

When the primer pair SecAfor1/SecArev3 was used in PCR, products of expected size (about 840 bp) were generated from approximately 90% of the 34 phytoplasma DNA samples tested, which included representative isolates from groups 16SrI through to 16SrXIII (with the exception of group 16SrVIII) and also 16SrXXII coconut lethal decline isolates. Some of the amplifications were weak, probably reflecting a low titre of phytoplasma DNA in these samples, or possibly poor primer binding, but when a semi-nested approach was used, in which phytoplasma DNA samples were first amplified with SecAfor1/SecArev3 and then reamplified with SecAfor2/SecArev3, the expected PCR product (approximately 480 bp) was readily obtained from all 34 phytoplasmas (Fig. 6.3). Subsequently, we have also shown that these primers work on 16SrXIV group phytoplasmas. Sequencing of these PCR products has confirmed that they all encode the SecA protein and that the region for the 16SrI group phytoplasmas is two amino acids longer than that from all the other phylogenetic groups.

Phylogenetic analysis based on the *secA* gene sequences has supported results obtained for other genes and provided improved resolution of groups and subgroups, particularly when compared with the 16S rRNA gene. The subgroups of 16SrI are clearly defined, and 16SrII clearly splits into a cluster that contains 16SrII-B and 16SrII-C isolates and a cluster that contains the 16SrII-D isolates. These results support the *in silico* results of Wei *et al.* (2007), who classified the 16SrII-B strains as '*Ca.* Phytoplasma aurantifolia' whilst classifying the 16SrII-D strains as '*Ca.* Phytoplasma australasiae'. In addition, the results using the *secA* gene have shown clear distinction between strains within the coconut lethal yellowing-type disease group. These are phytoplasmas that cause economically important lethal diseases of coconuts, all characterized by similar syndromes, which include premature fruit drop, floral necrosis, leaf discoloration and decline. Subgroup 16SrIV-A phytoplasmas are associated with lethal yellowing (LY) of coconut and other palm species in the Americas, whereas phytoplasmas that induce similar symptoms on coconut in Africa are referred to by other names, such as Cape St Paul wilt in Ghana, Awka in Nigeria and lethal decline in Tanzania. Strain differences were previously identified through 16S rRNA gene analysis, and Mpunami *et al.* (1999) also showed that it was possible to differentiate East African from West African coconut-associated phytoplasmas by selective amplification of 16S rRNA gene sequences during PCR or by RFLP analysis of 16S rRNA gene PCR products, such that they were classified as groups 16SrIV-B and 16SrIV-C, respectively. In the *in silico* work of Wei *et al.* (2007), the 16SrIV-C Nigerian coconut lethal decline phytoplasma (LDN) (Awka), which has an almost identical 16S rRNA gene sequence to the Cape St Paul wilt phytoplasma from Ghana, was allocated a distinct 16Sr group, 16SrXXII-A. The Hodgetts *et al.* (2008) analysis of the *secA* gene confirms the high degree of divergence between the different coconut phytoplasmas and supports this separation of the coconut lethal yellows diseases of the Americas from the coconut lethal decline diseases of Africa but also indicates that the Tanzanian lethal decline phytoplasma should be put in a separate group from the Nigerian and Ghanaian lethal declines and therefore probably in a new 16Sr group.

Other genes

A number of attempts have been made to identify other genes that could be used to design universal primers for phytoplasma amplification. Goodwin *et al.* (1994) used a plasmid that had been isolated from a 16SrI severe strain of western aster yellows (SAY) by CsCl ethidium bromide gradients to design primers which could amplify a 235 bp fragment of the plasmid DNA. These primers were shown to work on several other strains of AY, 16SrX pear decline, 16SrV EY and 16SrVI beet leafhopper-transmitted virescence from both plant and insect samples, and the PCR approach was at least 500 times more sensitive than previous tests using the plasmid DNA as a probe in dot blots. It was also possible to use restriction enzyme digests to differentiate between strains, and sequencing showed that there was around 93% homology between the EY and AY plasmid sequence. However, plasmids have not been found in all phytoplasma strains, the gene from which the sequence came has not been identified, and whether this technique and these primers would be universally applicable is unknown.

At about the same time, Jarausch *et al.* (1994) sequenced a clone that they derived from the AP and found that one of the open reading frames encoded a protein with significant homology to bacterial nitroreductases. The primers which they derived from this sequence worked successfully in PCR to amplify the same gene from other 16SrX phytoplasmas, and they were able to use RFLP analysis to assign the isolates to different subgroups. However, when the same primers were used on DNA from isolates belonging to groups 16SrI, III, V, VII and XII, no PCR products were obtained. Similarly, Streten and Gibb (2003) obtained random clones from 16SrII tomato big bud DNA and sequenced the clones to identify the putative genes located on them. Primers were then designed based on alignments of these sequences with sequences of equivalent genes from other phytoplasmas, mycoplasmas and bacteria. Four sets of primers were designed, for parts of the DNA polymerase beta II chain gene, the peptide chain release factor (RF-1) gene, the ATP-dependent RNA helicase gene and the *FtsH* gene, which encodes a protein involved in cell division. These primers were tested on three isolates of tomato big bud and two of the closely related 16SrII sweet potato little leaf, but none of the four primer sets were able to amplify from all five isolates. We have subsequently tested these primers on a range of other phytoplasmas and confirmed that they only amplify from selected 16SrII isolates and are therefore not universal phytoplasma primers. Wagner *et al.* (2001) used a different approach and designed primers based on known mycoplasma sequences for the RNase P gene, which encodes a protein essential for tRNA maturation. Following sequencing of the PCR product that they obtained from AP, they refined the primers and were able to get an improved PCR product for complete sequencing. However, there are no reports of these primers being used on any other phytoplasma groups.

A more circuitous route has been used to obtain DNA sequences and primers for immunodominant membrane proteins of phytoplasmas. These proteins are on the outer surface of the phytoplasmas and are believed to be involved in mediating the specific attachment of these bacteria to their insect vectors. They were first identified when monoclonal antibodies were prepared against 16SrII sweet potato witches'-broom-infected plant material that had been enriched for phytoplasma by using leaf midribs (Shen and Lin, 1993). These antibodies were shown to react to an 18.4 kDa antigenic protein in Western blots only from infected plants and were then used to screen a genomic library made from CsCl–bisbenzimide-enriched phytoplasma DNA (Yu *et al.*, 1998). Protein expression was induced from clones in this library using IPTG-saturated nitrocellulose filters, and clones that expressed the 18.4 kDa protein were identified from the library and sequenced. Using this approach, a 519 bp sequence that encoded the antigenic protein was identified, and primers were designed from this sequence and shown to amplify a PCR product of the correct size from the closely related peanut witches'-broom phytoplasma. However, these primers failed to amplify anything from the other phytoplasma-infected plants tested, which included 16SrI AY, 16SrV EY, 16SrVIII loofah witches'-broom and 16SrXI rice yellow dwarf. The same approach was used to obtain the AP immunodominant protein (IDP) (Berg *et al.*, 1999) and also that of Western X disease (Blomquist *et al.*, 2001), but sequencing of these genes showed that there was no significant homology between these, or between them and other phytoplasma IDP genes, and no primers were developed from this work for isolating immunodominant protein genes from other phytoplasmas.

For the 16SrI-B chlorante aster yellows (AY-C) phytoplasma, the gene was cloned by isolating the affinity-purified IDP on SDS-polyacrylamide gels, fragmenting the protein chemically and enzymatically, and sequencing the cleavage products. Degenerate primers were then designed, based on these amino acid sequences, and used to amplify the AY-C *amp* gene (Barbara *et al.*, 2002). Primers designed to this gene were then used to clone the equivalent gene from the 16SrI-C clover phyllody phytoplasma. However, whilst there were high levels of sequence similarity in the intergenic regions flanking the genes from the two isolates, and also in parts of the open reading frame, there were also regions, in particular the predicted large hydrophilic domain, where nucleic acid and amino acid conservation was low.

It is worth noting that the terminology for these major antigenic proteins can be confusing and that the terms antigenic membrane protein (amp), immunodominant membrane protein (imp) and immunodominant protein (idp) are sometimes being used for the same protein and sometimes for different ones. In OY, the *amp* gene (antigenic membrane protein) is located between the *GroEL* gene and the NAD^+ synthase gene, whilst in AY witches'-broom and '*Ca.* Phytoplasma australiense' the gene in the same location is referred to as the immunodominant membrane protein. However, in '*Ca.* Phytoplasma mali', there is no gene in an equivalent position in the genome, and the immunodominant protein that has been identified in this phytoplasma is between the *dnaD* gene and the *pyrG* gene. There is also a gene in the same position in the OY, AY witches'-broom and '*Ca.* Phytoplasma australiense' genomes, but in these cases it is referred to as encoding a hypothetical protein. Design of universal primers for phytoplasma immunogenic proteins is therefore unlikely, since they occur in at least two different locations in phytoplasma genomes and are very dissimilar in sequence. We have recently attempted to design primers from the sequences of genes flanking both of these major immunogenic proteins in the published phytoplasma genome sequences, to determine whether this could be a strategy for isolating these genes from other phytoplasmas, but as yet none of the primers that we have designed have been successful.

Based on phytoplasma genome sequences, we have also attempted to design primers from other genes. Some of these attempts have been to amplify the full-length gene sequence for specific genes from all phytoplasmas other than those already completely sequenced. To do this requires knowledge of the flanking genes so that primers can be designed from these to amplify into and through the gene of interest. However, recent studies have demonstrated that gene orders and synteny vary between phytoplasmas (Jomantiene *et al.*, 2007), including closely related strains such as AYWB and OY-M (Bai *et al.*, 2006), so it is therefore not possible, as yet, to predict which flanking genes to use for such a strategy to ensure success in amplifying complete genes from diverse groups of phytoplasmas. Instead, the current approach is to amplify parts of genes using primers designed for well-conserved regions following alignment of the particular gene sequences from 16SrI group phytoplasmas and the phylogenetically more-distant phytoplasmas such as 16SrXII '*Ca.* Phytoplasma australiense', 16SrX AP and any others that are available through past and current sequencing projects. However, even based on this strategy, it is often difficult to identify well-conserved sequences that are not too AT rich for good primer design or that result in such degenerate primers that artefacts are amplified.

Diagnostics Developments Based on Alternative Genes

Phytoplasma diagnostics has been routinely based on the 16S rRNA gene, using conventional PCR primers (Smart *et al.*, 1996; Firrao *et al.*, 2005), arrays (Nicolaisen and Bertaccini, 2007) and, most recently, real-time PCR (see Galetto and Marzachì, Chapter 1, this volume). However, there are sometimes problems with false positives, and in addition it is often difficult to diagnose the presence of mixed infections, since RFLP patterns can become complicated and difficult to interpret when overlapping patterns occur in the same track on a gel. As a result, primers based on alternative genes have been devised for conventional diagnostics, as already discussed in this chapter. In addition, alternative diagnostic methods based on these other gene sequences have been established, such as heteroduplex mobility assays (HMAs) (Wang and Hiruki, 2005), single-strand conformation polymorphisms (SSCP) (Music *et al.*, 2008), terminal restriction fragment length polymorphism (T-RFLP) (Hodgetts *et al.*, 2007) and real-time PCR (Wei *et al.*, 2004; Hren *et al.*, 2007).

The heteroduplex mobility assay is a technique in which PCR products from different samples are denatured and re-annealed prior to electrophoresis through acrylamide gels. Homoduplexes (annealing of complementary strands that match) and heteroduplexes (annealing of strands with mismatches) are then observed and variation between sequences can be detected. This technique has been developed for phytoplasmas, based on the 16S–23S spacer region, and was used to analyse variability between 62 phytoplasmas collected from North America, Europe and Asia (Wang and Hiruki, 2005). The results were useful for detecting subgroups and provided a rapid and sensitive test as an alternative to RFLP analysis.

SSCP is a similar method, in which PCR products are denatured to produce single-stranded DNA, which is then separated on polyacrylamide gels such that sequences with different primary structures fold differently and therefore migrate differently in the gels. Polymorphisms between sequences can then be observed, and, because the technique analyses the whole sequence rather than just the restriction endonuclease sites which are analysed in RFLP analyses, band shifts and polymorphisms between isolates are more readily observed. The technique has been used with primers for the 16S rRNA, *tuf* gene and *dnaB* gene (Music *et al.*, 2008) and has been shown to distinguish between 16SrXII isolates collected from Croatia and the FYR of Macedonia, which could not be separated into separate subgroups using conventional RFLP analysis. However, although the assay is easy to perform, conditions for separation of bands have to be established empirically for each set of fragments, making the technique quite time-consuming for routine assays.

T-RFLP has also been developed for identification of phytoplasmas in plants and for assigning them to phylogenetic groups (Hodgetts *et al.*, 2007). In this method, one of the PCR primers is labelled with a fluorescent tag, such that, following PCR and restriction enzyme digestion with an appropriate enzyme, it is the size of the terminal restriction fragment (TRF) containing the labelled primer, which can be analysed on an automated DNA sequencing machine. Hodgetts *et al.* (2007) have developed primers based on the 23S rRNA gene that can amplify simultaneously from the plant host chloroplast DNA and any phytoplasma present in a sample, so that, following restriction enzyme digestion, the chloroplast TRF acts as an internal control to show that PCR has worked, whilst any phytoplasma present will give a TRF of a specific size, depending on which group it belongs to. For example, if the chloroplast TRF and a phytoplasma TRF appear in a sample, the group to which the phytoplasma belongs can be ascertained. If the chloroplast TRF alone appears, the sample can be regarded as negative for phytoplasma, whilst, if no chloroplast TRF appears following PCR, it indicates that the DNA sample is unable to support PCR and presumably requires a clean-up step to remove PCR inhibitors prior to retesting. In addition, the method can also be used to monitor mixed infections if phytoplasmas from more than one taxonomic group are present in the same plant, and it can also be used as a semi-quantitative method for measuring fluxes in phytoplasma populations between samples.

A better automated, more accurate and sensitive method for quantifying phytoplasma levels is real-time PCR, as discussed by Galetto and Marzachì, Chapter 1, this volume. This method has been developed mainly based on the 16S rRNA gene, but primers based on other genes are also being used. For example, Wei *et al.* (2004) developed a specific real-time PCR assay to detect OY based on the *tuf* gene, whilst Hren *et al.* (2007) have developed a series of TaqMan® MGB assays with a specific assay for bois noir (BN) based on the *secY* gene and for flavescence doreé based on a random STOL genomic fragment. We have also developed a universal real-time PCR assay for phytoplasmas based on primers and probes from the 23S rRNA gene, along with specific assays for different groups and subgroups (Hodgetts *et al.*, 2009). What is perhaps more important in the future is to integrate these new and improved methods into rapid and easy to perform field-based detection and classification systems, so that material can be tested at the point of sampling without the need to return samples to the laboratory. Such approaches are already being developed for other pathogens to involve rapid on-site DNA extraction methods combined with portable real-time PCR platforms (Tomlinson *et al.*, 2005), or simpler diagnostic methods such as loop-mediated isothermal amplification (LAMP), which requires no specialized equipment (Tomlinson *et al.*, 2007). The aim must be to devise primers and tests that can be incorporated into such systems so that a sample in the field or where resources are limited, such as in developing countries, can be tested within minutes to identify not only whether a phytoplasma is present but also what taxonomic group it belongs to. It is probable that it will be genes other than 16S rRNA that provide the sequence specificity necessary for such advances.

Summary

The use of DNA sequences to build up phylogenetic trees is widespread and recognized as a valid approach for identifying taxonomic relationships between organisms. Early studies in most organisms have been based on well-conserved genes that are easy to analyse, particularly ribosomal RNA gene sequences. However, it has long been recognized that there are limitations to basing any classification system on a single, well-conserved gene, and in other organisms, such as the *Fungi*, taxonomic classifications have now moved on and are based on combined data sets for multiple genes, including some that encode proteins and are therefore less well conserved. Finding similar genes to improve phylogenetics in phytoplasmas has proven difficult because of the inability to culture the organisms, limited number of published sequences and problems in finding universal primers that can amplify a specified gene from a broad range of different phytoplasmas. However, semi-universal primers are now available for part of the *rp* operon, and universal primers for part of the *secA* gene, such that it is now possible to start combining data sets for different genes into a more complete and robust phylogenetic analysis of these organisms, resulting in better definition of groups and subgroups. With more sequence information becoming available, it is likely that universal primers for other genes will follow in the future, so that further data sets can be added, and already the results have led to improved diagnostic methods in the development of both new primers for conventional PCR and new primers and probes for real-time PCR, as discussed in Chapter 1, this volume. The aim in the future must be to build on these advances to develop rapid, simple, field-based diagnostic systems that can not only show that a phytoplasma is present in a sample but also identify the strain.

References

Angelini, E., Negrisolo, E., Clair, D., Borgo, M. and Boudon-Padieu, E. (2003) Phylogenetic relationships among Flavescence dorée strains and related phytoplasmas determined by heteroduplex mobility assay and sequence of ribosomal and non-ribosomal DNA. *Plant Pathology* 52, 663–672.

Anthony, R.M., Brown, T.J. and French, G.L. (2000) Rapid diagnosis of bacteremia by universal amplification of 23S ribosomal DNA followed by hybridization to an oligonucleotide array. *Journal of Clinical Microbiology* 38, 781–788.

Arnaud, G., Malembic-Maher, S., Salar, P., Bonnet, P., Maixner, M., Marcone, C., Boudon-Padieu, E. and Foissac, X. (2007) Multilocus sequence typing confirms the close genetic interrelatedness of three distinct flavescence dorée phytoplasma strain clusters and group 16SrV phytoplasmas infecting grapevine and alder in Europe. *Applied and Environmental Microbiology* 73, 4001–4010.

Bai, X., Zhang, J., Ewing, A., Miller, S.A., Radek, A.J., Shevchenko, D.V., Tsukerman, K., Walunas, T., Lapidus, A., Campbell, J.W. and Hogenhout, S.A. (2006) Living with genome instability: the adaptation of phytoplasmas to diverse environments of their insect and plant hosts. *Journal of Bacteriology* 188, 3682–3696.

Barbara, D.J., Morton, A., Clark, M.F. and Davies, D.L. (2002) Immunodominant membrane proteins from two phytoplasmas in the aster yellows clade (chlorante aster yellows and clover phyllody) are highly divergent in the major hydrophilic region. *Microbiology* 148, 157–167.

Berg, M. and Seemüller, E. (1999) Chromosomal organization and nucleotide sequence of the genes coding for the elongation factors G and Tu of the apple proliferation phytoplasma. *Gene* 226, 103–109.

Berg, M., Davies, D.L., Clark, M.F., Vetten, H.J., Maier, G., Marcone, C. and Seemüller, E. (1999) Isolation of the gene encoding an immunodominant membrane protein of the apple proliferation phytoplasma, and expression and characterization of the gene product. *Microbiology* 145, 1937–1943.

Blomquist, C.L., Barbara, D.J., Davies, D.L., Clark, M.F. and Kirkpatrick, B.C. (2001) An immunodominant membrane protein gene from the Western X-disease phytoplasma is distinct from those of other phytoplasmas. *Microbiology* 147, 571–580.

Daire, X., Clair, D., Reinert, W. and Boudon-Padieu, E. (1997) Detection and differentiation of grapevine yellows phytoplasmas belonging to the elm yellows group and to the stolbur subgroup by PCR amplification of non-ribosomal DNA. *European Journal of Plant Pathology* 103, 507–514.

De La Rue, S., Padovan, A. and Gibb, K. (2001) *Stylosanthes* is a host for several phytoplasmas, one of which shows unique 16S–23S intergenic spacer region heterogeneity. *Journal of Phytopathology* 149, 613–619.

Firrao, G., Gibb, K. and Streten, C. (2005) Short taxonomic guide to the genus '*Candidatus* phytoplasma'. *Journal of Plant Pathology* 87, 249–263.

Goodwin, P.H., Xue, B.G., Kuske, C.R. and Sears, M.K. (1994) Amplification of plasmid DNA to detect plant pathogenic mycoplasmalike organisms. *Annals of Applied Biology* 124, 27–36.

Gundersen, D.E., Lee, I.-M., Rehner, S.A., Davis, R.E. and Kingsbury, D.T. (1994) Phylogeny of mycoplasmalike organisms (Phytoplasmas): a basis for their classification. *Journal of Bacteriology* 176, 5244–5254.

Harrison, N.A., Womack, M. and Carpio, M.L. (2002) Detection and characterization of a lethal yellowing (16SrIV) group phytoplasma in Canary Island date palms affected by lethal decline in Texas. *Plant Disease* 86, 676–681.

Hodgetts, J., Ball, T., Boonham, N., Mumford, R. and Dickinson, M. (2007) Use of terminal restriction fragment length polymorphism (T-RFLP) for identification of phytoplasmas in plants. *Plant Pathology* 56, 357–365.

Hodgetts, J., Boonham, N., Mumford, R., Harrison, N. and Dickinson, M. (2008) Phytoplasma phylogenetics based on analysis of *secA* and 23S rRNA gene sequences for improved resolution of candidate species of '*Candidatus* Phytoplasma'. *International Journal of Systematic and Evolutionary Microbiology* 58, 1826–1837.

Hodgetts, J., Boonham, N., Mumford, R. and Dickinson, M. (2009) Panel of 23S rRNA gene-based real-time PCR assays for improved universal and group specific detection of phytoplasmas. *Applied and Environmental Microbiology* 75, 2945–2950.

Hren, M., Boben, J., Rotter, A., Kralj, P., Gruden, K. and Ravnikar, M. (2007) Real-time PCR detection systems for Flavescence doreé and Bois noir phytoplasmas in grapevine: comparison with conventional PCR detection and application in diagnostics. *Plant Pathology* 56, 785–796.

IRPCM Phytoplasma/Spiroplasma Working Team – Phytoplasma Taxonomy Group (2004) '*Candidatus* Phytoplasma', a taxon for the wall-less, non-helical prokaryotes that colonize plant phloem and insects. *International Journal of Systematic and Evolutionary Microbiology* 54, 1243–1255.

James, T.Y., Kauff, F., Schoch, C.L. *et al.* (2006) Restructuring the early evolution of *Fungi* using a six-gene phylogeny. *Nature* 443, 818–822.

Jarausch, W., Saillard, C., Dosba, F. and Bové, J.-M. (1994) Differentiation of mycoplasmalike organisms (MLOs) in European fruit trees by PCR using specific primers derived from the sequence of a chromosomal fragment of the apple proliferation MLO. *Applied and Environmental Microbiology* 60, 2916–2923.

Jomantiene, R., Zhao, Y. and Davis, R.E. (2007) Sequence-variable mosaics: composites of recurrent transposition characterizing the genomes of phylogenetically diverse phytoplasmas. *DNA and Cell Biology* 26, 557–564.

Kakizawa, S., Oshima, K., Kuboyama, T., Nishigawa, H., Jung, H.-Y., Sawayanagi, T., Tsuchizaki, T., Miyata, S.-I., Ugaki, M. and Namba, S. (2001) Cloning and expression analysis of phytoplasma protein translocation genes. *Molecular Plant–Microbe Interactions* 14, 1043–1050.

Keeling, P.J., Burger, G., Durnford, D.G., Lang, B.F., Lee, R.W., Pearlman, R.E., Roger, A.J. and Gray, M.W. (2005) The tree of eukaryotes. *Trends in Ecology and Evolution* 20, 670–676.

Kube M., Schneider, B., Kuhl, H., Dandekar, T., Heitmann, K., Migdoll, A.M., Reinhardt, R. and Seemüller, E. (2008) The linear chromosome of the plant-pathogenic mycoplasma '*Candidatus* Phytoplasma mali'. *BMC Genomics* 9, 306.

Lee, I.-M., Gundersen-Rindal, D.E., Davis, R.E. and Bartoszyk, I.M. (1998) Revised classification scheme of phytoplasmas based on RFLP analyses of 16S rRNA and ribosomal protein gene sequences. *International Journal of Systematic Bacteriology* 48, 1153–1169.

Lee, I.-M., Zhao, Y. and Bottner, K.D. (2006) SecY gene sequence analysis for finer differentiation of diverse strains in the aster yellows phytoplasma group. *Molecular and Cellular Probes* 20, 87–91.

Liefting, L.W., Andersen, M.T., Beever, R.E., Gardner, R.C. and Forster, R.L. (1996) Sequence heterogeneity in the two 16S rRNA genes of *Phormium* yellow leaf phytoplasma. *Applied and Environmental Microbiology* 62, 3133–3139.

Lim, P.-O. and Sears, B.B. (1991) DNA sequence of the ribosomal protein genes *rpl2* and *rps19* from a plant pathogenic mycoplasma-like organism. *FEMS Microbiology Letters* 84, 71–74.

Lim, P.-O. and Sears, B.B. (1992) Evolutionary relationships of a plant-pathogenic mycoplasmalike organism and *Acholeplasma laidlawii* deduced from two ribosomal protein gene sequences. *Journal of Bacteriology* 174, 2606–2611.

Marcone, C., Lee, I.-M., Davis, R.E., Ragozzino, A. and Seemüller, E. (2000) Classification of aster yellows-group phytoplasmas based on combined analyses of rRNA and *tuf* gene sequences. *International Journal of Systematic and Evolutionary Microbiology* 50, 1703–1713.

Martini, M., Lee, I.-M., Bottner, K.D., Zhao, Y., Botti, S., Bertaccini, A., Harrison, N.A., Carraro, L., Marcone, C., Khan, A.J. and Osler, R. (2007) Ribosomal protein gene-based phylogeny for finer differentiation and classification of phytoplasmas. *International Journal of Systematic and Evolutionary Microbiology* 57, 2037–2051.

Mpunami, A.A., Tymon, A., Jones, P. and Dickinson, M.J. (1999) Genetic diversity in the coconut lethal yellowing disease phytoplasmas of East Africa. *Plant Pathology* 48, 109–114.

Musić, M.Š., Krajačić, M. and Škorić, D. (2008) The use of SSCP analysis in the assessment of phytoplasma gene variability. *Journal of Microbiological Methods* 73, 69–72.

Nei, M. and Li, W.-H. (1979) Mathematical model for studying genetic variation in terms of restriction endonucleases. *Proceedings of the National Academy of Sciences of the USA* 76, 5269–5273.

Nicolaisen, M. and Bertaccini, A. (2007) An oligonucleotide microarray-based assay for identification of phytoplasma 16S ribosomal groups. *Plant Pathology* 56, 332–336.

Oshima, K., Kakizawa, S., Nishigawa, H., Jung, H.-Y., Wei, W., Suzuki, S., Arashida, R., Nakata, D., Miyata, S.-I., Ugaki, M. and Namba, S. (2004) Reductive evolution suggested from the complete genome sequence of a plant-pathogenic phytoplasma. *Nature Genetics* 36, 27–29.

Schneider, B., Gibb, K.S. and Seemüller, E. (1997) Sequence and RFLP analysis of the elongation factor Tu gene used in differentiation and classification of phytoplasmas. *Microbiology* 143, 3381–3389.

Shen, W.C. and Lin, C.P. (1993) Production of monoclonal antibodies against a mycoplasmalike organism associated with sweet potato witches' broom. *Phytopathology* 83, 671–675.

Smart, C.D., Schneider, B., Blomquist, C.L., Guerra, L.J., Harrison, N.A., Ahrens, U., Lorenz, K.-H., Seemüller, E. and Kirkpatrick, B.C. (1996) Phytoplasma-specific PCR primers based on sequence of the 16S–23S rRNA spacer region. *Applied and Environmental Microbiology* 62, 2988–2993.

Streten, C. and Gibb, K.S. (2003) Identification of genes in the tomato big bud phytoplasma and comparison to those in sweet potato little leaf-V4 phytoplasma. *Microbiology* 149, 1797–1805.

Streten, C. and Gibb, K.S. (2005) Genetic variation in '*Candidatus* Phytoplasma australiense'. *Plant Pathology* 54, 8–14.

Tomlinson, J.A., Boonham, N., Hughes, K.J.D., Griffin, R.L. and Barker, I. (2005) On-site DNA extraction and real-time PCR for detection of *Phytophthora ramorum* in the field. *Applied and Environmental Microbiology* 71, 6702–6710.

Tomlinson, J.A., Barker, I. and Boonham, N. (2007) Faster, simpler, more-specific methods for improved molecular detection of *Phytophthora ramorum* in the field. *Applied and Environmental Microbiology* 73, 4040–4047.

Toth, K.F., Harrison, N. and Sears, B.B. (1994) Phylogenetic relationships among members of the class *Mollicutes* deduced from rps3 gene sequences. *International Journal of Systematic and Evolutionary Microbiology* 44, 119–124.

Tran-Nguyen, L.T.T., Kube, M., Schneider, B., Reinhardt, R. and Gibb, K.S. (2008) Comparative genome analysis of '*Candidatus* Phytoplasma australiense' (subgroup *tuf*-Australia 1: *rp*-A) and '*Ca.* P. asteris' strains OY-M and AY-WB. *Journal of Bacteriology* 190, 3979–3991.

Wagner, M., Fingerhut, C., Gross, H.J. and Schön, A. (2001) The first phytoplasma RNaseP RNA provides new insights into the sequence requirements of this ribozyme. *Nucleic Acids Research* 29, 2661–2665.

Wang, K. and Hiruki, C. (2005) Distinctions between phytoplasmas at the subgroup level detected by heteroduplex mobility assay. *Plant Pathology* 54, 625–633.

Wang, K., Hiruki, C. and Yeh, F. (2003) Molecular evolution of phytoplasmas based on polymorphisms in the 16S rRNA genes and the 16/23S spacer regions. *Proceedings of the Japanese Academy Series B* 79, 155–162.

Wei, W., Kakizawa, S., Suzuki, S., Jung, H.Y., Nishigawa, H., Miyata, S., Oshima, K., Ugaki, M., Hibi, T. and Namba, S. (2004) *In planta* dynamic analysis of onion yellows phytoplasma using localized inoculation by insect transmission. *Phytopathology* 94, 244–250.

Wei, W., Davis, R.E., Lee, I.-M. and Zhao, Y. (2007) Computer-simulated RFLP analysis of 16S rRNA genes: identification of ten new phytoplasma groups. *International Journal of Systematics and Evolutionary Microbiology* 57, 1855–1867.

Wei, W., Lee, I.-M., Davis, R.E., Suo, X. and Zhao, Y. (2008) Automated RFLP pattern comparison and similarity coefficient calculation for rapid delineation of new and distinct phytoplasma 16Sr subgroup lineages. *International Journal of Systematic and Evolutionary Microbiology* 58, 2368–2377.

Yu, Y.-L., Yeh, K.-W. and Lin, C.-P. (1998) An antigenic protein gene of a phytoplasma associated with sweet potato witches' broom. *Microbiology* 144, 1257–1262.

7 Movement of Phytoplasmas and the Development of Disease in the Plant

Carmine Marcone

Università degli Studi di Salerno, Italy

Introduction

Plant-pathogenic phytoplasmas are wall-less, unculturable bacteria of the class *Mollicutes* with a small genome size, which ranges from 530 to 1350 kilobases (Marcone *et al.*, 1999). In diseased plants, phytoplasmas reside almost exclusively in the phloem sieve tube elements, to which they are introduced by phloem-feeding homopteran insects, mainly leafhoppers (Cicadellidae) and planthoppers (Fulgoromorpha), and less frequently psyllids (Psyllidae) (Weintraub and Beanland, 2006). Phytoplasmas can also have access to phloem sieve tube elements through the parasitic plant dodder (*Cuscuta* spp.). Once phytoplasmas have entered the phloem sieve tube elements, they spread systemically throughout the plant by passing through phloem sieve plate pores (Fig. 7.1).

Occasionally, a few phloem parenchyma cells adjacent to sieve tubes are also invaded. Phytoplasmas are associated with a variety of diseases in more than a thousand plant species worldwide (McCoy *et al.*, 1989). Some of these diseases, especially those of woody plants, are lethal. The list of plants and insects known to harbour phytoplasmas is continuously increasing, as is the number of taxonomically characterized phytoplasma strains (Seemüller *et al.*, 1998a, 2002; Lee *et al.*, 2000). In contrast to progress made in detection, differentiation and phylogenetic classification of phytoplasmas, very little is known about the mechanisms by which phytoplasmas induce disease in plants and the reason for different reactions of the host plants to phytoplasmal infections. Knowledge of the movement of phytoplasmas within the host plant and their final distribution in various organs is usually essential for understanding the phytoplasma–plant host interactions.

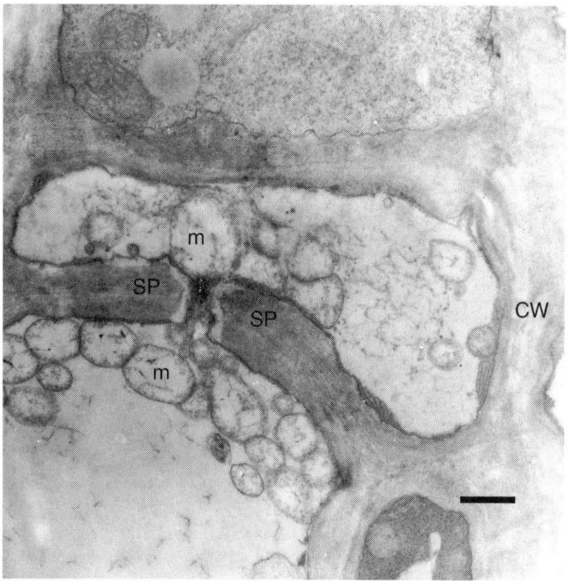

Fig. 7.1. Transmission electron micrograph showing phytoplasma bodies passing through phloem sieve plate pores. Some phytoplasmas are attached to the inner surface of the sieve tube plasma membrane. CW, cell wall; m, phytoplasma bodies; SP, sieve plate. Bar represents 700 nm.

Disease Induction

Since phytoplasmas live and multiply in functional phloem sieve tube elements, the main effect of phytoplasmal infections apparently is the impairment of the sieve tube function. Several studies have shown that inhibition of phloem transport occurs in phytoplasma-infected plants, which, in turn, leads to an accumulation of abnormal amounts of carbohydrates in source leaves, i.e. mature leaves, and a marked reduction of these essential energy-storage compounds in sink organs, i.e. young leaves and roots (Catlin *et al.*, 1975; Braun and Sinclair, 1978; Kartte and Seemüller, 1991a; Lepka *et al.*, 1999; Guthrie *et al.*, 2001; Maust *et al.*, 2003). Changes in photosynthate translocation along with other impaired physiological functions, including reduced photosynthesis, stomatal conductance and root respiration, altered secondary metabolism and disturbed plant hormone balance, possibly mediated by phloem dysfunction, could account for symptoms exhibited by infected plants (McCoy, 1979; León *et al.*, 1996; Lepka *et al.*, 1999; Tan and Whitlow, 2001; Maust *et al.*, 2003; Choi *et al.*, 2004). However, the exact mechanisms involved in symptom development or the genes that control these events are still poorly understood. Recent studies on the effects of phytoplasmal infections on host plant gene expression have shown that symptoms of flower abnormalities occurring in tomato plants infected by the stolbur (STOL) phytoplasma are associated with deregulations of key floral development genes

(Pracros *et al.*, 2006), whereas, in free-branching-affected poinsettia (*Euphorbia pulcherrima*) plants, genes involved in phytohormone activity, which are upregulated upon phytoplasmal infections, may play a major role in symptom expression (Nicolaisen and Horvath, 2008). Plants infected by phytoplasmas exhibit a wide range of specific and non-specific symptoms. Symptoms of diseased plants may vary with the phytoplasma, host plant, stage of the disease, age of the plant at the time of infection and environmental conditions (McCoy, 1979; McCoy *et al.*, 1989; Lee *et al.*, 2000; Seemüller *et al.*, 2002). Specific symptoms include flower discolorations and distortions such as virescence, phyllody, big bud, flower proliferation and other flower abnormalities – all resulting in sterility, witches'-brooms, rosetting, internode elongation and etiolation, shortened internodes, enlarged stipules, off-season growth and brown discoloration of phloem tissue (Plate 2).

Less specific and non-specific symptoms, which are most often common in woody plants, include foliar yellowing and reddening, small leaves, leaf roll, leaf curl, vein clearing, vein enlargement, vein necrosis, premature autumn coloration, premature defoliation, undersized fruits, poor terminal growth, sparse foliage, dieback, stunting of overall plant growth and decline. In rare instances, phytoplasma-infected plants are fully non-symptomatic over their lifespan; a temporary or permanent remission of symptoms may also occur.

Colonization Behaviour of Phytoplasmas in Plants and Relationship to Symptom Expression

Phytoplasmas differentially colonize plants, depending on the pathogen type, properties of the host and its specific reactions. Movement, distribution and multiplication of phytoplasmas in plants and relationship to symptom expression have been examined in both herbaceous and perennial woody hosts infected by phylogenetically different phytoplasmas, using fluorescence and electron microscopy, bioimaging, histological, serological, DNA hybridization, polymerase chain reaction (PCR) and quantitative real-time PCR assays, and transmission experiments (Seemüller *et al.*, 1984a; Douglas, 1986; Kuske and Kirkpatrick, 1992; Lherminier *et al.*, 1994; Siddique *et al.*, 1998; Constable *et al.*, 2003; Christensen *et al.*, 2004; Wei *et al.*, 2004; Saracco *et al.*, 2006).

Herbaceous hosts

Work by Kuske and Kirkpatrick (1992) has shown that two strains of the aster yellows (AY) phytoplasma '*Ca.* Phytoplasma asteris', the severe (SAY) and dwarf (DAY) strains, had similar colonization patterns in the experimental host *Catharanthus roseus* (periwinkle), following graft inoculation, over a 10-week period of observation. Both strains moved from the grafted shoots into ungrafted shoots that originated from the main stem below the grafted shoots, approximately 1 week after the graft unions had healed and then spread systemically throughout the plant. Strain SAY colonized periwinkle more rapidly than strain DAY, and strain SAY was detected about 1 week earlier than the other. Moreover, periwinkle plants infected by SAY showed symptoms 1 week earlier than those infected by DAY. By 4 weeks after the plants were grafted, the titre of SAY in the various plant parts was higher than that of DAY, and the symptoms induced by SAY were more pronounced. However, SAY titre declined after 4 weeks from inoculation, whereas the titre of DAY continued to increase, and at the end of the observation period was higher than that of SAY. Also, at the end of the observation period, symptoms on the plants infected by DAY were not as pronounced as those infected by SAY. For both strains, titres were highest in actively growing meristematic regions, such as ungrafted shoots, with symptoms arising from the main stem below the grafted shoots and side shoots that emerged from the grafted shoots, but lowest in the roots. Distribution and titres of phytoplasmal infections were clearly correlated to symptom expression in the aerial parts.

A similar colonization pattern was observed in strawberry plants naturally infected by the clover phyllody phytoplasma, in which phytoplasma titres were highest in pedicels and receptacles with symptoms, followed by sepals, petals and leaves. However, the clover phyllody agent was not detectable in strawberry roots (Clark *et al.*, 1983). In celery (*Apium graveolens*) plants experimentally inoculated with the X-disease phytoplasma using the leafhopper vector *Colladonus montanus*, phytoplasmal infections were first detected in the roots (Kirkpatrick, 1991). However, in plants with full symptoms, the phytoplasma concentration was highest in the shoot apex, followed by youngest, severely affected leaves. The X-disease phytoplasma was also present, but in much lower concentration, in older, completely chlorotic leaves.

Wei *et al.* (2004) analysed the dynamic of onion yellows (OY) phytoplasma in garland chrysanthemum (*Chrysanthemum coronarium*) using a localized inoculation technique and nested PCR, real-time PCR and immunohistochemical assays. Following localized inoculation of a leaf of garland chrysanthemum by the leafhopper vector *Macrosteles striifrons*, the OY agent spread sequentially from the inoculated leaf to the main stem, root, upper leaves and then lower leaves. The OY phytoplasma systemically colonized the plant by 21 days after inoculation and its concentration increased approximately sixfold per week from 14 to 28 days after inoculation. The phytoplasma concentration in the roots was higher than in inoculated leaves. Within the aerial parts, the OY agent was most abundant in inoculated leaves and shoot apex, followed by youngest leaves and lower leaves. Furthermore, at 28 days after inoculation, characteristic internal symptoms such as phloem necrosis in the root and phloem hyperplasia in the stem, as well as external symptoms such as stunted growth, were apparent. Similar data were obtained by Saracco *et al.* (2006), who investigated the colonization pattern of chrysanthemum yellows (CY) phytoplasma in daisy (*Chrysanthemum carinatum*). Following localized inoculation of either a basal or an apical leaf by the leafhopper vector *Macrosteles quadripunctulatus*, the CY phytoplasma colonized and multiplied to high numbers in roots and young, apical leaves rather than in the old, basal leaves. However, it was always more concentrated in roots than in aerial parts (Saracco *et al.*, 2006).

Movement and distribution of the flavescence dorée phytoplasma within the experimental host *Vicia faba* have been described by Lherminier *et al.* (1994). This work has shown that, although the FD phytoplasma was inoculated into the aerial parts by the leafhopper *Euscelidius variegatus*, the FD agent was first detected in roots, where it multiplied to a high extent prior to symptom development. Subsequently, the FD phytoplasma invaded collar and axillary shoots, emerging about 20 days after inoculation from the basal part of the main stem, and then moved acropetally along the main stem. As soon as symptoms became apparent, FD phytoplasma could be detected in all parts of the inoculated plant. However, the highest phytoplasma concentrations were recorded in the upper parts of basal axillary shoots rather than in the roots. In all of the colonization patterns mentioned, phytoplasma movement was directed towards actively growing regions of the plant with symptoms, which acted as metabolic sinks, to which nutrients are directed through the phloem conduit. In contrast, in free-branching-affected poinsettia plants, the infecting phytoplasma, the poinsetta branch-inducing (PoiBI) agent, which is a member of the X-disease phytoplasma group or 16SrIII group, preferentially colonized and accumulated in source leaves and, to a lesser extent, in petioles of source leaves and stems (Christensen *et al.*, 2004). PoiBI phytoplasma did not colonize or was present in very low numbers in sink organs such as roots and young leaves. Moreover, the extent of PoiBI phytoplasma invasion was correlated with symptom severity.

Woody hosts

For a number of deciduous woody hosts it has been shown that phytoplasma colonization of the aerial parts of infected plants is subject to seasonal fluctuation. In apple (*Malus* spp.) and pear (*Pyrus* spp.) trees affected by apple proliferation (AP) and pear decline (PD) diseases respectively, colonization patterns of the causal agents, '*Ca.* Phytoplasma mali' and '*Ca.* Phytoplasma pyri', have been monitored over a period of several years with fluorescence microscopy using the DNA dye 4´-6-diamidino-2-phenylindole (DAPI) test and periodic transmission grafting (Schaper and Seemüller, 1982; Seemüller *et al.*, 1984a, b). Since phytoplasmas depend on functional sieve tubes and because the sieve tubes in the above-ground parts of apple and pear trees degenerate in late autumn and early winter, AP and PD phytoplasmas are, in almost all cases, eliminated in the aerial parts during winter. In contrast, they persist in the roots, where intact sieve tubes are present throughout the year. From the roots, both pathogens may recolonize the aerial parts in spring when new phloem is being formed. This recolonization was studied by Schaper and Seemüller (1984). Over a period, from the middle of March to the middle of June, a narrow strip of bark was removed at regular intervals all around the trunks or scaffold limbs of orchard apple and pear trees that were colonized in the roots. The girdling interrupted the continuity of the phloem, which prevented the translocation of phytoplasmas above the girdling line. The study revealed that recolonization started as soon as newly differentiated sieve tubes were present. This point was usually reached in March. The upward movement of AP and PD phytoplasmas was rather slow, reaching a velocity of 7.5–30 mm per day. However, the mechanism of this movement is still unknown. Recolonization usually occurs annually during the first few years of the diseases. Later, the aerial parts may be only partly or weakly colonized or may not be colonized at all. Because expression of symptoms depends on the presence of the invading phytoplasmas, trees intensively colonized in the aerial parts develop AP-specific symptoms, whereas those only partially, weakly or not colonized develop mild or no symptoms. Therefore, the variation in symptom expression over the years is related to seasonal fluctuation of the phytoplasma population in the aerial parts of infected trees.

The root system of infected trees remains colonized for the life of trees. The colonization behaviour described is typical for trees grown on established rootstocks based on *Malus* × *domestica* (Schaper and Seemüller, 1982, 1984; Seemüller *et al.*, 1984a, b; Carraro *et al.*, 2004). Extensive studies on AP resistance have shown that diseased apple trees worked on experimental *M. sieboldii*-based AP-resistant apomictic rootstocks either never developed symptoms or, only rarely, developed temporary, mild symptoms. Such trees were not colonized in the aerial parts and harboured extremely low phytoplasma numbers in the roots. Thus, the low starting concentration in the roots and the unsuitable host properties of *M. sieboldii*-based genotypes may impair the spread of the pathogen from the roots into the scion during recolonization of the stem in spring (Kartte and Seemüller, 1991b; Seemüller *et al.*, 2008). Similarly, pear trees on clonal quince (*Cydonia oblonga*) rootstocks are little affected by the PD disease, owing the poor host properties of quince. In this rootstock, the PD phytoplasma occurs in a very low titre, which has a negative effect on the recolonization of the aerial parts in spring, resulting in no or mild symptoms (Seemüller *et al.*, 1986). The degeneration of the phytoplasma population during winter in above-ground parts of trees, the overwintering in the root system and the reinvasion of the stem in spring seems to be a general phenomenon for many deciduous woody plants. It has been described for elm yellows (EY), mulberry dwarf and paulownia witches'-broom diseases affecting the respective hosts (for references see Schaper and Seemüller, 1984). Braun and Sinclair (1976) were the first to relate the conditions of the phloem to the survival of the causal agent. They found no functional sieve tubes in the stem of EY-affected elm (*Ulmus*) trees in winter but did find intact sieve tubes in roots. Thus, they concluded that the EY phytoplasma ('*Ca.* Phytoplasma ulmi') overwinters in the roots and recolonizes the stem in spring.

The X-disease phytoplasma, which infects stone fruits in North America, overwinters in the roots and in a low percentage of buds. From these sites, spread of the pathogen throughout the tree may take place in spring (for review see Kirkpatrick, 1991). Douglas (1986), who monitored the distribution of X-disease phytoplasma in diseased peach (*Prunus persica*) and chokecherry (*Pr. virginiana*) trees during two growing seasons by DAPI fluorescence, found a good correlation between symptom severity and extent of phytoplasma invasion. However, in chokecherry trees, invasion occurred earlier and was more extensive than in peach trees. Also, the pathogen was more evenly distributed in chokecherry than in peach. Epidemiological studies conducted in California revealed that in spring in sweet cherry (*Prunus avium*) trees, the X-disease phytoplasma initially colonized and multiplied in the fruit peduncles rather than in leaves, reaching its highest concentration in peduncles when fruits were mature (Kirkpatrick, 1991). Later, after ripening and senescence of fruits, the pathogen persisted and continued to increase in leaves until late autumn.

The European stone fruit yellows (ESFY) phytoplasma ('*Ca.* Phytoplasma prunorum'), which is the only confirmed phytoplasmal pathogen causing diseases of *Prunus* species in Europe, can persist in the aerial parts of diseased trees during the dormant season, whereas it is always present at root level throughout the year (Seemüller *et al.*, 1998b; Jarausch *et al.*, 1999). However, monitoring of the ESFY phytoplasma population by DAPI fluorescence and PCR assays showed that the pathogen slowly colonized the leaves in spring, giving a colonization pattern like AP and PD phytoplasmas. Systemic colonization of diseased trees occurred from July throughout late autumn. In susceptible genotypes, specific symptoms are highly correlated with the presence of the pathogen (Jarausch *et al.*, 1999). Work by Kison and Seemüller (2001), who examined the response of many established and experimental *Prunus* rootstocks to ESFY phytoplasma infections using graft-inoculation, PCR assays and DAPI fluorescence tests, has shown that phytoplasmal infections occurred in all inoculated *Prunus* genotypes, irrespective of symptom expression. The colonization density was lower in little-affected or unaffected rootstocks than in susceptible rootstocks. There were also differences between colonization of the roots and the aerial parts of rootstocks. Usually phytoplasma numbers were higher in the roots than in the stem of the same genotype. In some rootstocks, infections were not found in the stem but were present in the roots. Colonization appeared to be persistent over the observation period of 5–8 years, even in the least affected rootstocks, such as *Pr. domestica* stocks Achermann's, Brompton and P 2175, and *Pr. cerasifera* stock Myrabi.

Siddique *et al.* (1998), working with Australian papaya dieback disease, monitored, through PCR assays, the distribution of the presumed causal agent, i.e. '*Ca.* Phytoplasma australiense', within different parts of diseased papaya plants at progressive stages of disease development. In the early stages of the disease, the papaya-infecting phytoplasma proved to be present in expanding leaves, flowers and the upper part of the stem. As the disease progressed and symptoms became more pronounced, the pathogen occurred in the mid–lower part of the stem and even in roots. However, no phytoplasmal infections occurred in symptomatic mature leaves in all stages of disease development. Also, the pathogen occurred in very low numbers throughout the plant and its concentration decreased as the disease progressed.

Similar distribution patterns have been observed in coconut palms affected by the lethal yellowing (LY) disease. In this case, the colonizing phytoplasma occurs mainly in sink plant parts, such as the apical meristem, immature leaves, root meristem and inflorescences, but is rarely detectable in mature leaves with symptoms (León *et al.*, 1996; Maust *et al.*, 2003).

Distribution and persistence of phytoplasmas in grapevine plants affected by the Australian grapevine yellows (AGY) disease have been studied by Constable *et al.* (2003) using PCR technology over a three-year observation period. AGY disease is known to be associated with two phylogenetically different phytoplasmas, namely '*Ca.* Phytoplasma australiense' and the tomato big bud agent. The AGY phytoplasmas were detected in shoots, cordons, trunks and roots of diseased grapevine plants throughout each year and thus they appear to infect Australian grapevines persistently from year to year. However, the detection frequency within the different plant organs differed considerably. In the roots, the detection frequency was, in comparison to aerial parts, extremely low throughout the observation period, thus indicating a poor colonization of the root system. The highest detection rates were recorded in the aerial parts in July or October of each year of investigation. Also, there was a good correlation between detection frequency and symptom expression. AGY phytoplasmas were more frequently detected in symptomatic shoots than in non-symptomatic shoots taken from the same plant. In a few cases, phytoplasmal infections were detected in non-symptomatic plants.

Alder yellows (ALY) phytoplasma, which is known to affect several *Alnus* (alder) species in Europe, persists in the aerial parts of trees throughout the year rather than in the roots. In this respect the colonization differs significantly from that of AP and PD phytoplasmas. Of the several alder trees examined by Lederer and Seemüller (1991) using DAPI fluorescence, ALY infections proved to be detectable in all trees older than approximately five years, irrespective of symptom expression. About 80% of infected trees were non-symptomatic for ALY, whereas only 20% showed ALY symptoms. Usually, healthy-appearing trees were more heavily colonized than trees with symptoms. The phytoplasma population was always higher in petioles and young twigs than in several-year-old branches, trunks and roots. Colonization was characterized by an uneven phytoplasma distribution in adjacent sieve tubes, with some tubes packed by phytoplasmas while others contained a considerably lower population (Plate 3).

Possible Factors Involved in Disease Development

Strain virulence

Symptom development also depends on the virulence of the infecting phytoplasma strain within a given taxon. Great differences in virulence have been observed among strains of several phytoplasmas, including AP, ESFY, ash yellows (AshY) ('*Ca.* Phytoplasma fraxini') and ALY agents. In a recent study by Seemüller and Schneider (2007), the virulence of 24 strains of the AP phytoplasma was examined by graft-inoculating 'Golden Delicious' apple trees worked on to M 11 rootstock and monitoring symptom development over a 12-year period. Based on symptomatology, the strains were classified into three virulence groups defined as: (i) not or mildly virulent; (ii) moderately virulent; and (iii) severely virulent. Strains of the first group did not induce symptoms or induced very mild symptoms, which consisted of slight to moderate foliar reddening and yellowing. However, they never induced specific AP symptoms, such as witches'-brooms and undersized fruits, and thus the performance of infected trees was similar to that of healthy trees. Strains of the other groups caused symptoms that ranged from moderate to severe and included the typical AP symptoms. A similar wide range of virulence was observed in previous work, in which trees of different *Prunus* genotypes were inoculated with 20 strains of the ESFY phytoplasma (Kison and Seemüller, 2001). While some strains were nearly avirulent or weakly virulent, inducing only mild foliar symptoms and slightly reduced vigour but no mortality, others were highly virulent and caused severe symptoms and a high mortality rate of affected trees. Differences in strain virulence are also known from the AshY phytoplasma. Different strains of this pathogen caused significantly different degrees of growth suppression and loss of foliar greenness, ranging from slight or imperceptible to severe, in graft-inoculated green ash (*Fraxinus pennsylvanica*) and periwinkle plants (Sinclair and Griffiths, 2000). As mentioned above, in nature only a minority of alder trees infected by the ALY agent develop symptoms. By graft-inoculation of healthy alder seedlings with scion wood from differently affected and non-symptomatic trees and observation of disease development over a 5-year period, Berges and Seemüller (2002) provided evidence that the ALY phytoplasma is pathogenic to alder and may induce severe symptoms, but avirulent strains occurring within this taxon account for the latent infections which are widespread in Europe.

Strain interference

Interference between strains of the same taxon has been described for several phytoplasma–plant host combinations. Studies conducted by Sinclair and Griffiths (2000) have revealed that, in periwinkle plants which were co-inoculated with two strains of the AshY phytoplasma that greatly differed in aggressiveness, the most aggressive strain appeared sooner and more frequently than the less aggressive strain in leaves located at a distance from the inoculation sites. Thus, aggressiveness was associated with either a higher movement or a higher multiplication rate than that achieved by the less aggressive strain. However, when either strain was inoculated 11 weeks before the other into the same plant, only the initial strain could be detected after a further 12 weeks of incubation. Therefore, the initial strain or its effect on the host may have interfered with long-distance movement or multiplication of the second strain. A concept of pre-emptive dominance has been proposed by the authors to explain the continuing dominance of the first strain that colonized a plant, regardless of its aggressiveness. Interaction among three strains of the AY phytoplasma has been reported by Freitag (1964). When tobacco (*Nicotiana rustica* var. *humilis*) and plantain (*Plantago major*) plants were inoculated first with either DAY or SAY strains and 1 week later with the Tulelake (TLAY) strain, the plants developed mainly symptoms of the first strains. The results indicated a cross-protection reaction. The strain that was inoculated first nearly always predominated and prevented development of symptoms of the challenging strain. Plants inoculated simultaneously with DAY and SAY strains usually developed symptoms of only one strain. However, some plants first showed symptoms of one strain and subsequently also symptoms of the other strain. In plantain plants inoculated first with the TLAY strain and then challenged by the DAY or SAY strain, the initial strain was displaced and symptoms of the challenging strain developed. The most interesting results were obtained when tobacco plants were first inoculated with the TLAY strain and then, a week later, by either the DAY or SAY strains. In these instances, the plants first developed pronounced vein-clearing symptoms, indicative of TLAY strain infections, and then recovered completely. This was interpreted as a phenomenon of mutual suppression, which resulted in neutralization of the two competing strains. It is conceivable that, in nature, interaction amongst strains of the same taxon occurring within a given plant host may account for either the lack of symptom expression or mild symptom expression. For instance, a graft-transmissible agent that mediated cross-protection has been identified in apricot. This agent most probably is an avirulent strain of the ESFY phytoplasma, which protected the plant from subsequent infection by severe strains (for references see Berges and Seemüller, 2002).

Phytoplasma concentration

There is increasing evidence that phytoplasma concentration in infected plants differs greatly. High-titre hosts are periwinkle and other herbaceous plants, including lettuce, celery, tobacco and several *Brassica* spp. Low numbers are known from woody hosts, in which phytoplasma concentration is often below the detection level of microscopical methods (Berges *et al.*, 2000). However, in some genera of woody plants both low- and high-titre hosts are known to occur. Since phytoplasma concentration is usually regarded as an important pathogenicity factor, it is possible that different mechanisms of pathogenicity exist in high- and low-titre hosts. Also, plants differing in host suitability, and thus phytoplasma titre, may respond very differently to phytoplasmal infections. Previous studies have shown that AP phytoplasma-infected apple trees on *M.* × *domestica*, *M. silvestris* and *M. domestica* × *baccata* rootstocks have high phytoplasma titres, a high frequency of witches'-brooms as a specific symptom (especially for trees on *M.* × *domestica*) and a low mortality rate. In these taxa, both phloem conditions and starch content in the roots differ only slightly from those of healthy trees. Thus, sieve tubes are rather tolerant of infection and have good host properties, allowing multiplication and spread of the pathogen. In contrast, in genotypes such as *M.*

tschonoskii, *M. kansuensis* and *M. sargentii*, phytoplasma titre is extremely low, only non-specific yellowing symptoms develop and a high mortality rate occurs. At root level, these taxa are characterized by extensive phloem necrosis and depletion of starch. Their sieve tubes are so sensitive to phytoplasma infections that they collapse before the pathogen reaches a high titre (Kartte and Seemüller, 1991a, b). As mentioned above, phytoplasma titre is very low in AP-resistant, *M. sieboldii*-based apomictic rootstocks. In these genotypes, the AP phytoplasma concentration proved to be 100 to 5000 times lower than that occurring in susceptible *M.* × *domestica*-based rootstocks. Also, other apomictic rootstocks, which were mainly derived from *M. sargentii*, have been shown to have a very low AP phytoplasma titre. However, these proved to be very susceptible, much more than the *M.* × *domestica*-based rootstocks. It seems that host suitability per se, as expressed in phytoplasma titre, is obviously not the only defining factor for disease development. Instead, the pathogenic effect of phytoplasma infection may be of a qualitative rather than a quantitative nature (Seemüller *et al.*, 2008, and references cited therein). According to Seemüller and Schneider (2007), phytoplasma concentration in AP-affected apple trees is not markedly affected by strain virulence. These authors have reported that the three virulence groups identified among strains of the AP phytoplasma have similar phytoplasma concentrations. Thus, avirulent and mild strains have the same fitness as severe strains for multiplication and spread in apple trees.

A wide range of phytoplasma concentrations is also known to occur in EY-infected species of the genus *Ulmus*. These species differ greatly in their response to EY-phytoplasma infections. North American species such as *U. americana* (American elm) and *U. rubra* (red elm) are low-titre hosts. The affected trees show symptoms of foliar yellowing and extensive phloem necrosis in the roots and stem, and usually die within one or a few years after appearance of the foliar symptoms. In contrast, European and Asian species, including *U. minor* (European field elm), *U. laevis* (European white elm), *U. parvifolia* (Chinese elm) and *U. pumila* (Siberian elm), are suitable hosts for the EY pathogen and allow its multiplication to relatively high levels. The affected trees are primarily characterized by witches'-brooms as a specific symptom, do not show phloem necrosis and are less prone to decline. Eurasian elm genotypes that proved to be tolerant to EY phytoplasma infections have also been identified (Braun and Sinclair, 1976; Sinclair *et al.*, 2000).

Extremely low phytoplasma numbers, which could only be detected using the highly sensitive nested PCR assays, have been observed in several naturally infected plants showing non-specific, mild or no symptoms (Berges *et al.*, 2000). However, the pathological and epidemiological significance of these phytoplasmal infections is unclear and needs to be investigated through pathological and vector transmission studies.

Toxins

In some phytoplasma diseases, such as Australian papaya dieback, LY of coconut palms and rice yellow dwarf, because of poor correlation between phytoplasma presence and phloem aberrations or external symptoms occurring in some parts of infected plants, a long-distance effect of phytoplasmal infections through a toxic metabolite has been hypothesized (León *et al.*, 1996; Siddique *et al.*, 1998; Guthrie *et al.*, 2001; Tan and Whitlow, 2001). Also, in PD-affected pear trees grown on the highly susceptible oriental rootstocks *Pyrus pyrifolia* and *Py. ussuriensis*, severe sieve tube necrosis immediately below the bud union, resulting in the formation of replacement phloem, occurs. According to Schneider (1977), a substance produced by the causal phytoplasma, mainly in the leaves and translocated down the trunk, is toxic to the rootstock sieve tubes but not to the sieve tubes of scions. The author also reported that the *Py. communis* seedling indicators 'Variolosa', 'Magness' and 'Precocious' developed specific foliar symptoms, such as browning and enlargement of midribs and major lateral veins, upon infection with the PD phytoplasma. However, the PD phytoplasma concentration proved to be highest in the small minor veins, where phloem aberrations did not occur, whereas phytoplasma numbers were too low or undetectable in the coarse veins. From this finding it was suggested that toxic metabolites are produced or induced by the PD pathogen in the small minor veins and are then translocated to other sites, such as midribs and major lateral leaf vein sieve tubes and other vein-encasing tissues, where they cause necrosis. Histopathological studies have shown that, in EY-affected American elm trees, the occurrence of pathological phloem, which consists of abnormal callose deposition followed by sieve tube necrosis, collapse of sieve elements and companion cells, hyperactivity of the cambium and formation of excessive replacement phloem, is apparently induced at a distance from the phytoplasma-containing sieve elements (Braun and Sinclair, 1976). Therefore, it is likely that a phytoplasma-derived or phytoplasma-induced metabolite is involved in the pathogenic action of the EY agent. In the system X-disease phytoplasma–peach, the suddenness and synchronicity with which diseased peach trees developed X-disease symptoms, irrespective of X-disease phytoplasma titre, led to the hypothesis that symptom expression might be related to complex interactions involving metabolically active compounds such as toxins (Douglas, 1986). However, studies on detection and characterization of toxic substances in phytoplasma-infected plants have not been reported nor have phytoplasma genes showing significant sequence similarity to typical pathogenicity genes, which are related to the synthesis or secretion of toxins in walled plant-pathogenic bacteria, been identified.

Hormone imbalance

The extensive abnormalities in plant growth and development caused by phytoplasmas are suggestive of a profound disturbance in plant hormone balance. Several studies have shown hormone imbalance in phytoplasma-infected plants. For instance, in diseased periwinkle plants a decrease of cytokinin levels in mature leaves and roots and an increase in flowers were observed, whereas in LY-affected coconut palms an increase in abscisic acid and ethylene concentrations occurred in leaves. Also, infections by the AY phytoplasma of micropropagated periwinkle plants proved to be associated with a marked decrease of the endogenous auxin levels (for reviews see Kirkpatrick, 1991; León *et al.*, 1996; Lee *et al.*, 2000; Tan and Whitlow, 2001). However, it is not known whether phytoplasmas synthesize plant growth regulators, like some walled plant-pathogenic bacteria, or if phytoplasmas change the natural levels of one or more of the endogenous plant hormones. It has been shown that several plant species that normally require short photoperiods or cold temperature to induce flowering flowered when grown under non-inductive environmental conditions upon infection with the beet leafhopper-transmitted virescent (BLTV) phytoplasma. Exogenous applications of gibberellic acid (GA) also induced similar flowering responses in these plants, whereas inhibitors of GA biosynthesis prevented the BLTV phytoplasma-induced host response. Thus, it was suggested that the BLTV agent either may produce a compound with GA-like activity or alters normal endogenous GA levels in the plant (Kirkpatrick, 1991). Moreover, some LY symptoms in coconut palms, such as nut fall and leaf senescence, were mimicked by treating healthy plants with an ethylene-releasing agent (León *et al.*, 1996). This finding further supports the involvement of phytoplasmas in plant hormone imbalance. However, work by Smart and Kirkpatrick (for references see Lee *et al.*, 2000) revealed that *Arabidopsis thaliana* mutants that were insensitive to changes in abscisic acid, auxin, ethylene or gibberellin concentrations showed typical virescence and phyllody symptoms following infection with the AY phytoplasma. From this finding the authors concluded that changes in the levels of these phytohormones may not be responsible for the symptom expression. Recent studies have shown that symptoms of flower abnormalities in STOL-infected tomato plants are associated with deregulations of key floral development genes, whereas in free-branching-affected poinsettia plants genes involved in phytohormone activity, which are upregulated upon phytoplasmal infections, may play a major role in symptom expression (see above under 'Disease Induction'). Further work, mainly at a molecular level, is needed in order to get firm data on the role of phytoplasmal infections in plant hormone imbalance and its relationships to symptom expression.

Attachment to host cell membrane

Attachment to the host membrane is a characteristic feature of many human and animal pathogenic mollicutes and is considered to be an important requirement for pathogenicity. This attachment is mediated by exposed surface protein adhesins, which in some species are organized in specialized structures (tips). Several studies have shown that in most phytoplasmas a subset of membrane proteins, referred to as immunodominant membrane proteins (IDPs), constitutes a major portion of the total cellular membrane proteins (reviewed in Hogenhout *et al.*, 2008). Genes encoding IDPs have been isolated from several phytoplasmas, and these proteins have been

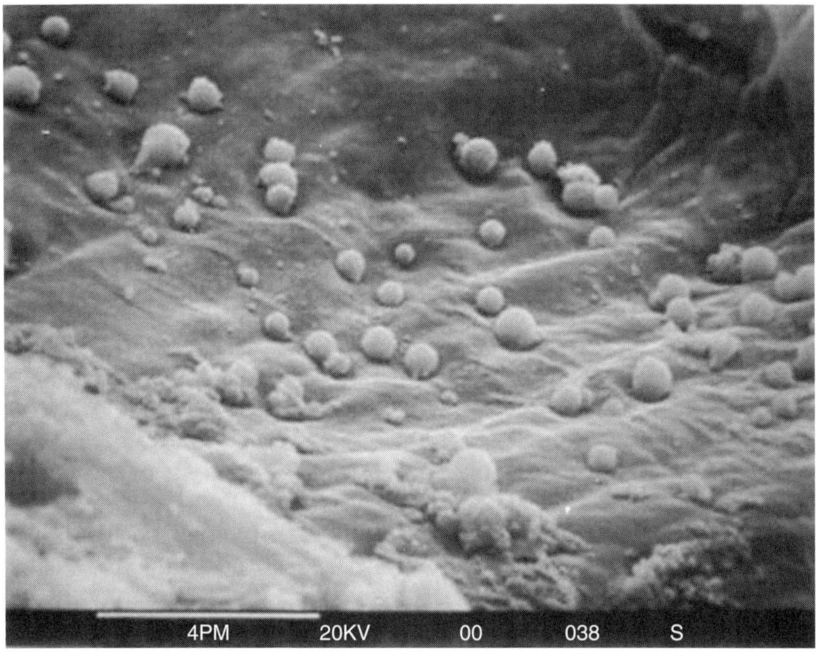

Fig. 7.2. Scanning electron micrograph of a phloem sieve tube of phytoplasma-infected plants, showing phytoplasmas attached to the sieve tube plasma membrane.

classified into three distinct groups on the basis of their type of interaction with the lipid bilayer. All IDPs have a central hydrophilic region, which may be located on the external surface of the phytoplasma cell membrane, and one or two hydrophobic transmembrane domains. Because of their location and reasonable abundance, IDPs are good candidates for phytoplasma–host interactions. They apparently function as adhesins, allowing phytoplasma attachment to host cells. Transmission and scanning electron microscope observations carried out in several laboratories have provided evidence that the parasitizing phytoplasmas are attached to the inner surface of the sieve tube plasma membrane (Figs 7.1 and 7.2). In this respect, phytoplasmas bear a striking resemblance to the adhering culturable mycoplasmas (Marcone and Ragozzino, 1996).

Other pathogenicity determinants

Comparative analysis of genomes of the few phytoplasmas that have been completely sequenced to date revealed the presence of several potential virulence factors. These include *hlyC* and *tlyC* genes, which encode haemolysins, i.e. membrane-damaging agents that serve as important virulence factors, mainly for Gram-positive human pathogens. Also, genes encoding the protein export and targeting components of the Sec-dependent pathway have been identified in phytoplasma genomes. This finding indicates the presence of a functional protein translocation system in phytoplasmas. Thus, virulence-related proteins might be transported from the phytoplasma cytoplasm to the phytoplasma cell surface or to host cytoplasm through the Sec pathway and may influence pathogenicity, as reported for *Streptococcus pyogenes*. Another possible virulence factor is the *hflB* (or *ftsH*) gene, which encodes an essential ATP-dependent metalloprotease anchored in the cell membrane. This protease may function by degrading host proteins into amino acids for uptake as essential compounds or by degrading proteins produced by the host cell as a plant defence reaction. Nucleases, which are reported as potential virulence determinants in culturable mycoplasmas, may also be involved in phytoplasma pathogenicity. It has been shown that the severe strain of the OY phytoplasma has two sets of certain glycolytic genes, whereas the mild strain of the same pathogen carries only one set. From this finding it was suggested that the difference accounts for the greater aggressiveness of the severe strain, which is due to a depletion of sugar in the phloem sap, and the resulting higher multiplication rate of the pathogen (Hogenhout *et al.*, 2008).

Summary

Plant-pathogenic phytoplasmas systemically colonize the plants by passing through phloem sieve plate pores. Since they live and multiply in functional phloem sieve tube elements, the main effect of phytoplasmal infections is, apparently, the impairment of the sieve tube function. However, the exact mechanisms involved in symptom development or the genes that control these events are still poorly understood. Knowledge of the movement of phytoplasmas within the host plant and their final distribution in various organs is usually essential for understanding the phytoplasma–plant host interactions. For most phytoplasmal diseases for which the colonization behaviour of the causal agent(s) is known, expression of symptoms depends on the presence of the invading phytoplasma(s). However, in some diseases, because of poor correlation between phytoplasma presence and phloem aberrations or external symptoms occurring in some parts of infected plants, a long-distance effect of phytoplasmal infections is hypothesized. Several factors which may account for disease development include strain virulence, strain interference, phytoplasma concentration, toxins, plant hormone imbalance and attachment of phytoplasmas to host cell membrane. Also, a number of other putative pathogenicity factors are known from the complete genome sequences of the few phytoplasmas that have been determined to date. Sequence comparisons of the entire genomes of several phytoplasmas, including strains within a given taxon that differ greatly in aggressiveness, will provide insights into the largely unknown phytoplasma pathology.

References

Berges, R. and Seemüller, E. (2002) Impact of phytoplasma infection of common alder (*Alnus glutinosa*) depends on strain virulence. *Forest Pathology* 32, 357–363.

Berges, R., Rott, M. and Seemüller, E. (2000) Range of phytoplasma concentrations in various plant hosts as determined by competitive polymerase chain reaction. *Phytopathology* 90, 1145–1152.

Braun, E.J. and Sinclair, W.A. (1976) Histopathology of phloem necrosis in *Ulmus americana*. *Phytopathology* 66, 598–607.

Braun, E.J. and Sinclair, W.A. (1978) Translocation in phloem necrosis-diseased American elm seedlings. *Phytopathology* 68, 1733–1737.

Carraro, L., Ermacora, P., Loi, N. and Osler, R. (2004) The recovery phenomenon in apple proliferation-infected trees. *Journal of Plant Pathology* 86, 141–146.

Catlin, P.B., Olsson, E.A. and Beutel, J.A. (1975) Reduced translocation of carbon and nitrogen from leaves with symptoms of pear curl. *Journal of the American Society for Horticultural Science* 100, 184–187.

Choi, Y.H., Tapias, E.C., Kim, H.K., Lefeber, A.W.M., Erkelens, C., Verhoeven, J.Th.J., Brzin, J., Zel, J. and Verpoorte, R. (2004) Metabolic discrimination of *Catharanthus roseus* leaves infected by phytoplasma using ^{1}H-NMR spectroscopy and multivariate data analysis. *Plant Physiology* 135, 2398–2410.

Christensen, N.M., Nicolaisen, M., Hansen, M. and Schulz, A. (2004) Distribution of phytoplasmas in infected plants as revealed by real-time PCR and bioimaging. *Molecular Plant–Microbe Interactions* 17, 1175–1184.

Clark, M.F., Barbara, D.J. and Davies, D.L. (1983) Production and characteristics of antisera to *Spiroplasma citri* and clover phyllody-associated antigens derived from plants. *Annals of Applied Biology* 103, 251–259.

Constable, F.E., Gibb, K.S. and Symons, R.H. (2003) Seasonal distribution of phytoplasmas in Australian grapevines. *Plant Pathology* 52, 267–276.

Douglas, S.M. (1986) Detection of mycoplasmalike organisms in peach and chokecherry with X-disease by fluorescence microscopy. *Phytopathology* 76, 784–787.

Freitag, J.H. (1964) Interaction and mutual suppression among three strains of aster yellows virus. *Virology* 24, 401–413.

Guthrie, J.N., Walsh, K.B., Scott, P.T. and Rasmussen, T.S. (2001) The phytopathology of Australian papaya dieback: a proposed role for the phytoplasma. *Physiological and Molecular Plant Pathology* 58, 23–30.

Hogenhout, S.A., Oshima, K., Ammar, E.-D., Kakizawa, S., Kingdom, H.N. and Namba, S. (2008) Phytoplasmas: bacteria that manipulate plants and insects. *Molecular Plant Pathology* 9, 403–423.

Jarausch, W., Lansac, M. and Dosba, F. (1999) Seasonal colonization pattern of European stone fruit yellows phytoplasmas in different *Prunus* species detected by specific PCR. *Journal of Phytopathology* 147, 47–54.

Kartte, S. and Seemüller, E. (1991a) Histopathology of apple proliferation in *Malus* taxa and hybrids of different susceptibility. *Journal of Phytopathology* 131, 149–160.

Kartte, S. and Seemüller, E. (1991b) Susceptibility of grafted *Malus* taxa and hybrids to apple proliferation disease. *Journal of Phytopathology* 131, 137–148.

Kirkpatrick, B.C. (1991) Mycoplasma-like organisms – plant and invertebrate pathogens. In: Balows, A., Trüper, H.G., Dworkin, M., Harder, W. and Schleifer K.-S. (eds) *The Prokaryotes*, Vol. III. Springer-Verlag, New York, pp. 4050–4067.

Kison, H. and Seemüller, E. (2001) Differences in strain virulence of the European stone fruit yellows phytoplasma and susceptibility of stone fruit trees on various rootstocks to this pathogen. *Journal of Phytopathology* 149, 533–541.

Kuske, C.R. and Kirkpatrick, B.C. (1992) Distribution and multiplication of western aster yellows mycoplasmalike organisms in *Catharanthus roseus* as determined by DNA hybridization analysis. *Phytopathology* 82, 457–462.

Lederer, W. and Seemüller, E. (1991) Occurrence of mycoplasma-like organisms in diseased and non-symptomatic alder trees (*Alnus* spp.). *European Journal of Forest Pathology* 21, 90–96.

Lee, I.-M., Davis, R.E. and Gundersen-Rindal, D.E. (2000) Phytoplasma: phytopathogenic mollicutes. *Annual Review of Microbiology* 54, 221–255.

León, R., Santamaría, J.M., Alpizar, L., Escamilla, J.A. and Oropeza, C. (1996) Physiological and biochemical changes in shoots of coconut palms affected by lethal yellowing. *New Phytologist* 134, 227–234.

Lepka, P., Stitt, M., Moll, E. and Seemüller, E. (1999) Effect of phytoplasmal infection on concentration and traslocation of carbohydrates and amino acids in periwinkle and tobacco. *Physiological and Molecular Plant Pathology* 55, 59–68.

Lherminier, J., Courtois, M. and Caudwell, A. (1994) Determination of the distribution and multiplication sites of Flavescence dorée mycoplasma-like organisms in the host plant *Vicia faba* by ELISA and immunocytochemistry. *Physiological and Molecular Plant Pathology* 45, 125–138.

Marcone, C. and Ragozzino, A. (1996) Comparative ultrastructural studies on genetically different phytoplasmas using scanning electron microscopy. *Petria* 6(2), 125–136.

Marcone, C., Neimark, H., Ragozzino, A., Lauer, U. and Seemüller, E. (1999) Chromosome sizes of phytoplasmas composing major phylogenetic groups and subgroups. *Phytopathology* 89, 805–810.

Maust, B.E., Espadas, F., Talavera, C., Aguilar, M., Santamaría, J.M. and Oropeza, C. (2003) Changes in carbohydrate metabolism in coconut palms infected with the lethal yellowing phytoplasma. *Phytopathology* 93, 976–981.

McCoy, R.E. (1979) Mycoplasmas and yellows diseases. In: Whitcomb, R.F. and Tully, J.G. (eds) *The Mycoplasmas*, Vol. III. Academic Press, New York, pp. 229–264.

McCoy, R.E., Caudwell, A., Chang, C.J., Chen, T.A., Chiykowski, L.N., Cousin, M.T., Dale, J.L., de Leeuw, G.T.N., Golino, D.A., Hackett, K.J., Kirkpatrick, B.C., Marwitz, R., Petzold, H., Sinha, R.C., Sugiura, M., Whitcomb, R.F., Yang, I.L., Zhu, B.M. and Seemüller, E. (1989) Plant diseases associated with mycoplasma-like organisms. In: Whitcomb, R.F. and Tully, J.G. (eds) *The Mycoplasmas*, Vol. V. Academic Press, San Diego, California, pp. 545–640.

Nicolaisen, M. and Horvath, D.P. (2008) A branch-inducing phytoplasma in *Euphorbia pulcherrima* is associated with changes in expression of host genes. *Journal of Phytopathology* 156, 403–407.

Pracros, P., Renaudin, J., Eveillard, S., Mouras, A. and Hernould, M. (2006) Tomato flower abnormalities induced by stolbur phytoplasma infection are associated with changes of expression of floral development genes. *Molecular Plant–Microbe Interactions* 19, 62–68.

Saracco, P., Bosco, D., Veratti, F. and Marzachì, C. (2006) Quantification over time of chrysanthemum yellows phytoplasma (16Sr-I) in leaves and roots of the host plant *Chrysanthemum carinatum* (Schousboe) following inoculation with its insect vector. *Physiological and Molecular Plant Pathology* 67, 212–219.

Schaper, U. and Seemüller, E. (1982) Condition of the phloem and persistence of mycoplasmalike organisms associated with apple proliferation and pear decline. *Phytopathology* 72, 736–742.

Schaper, U. and Seemüller, E. (1984) Recolonization of the stem of apple proliferation and pear decline-diseased trees by the causal organisms in spring. *Zeitschrift für Pflanzenkrankheiten und Pflanzenschutz* 91, 608–613.

Schneider, H. (1977) Indicator hosts for pear decline: symptomatology, histopathology, and distribution of mycoplasmalike organisms in leaf veins. *Phytopathology* 67, 592–601.

Seemüller, E. and Schneider, B. (2007) Differences in virulence and genomic features of strains of '*Candidatus* Phytoplasma mali', the apple proliferation agent. *Phytopathology* 97, 964–970.

Seemüller, E., Kunze, L. and Schaper, U. (1984a) Colonization behavior of MLO, and symptom expression of proliferation-diseased apple trees and decline-diseased pear trees over a period of several years. *Zeitschrift für Pflanzenkrankheiten und Pflanzenschutz* 91, 525–532.

Seemüller, E., Schaper, U. and Zimbelmann, F. (1984b) Seasonal variation in the colonization patterns of mycoplasmalike organisms associated with apple proliferation and pear decline. *Zeitschrift für Pflanzenkrankheiten und Pflanzenschutz* 91, 371–382.

Seemüller, E., Schaper, U. and Kunze, L. (1986) Effect of pear decline on pear trees on 'Quince A' and *Pyrus communis* seedling rootstocks. *Zeitschrift für Pflanzenkrankheiten und Pflanzenschutz* 93, 44–50.

Seemüller, E., Marcone, C., Lauer, U., Ragozzino, A. and Göschl, M. (1998a) Current status of molecular classification of the phytoplasmas. *Journal of Plant Pathology* 80, 3–26.

Seemüller, E., Stolz, H. and Kison, H. (1998b) Persistence of the European stone fruit yellows phytoplasma in aerial parts of *Prunus* taxa during the dormant season. *Journal of Phytopathology* 146, 407–410.

Seemüller, E., Garnier, M. and Schneider, B. (2002) Mycoplasmas of plants and insects. In: Razin, S. and Hermann, R. (eds) *Molecular Biology and Pathogenicity of Mycoplasmas*. Kluwer Academic/Plenum Publishers, New York, pp. 91–115.

Seemüller, E., Moll, E. and Schneider, B. (2008) Apple proliferation resistance of *Malus sieboldii*-based rootstocks in comparison to rootstocks derived from other *Malus* species. *European Journal of Plant Pathology* 121, 109–119.

Siddique, A.B.M., Guthrie, J.N., Walsh, K.B., White, D.T. and Scott, P.T. (1998) Histopathology and within-plant distribution of the phytoplasma associated with Australian papaya dieback. *Plant Disease* 82, 1112–1120.

Sinclair, W.A. and Griffiths, H.M. (2000) Variation in aggressiveness of ash yellows phytoplasmas. *Plant Disease* 84, 282–288.

Sinclair, W.A., Townsend, A.M, Griffiths, H.M. and Whitlow, T.H. (2000) Responses of six Eurasian *Ulmus* cultivars to a North American elm yellows phytoplasma. *Plant Disease* 84, 1266–1270.

Tan, P.Y. and Whitlow, T. (2001) Physiological responses of *Catharanthus roseus* (periwinkle) to ash yellows phytoplasmal infection. *New Phytologist* 150, 757–769.

Wei, W., Kakizawa, S., Suzuki, S., Jung, H.-Y., Nishigawa, H., Miyata, S.-I., Oshima, K., Ugaki, M., Hibi, T. and Namba, S. (2004) In planta dynamic analysis of onion yellows phytoplasma using localized inoculation by insect transmission. *Phytopathology* 94, 244–250.

Weintraub, P.G. and Beanland, L. (2006) Insect vectors of phytoplasmas. *Annual Review of Entomology* 51, 91–111.

8 Biochemical Changes in Plants Infected by Phytoplasmas

Rita Musetti

Università di Udine, Italy

Introduction

Phytoplasmas are plant-pathogenic prokaryotes belonging to the class *Mollicutes*, a group of wall-less microorganisms phylogenetically related to low G+C, Gram-positive bacteria (Weisburg *et al.*, 1989). Plant-pathogenic mollicutes have been associated with several hundred diseases affecting economically important crops, such as ornamentals, vegetables, fruit trees and grapevine (Lee *et al.*, 2000). However, the mechanisms by which plant mollicutes induce diseases are still to be deciphered. In host plants, bacteria are restricted to the phloem sieve tubes and are transmitted between plants by phloem-sap-feeding leafhoppers or psyllids in a persistent manner. Phytoplasmas (Figs 8.1 and 8.2) are obligate parasites inducing characteristic symptoms in host plants, such as low growth rate, stunting, yellowing or reddening of the leaves, reduced leaf size, shortening of internodes and loss of apical dominance. These lead to reduced yields, proliferation of shoots or roots, witches'-brooming, general decline and, sometimes, death of the plant. Several symptoms affect flowers, including virescence, phyllody and sterility. Symptom appearance is preceded by cellular modifications, such as callose deposition near sieve plates and plasmodesmata (Fig. 8.1), starch accumulation in the chloroplasts and their disorganization, and phloem necrosis (Musetti, 2006).

Phytoplasma diseases are classified as 'auxonic diseases', indicating a possible interaction with the hormonal balance of the host (Chang, 1998; Pertot *et al.*, 1998), but physiological relationships between phytoplasma and host plant have remained largely undiscovered. Recent developments have improved our knowledge of the effect of phytoplasma infection on host secondary metabolites, mainly in herbaceous host plants (Musetti *et al.*, 1999, 2000; Tan and Whitlow, 2001; Choi *et al.*, 2004), but the literature available is still scarce regarding the physiology of phytoplasma infections in fruit crops (Lepka *et al.*, 1999; Musetti *et al.*, 2004, 2005) and grapevine (Bertamini *et al.*,

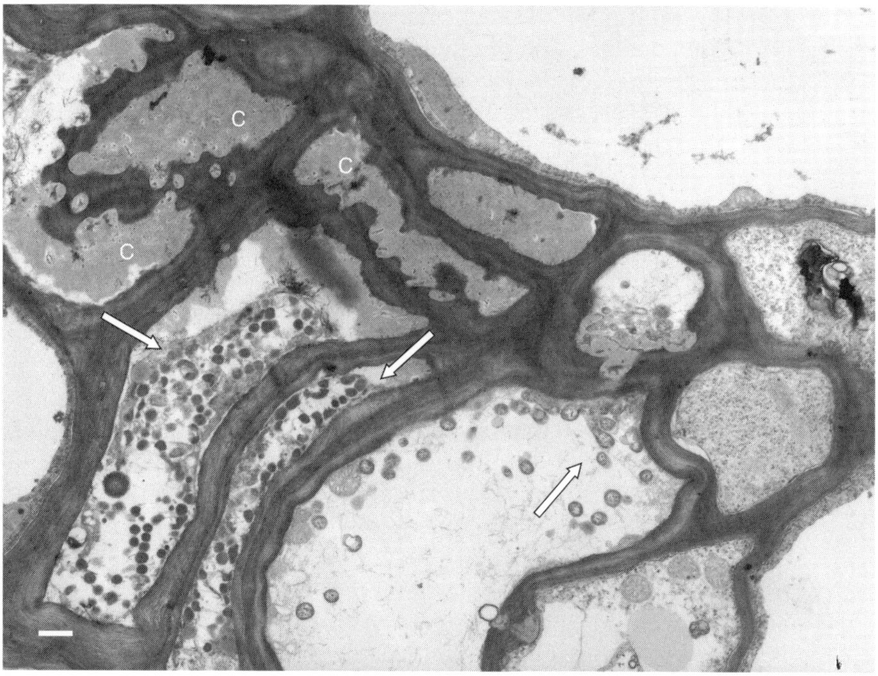

Fig. 8.1. Transmission electron micrograph of phytoplasmas (arrowed) in sieve tubes of the host plant. Sieve plates are filled by callose (C). Bar = 0.5 μm.

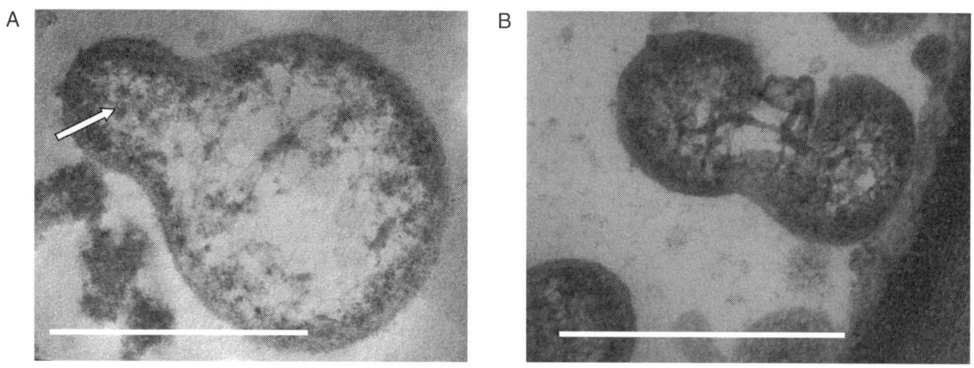

Fig. 8.2. High magnification of phytoplasma cells. In Fig. 8.2A, note the bud-like structure (arrow); in Fig. 8.2B a phytoplasma cell appears to be dividing by scission. Bars = 0.5 μm.

2002b; Musetti *et al.*, 2007). Phytoplasma infection can lead to the production of defence proteins, increase in phenolic compounds and involvement of important signal molecules such as Ca^{2+}, H_2O_2 and salicylic acid (Musetti and Favali, 2003; Musetti *et al.*, 2005).

The molecular mechanisms of pathogenicity are poorly understood. However, deregulation of plant genes involved in phytoplasma–plant interactions has been evaluated using differential display of mRNA, RT-PCR and microarray analysis.

Hormones and Growth Regulators

Phytoplasma-associated disorders are classified among the auxonic diseases of plants. The most typical symptoms of phytoplasma diseases, such as premature bud opening, out of season flowering, phyllody, proliferation and other growth aberrations, indicate perturbations in plant hormonal balance. However, few reports have been published to support this hypothesis. Pertot *et al.* (1998) reported that in phytoplasma-infected micropropagated periwinkle shoots the level of indole-3-acetic acid (IAA) increased ten times compared with the controls. The fact that exogenous addition of IAA or other auxin-like substances to healthy plants does not induce symptoms similar to those caused by phytoplasma infection supports the hypothesis that increasing levels of IAA in phytoplasma-infected tissues could be a non-specific or secondary response to a stress condition (Pertot *et al.*, 1998). Perica (2008) found that prolonged auxin treatment or indole-3-butyric acid (IBA) induces recovery in phytoplasma-infected periwinkle, suggesting that an anti-phytoplasmal activity of auxins could be involved. The possible molecular basis of this phenomenon has not been ascertained.

With regard to cytokinins, Kesumawati *et al.* (2006) reported on the role of this hormone in the phyllody symptom development in *Hydrangea macrophylla*. The development of phytoplasma-infected green flowers was correlated with the accumulation of high levels of cytokinin in the floral parts (Davey *et al.*, 1981). In fact, cytokinins play a role in chloroplast development and prevent chlorophyll breakdown.

Nicolaisen and Horvath (2008) studied gene expression in *Euphorbia pulcherrima* infected with a branch-inducing phytoplasma. Using a differential display of cDNA-polymerase chain reaction products, they identified differentially expressed genes after phytoplasma infection. Three genes showing high homology with genes involved in the biosynthesis or the perception of plant hormones or other growth regulators, such as cytokinin, expansin and gibberellins, were shown to be upregulated. Smart and Kirkpatrick (1996) and He *et al.* (1998) revealed that differential gene expression (including protein kinase genes) may be responsible for the induction of virescence and phyllody in plants during early stages of infection with aster yellows.

Tomato plants infected with stolbur phytoplasma (STOL) also show different floral abnormalities associated with early changes in the expression of key flower development genes. Three genes involved in meristem development (*LeWUSCHEL* and *LeCLAVATA1*) and organ identity (*LeDEFICIENS*) were found to be downregulated, whereas *FALSIFLORA*, a gene controlling the identity of the inflorescence meristem, was shown to be upregulated. Gene *TAG1*, which regulates stamen and carpel identities, was upregulated at the early stages and downregulated at the late stages of infection. However, tissue-specific expression of *TAG1* was not affected by STOL infection (Pracros *et al.*, 2006).

Because phytoplasma cells have never been detected in apical meristems, the signal inducing the misregulation of flower development genes is probably a long-distance signal. It has been hypothesized that this signal could be represented by sugars that influence the expression of genes controlling floral transition. Interestingly, phytoplasmas affect phloem function, impairing carbohydrate translocation and subsequently causing the accumulation of soluble sugars in source leaves (Lepka *et al.*, 1999). Accumulation of sugars has also been observed in the leaves of periwinkle infected by the culturable mollicute *Spiroplasma citri* (Andrè *et al.*, 2005) Abscisic acid (ABA) and ethylene increase in phytoplasma-infected plants and their involvement in leaf senescence processes associated with leaf yellowing has been hypothesized. Jagoueix-Eveillard *et al.* (2001) suggested that a gene coding for a putative sterol C-methyl-transferase, an enzyme involved in phytosterol biosynthesis, could play a role in yellows and stunting symptom development. This gene has been found to be downregulated in *Catharanthus roseus* plants infected by different kinds of *Mollicutes*, such as *S. citri* or STOL, both of which induce stunting and internode shortening.

Callose and Carbohydrate Metabolism

Histological studies on several plant species affected with phytoplasmas showed that the first detectable anatomical aberration is an abnormal deposition of callose on the plates of sieve tubes, followed by the collapse of these elements and the companion cells. Depending on sensitivity of the host, a small or large portion of the phloem becomes necrotic. Callose accumulation is a non-specific response of plants to pathogen attack and to wounds.

Callose synthesis and aggregation of P-protein filaments in the phloem to form plugs are Ca^{2+}-dependent phenomena (Köhle *et al.*, 1985; Knoblauch *et al.*, 2001) regulated by Ca^{2+} flux into the phloem, and they are among the early key events leading to the formation of physical barriers that might prevent the movement of phytoplasmas *in planta* (Musetti *et al.*, 2008). It has been reported that, in apple plants, infection by the apple proliferation phytoplasma is associated with a decrease in cytosolic Ca^{2+} in the phloem (Musetti *et al.*, 2008).

Lepka *et al.* (1999) described the effect of phytoplasma infection on concentration and translocation of carbohydrates in periwinkle and tobacco plants. They found higher levels of reducing sugars and sucrose in source leaves of infected plants than in healthy ones. In roots, concentration of sugars was low and seemed not to be affected by the phytoplasma infection. Sucrose levels appeared to be similar to those of healthy plants, but variations, depending on the virulence of the phytoplasma isolate and the host/phytoplasma association, were reported. Phytoplasma infection led to a significant increase of starch in source leaves and a decrease in sink leaves and roots when compared with healthy plants.

These data support those obtained from fruit trees and woody plants (Braun and Sinclair, 1976; Kartte and Seemüller, 1991), which showed that phytoplasma infection led to the accumulation of carbohydrates in mature leaves and decreased starch in the roots. These data are consistent with ultrastructural observations reporting starch accumulation in chloroplasts associated with a severe disorganization of thylakoids and a reduction in chlorophyll content (Musetti, 2006). Accumulation of carbohydrates, also reported in coconut palms affected by lethal yellowing (Maust *et al.*, 2003) and in maize plants affected by maize bushy stunt (Junqueira *et al.*, 2004), is considered a secondary effect of infection and can be explained through an inhibition of phloem transport (Maust *et al.*, 2003). As a consequence, photosynthetic products accumulate in chloroplasts, inhibiting photosynthesis and reducing the supply of sugars from source leaves to roots.

It has been demonstrated that some *Mollicutes* can repress genes involved in sugar transport, such as transketolase (Jagoueix-Eveillard *et al.*, 2001). Inhibition of these genes might also be responsible for the repression of the genes involved in photosynthesis. This gene deregulation can explain yellows symptoms induced in host plants by spiroplasmas, phytoplasmas such as STOL and other phloem-restricted pathogens.

The activity of the four main enzymes implicated in sugar partition in plants – sucrose synthase, cell wall invertase, neutral invertase and vacuolar invertase – has been investigated in STOL-infected periwinkle and tomato (Machenaud *et al.*, 2007). Only neutral invertase showed an increased activity in infected plants, but no significant variation in the regulation of corresponding genes has been observed. A role for neutral invertase in providing glucose and/or fructose for phytoplasma growth has been hypothesized.

Chlorophyll and Photosynthetic Activity

As reported above, carbohydrate metabolism variation in plants that are affected by phytoplasmas correlates with a marked reduction of total chlorophyll (Chl) content due to the decrease of both Chl *a* and Chl *b* in leaves. A decrease in photosynthetic pigments has been observed in maize plants infected with maize bushy stunt (Junqueira *et al.*, 2004), apples infected with apple proliferation and grapevine infected with the bois noir phytoplasma. This is probably the result of enhanced chlorophyllase activity in infected leaves (Bertamini *et al.*, 2002b).

The loss of chlorophyll is usually accompanied by a general sugar-mediated repression of genes involved in photosynthesis (Krapp *et al.*, 1993). It has been demonstrated that STOL induced a downregulation of genes involved in photosynthesis (Jagoueix-Eveillard *et al.*, 2001), and it has been suggested that phytoplasmas have a role in the inhibition of chlorophyll biosynthesis in plant host leaves (Bertamini *et al.*, 2002a).

The decline of photosynthesis is the result of phytoplasma infection on photosynthetic electron transport and enzymatic activities. Phytoplasma diseases induce a marked loss in the photosynthetic whole chain (mainly affecting photosystem II activity), due to the loss of several thylakoid membrane proteins and to the reduction of leaf soluble proteins, mainly ribulose-1,5-bisphosphate carboxylase (RuBPC). These changes are similar to those induced by leaf ageing, so an interference of phytoplasmas with plant hormones that regulate senescence processes in leaf tissues could be hypothesized.

Changes in photosynthetic activity are connected with several physiological parameters, such as stomatal conductance. In coconut palms affected by lethal yellowing (LY), stomatal conductance was shown to decrease progressively as the disease developed (Leon *et al.*, 1996). Since altered stomatal behaviour might affect gas exchange and related processes in the plant, such as water movement and photosynthesis, these results could support the hypothesis that LY-induced stomatal closure is central to the development of yellows symptoms in coconut palms. Leaf yellowing occurred simultaneously with a decrease in photosynthetic rates and a decrease in protein, chlorophyll and carotenoid content.

Based on these biochemical changes, the hypothesis that LY-associated leaf yellowing is part of a leaf senescence process has been proposed. In LY-affected palm trees, the concentration of abscisic acid (ABA) in the leaves, as well as the capacity for leaf tissue to form ethylene, increased, suggesting the probable involvement of these hormones. Hormonal imbalance might be related to LY symptom development, at least with respect to nutfall and leaf senescence (Leon *et al.*, 1996).

In *C. roseus* infected with ash yellows, the reduction of stomatal conductance was not associated with changes in ABA concentration (Tan and Whitlow, 2001) and could arise from stomatal closures due to sucrose accumulation as well as reductions in stomatal frequency and guard cell size during leaf ontogeny in a systemically infected plant. Reduced photosynthesis in ash yellows-infected *C. roseus* appears to be the result of reduced carboxylation capacity of Rubisco and a reduced regeneration rate of RuBP, indicating a downregulation of the Calvin cycle.

Amino Acid Transport

Phytoplasma infection affects amino acid transport in host plants. High amino acid content was found in source and sink leaves of ash yellows-infected *C. roseus* and in source leaves of apple proliferation-infected tobacco (Lepka *et al.*, 1999). Amino acid accumulation depends on the limitation of transport in the phloem, which is severely reduced in the phytoplasma-infected plants. Inhibition of amino acid transport produces negative effects on plant growth, contributing to a reduction in plant size. In fact, amino acids provide the quickest system for a plant to incorporate inorganic nitrogen, an essential element for growth. Several tissues, including developing leaves, meristems and reproductive organs, rely on the importation of amino acids to support growth and development. Moreover, a key role of amino acids in leaf senescence mechanisms has been suggested (Lepka *et al.*, 1999), and senescence-like symptoms are very common in phytoplasma diseases.

In agreement with these biochemical changes, mRNA differential display analyses revealed that, in apricot tissues infected by European stone fruit yellows (ESFY) phytoplasmas, a homologue of an amino acid transporter was downregulated (Carginale *et al.*, 2004).

Protein Content

Usually plants infected by pathogens show a high protein content, which could be due to both the activation of the host defence mechanisms and the pathogen attack mechanisms (Agrios, 1997). Amongst the proteins produced in the host plants there are pathogenesis-related proteins (PR-proteins), including peroxidases, chitinases and β-1,3-glucanases.

An increase in the total amount of proteins has been found in maize bushy stunt phytoplasma-infected maize plants (Junqueira *et al.*, 2004). In particular, resistant hybrids accumulate a higher protein content than susceptible ones, supporting the hypothesis that accumulation of PR-proteins contributes to the increase of total proteins in infected tissues.

Zhong and Shen (2004) found that six soluble proteins, accumulated specifically in *Chrysanthemum coronarium* infected with phytoplasmas, belonged to the 16SrI and 16SrIII groups. These proteins shared high sequence similarities with the N-terminal amino acid sequence of thaumatin-like or osmotin-like proteins, which are PR-5 protein group members.

Contradictory results have been obtained in maize plants infected with different *Mollicutes*, in tomato plants affected by STOL (Favali *et al.*, 2001), in grapevine affected by bois noir (Bertamini *et al.*, 2002b) or flavescence dorée (Musetti *et al.*, 2007) and in apple trees affected by apple proliferation (Bertamini *et al.*, 2002a), where a decrease in total soluble proteins has been observed. A possible explanation for this difference could be due to the fact that extremely susceptible plants were used in these last experiments.

Phenolics

It has often been observed that certain common phenolic substances are toxic to pathogens and accumulate in plants after infection, especially in resistant varieties (Agrios, 1997). Different phenolic substances appear concurrently in the same diseased tissue, showing a synergetic activity against the pathogens. Total polyphenol content has been determined in phytoplasma-infected apples and plums (Musetti *et al.*, 2000). For both plants, analyses performed in spring after flowering, and repeated in summer, revealed a threefold higher polyphenol content in the infected tissues as compared with healthy ones.

Accumulation of phenolic substances in infected plants has also been reported in *Zea mays* (Junqueira *et al.*, 2004), where the presence of phytoplasma triggered an increase in phenolic compounds.

H-NMR spectroscopy has been used to analyse *C. roseus* leaf tissues, where it revealed that the metabolites related to the biosynthesis of phenylpropanoids are present in higher amounts in phytoplasma-infected plants (Choi *et al.*, 2004). Chlorogenic acid and polyphenols (gallic acid derivatives) show a relatively higher abundance in infected leaves. In particular, chlorogenic acid was found to be increased by two to four times in phytoplasma-infected *C. roseus*, and the same is true of gallotannin.

Chlorogenic acid content was also doubled in dried flowering tops of *Hypericum perforatum* infected with ash yellows phytoplasmas (Bruni *et al.*, 2005). A different trend was shown by flavonoids, which decreased significantly, revealing that phytoplasma infection causes variations in the secondary metabolism of challenged tissues, inhibiting the flavonoid biosynthetic pathway and increasing the biosynthesis of caffeic and cinnamic derivatives.

Recently, Romanazzi *et al.* (2007) reported that grapevines, both affected by bois noir and recovered from the disease, showed an upregulation of phenylalanine ammonia-lyase (PAL), as compared with healthy plants. These data suggest that polyphenols are involved in the defence against phytoplasmas.

Polyamines

Polyamines (putrescine, spermidine and spermine) are nitrogen-containing compounds of low molecular weight and polycationic nature, found in bacteria, animal and plant cells (Bagni and Torrigiani, 1992), where they sustain cell division and differentiation.

In plants, polyamines play a major role in several stress-related processes (Martin-Tanguy, 1997). In infected plant tissues, they accumulate in response to diseases limiting pathogen spread (viruses in particular) in host tissues (Martin-Tanguy, 1997). Therefore a role for these compounds in resistance has been suggested.

Experiments on polyamine evaluation in phytoplasma infections in periwinkle plants have been carried out, with the aim of comparing the polyamine levels in healthy and phytoplasma-infected micropropagated shoots of *C. roseus*. This was done so as to determine a possible involvement of these compounds in supporting the plant defence mechanism or its possible recovery from the phytoplasma infection (Musetti *et al.*, 1999). The results of this work showed that the plant response to the infection in terms of polyamines is mainly localized in leaves. In fact, in phytoplasma-infected micropropagated material, endogenous free putrescine accumulation is at remarkably high levels in leaves when compared with the stem, where the level is the same as in healthy controls. As indicated, phytoplasmas are mainly distributed in the leaves, where photoassimilates are produced. Concomitant with the increase in putrescine levels, the levels of major polyamines decrease, indicating a possible shift in the polyamine biosynthetic pathway in infected leaves towards the polyamine precursor putrescine (Roustan *et al.*, 1992).

This shift could result from inhibition of S-adenosylmethionine (SAM) decarboxylase activity, which could also lead to accumulation of SAM, which is the common precursor of ethylene and the major polyamines. The senescence processes induced by phytoplasmas could therefore be mediated by ethylene synthesis, as observed in several plant diseases (Ricci *et al.*, 1988). Moreover, a direct correlation between ethylene synthesis and conjugated putrescine accumulation has also been found in cultured, 'senescing' tobacco thin layers (Scaramagli *et al.*, 1995).

In the phytoplasma infection, the response is evident from the appearance and accumulation in leaves of large amounts of free and conjugated polyamines, the latter being bound to cell wall polymers. Conjugated polyamines play a role in host–pathogen interactions by means of the phenolic group of the molecule, which, binding in the cell wall network (hemicelluloses, pectins and lignins), is responsible for the protective effect against the pathogen (Martin-Tanguy, 1997; Torrigiani *et al.*, 1997).

Other Secondary Metabolites

As already mentioned, plant secondary metabolites can be altered in response to the activity of pathogens, including phytoplasmas. Various phytoplasmas have been shown to affect medicinal plants, of which *C. roseus* represents the most common test-plant for investigating phytoplasma/host relationships.

The metabolic products and physiological responses of *C. roseus* implicated in the pathogenicity of phytoplasmas have been explored by Choi *et al.* (2004) and Favali *et al.* (2004). *C. roseus* produces about 150 alkaloids. It is known that the synthesis of alkaloids can be influenced by diseases, stress factors or elicitors (Van Der Heijden *et al.*, 1989). Therefore alkaloid metabolism has been investigated in different tissues of both healthy and phytoplasma-infected periwinkles (Favali *et al.*, 2004). In these studies, alkaloids from periwinkle plants and *in vitro*-cultured explants have been characterized using reversed-phase HPLC. Alkaloids, in particular vindoline, ajmalicine, serpentine, vinblastine and vincristine, were measured at higher concentration in the infected plants than in the controls (Fig. 8.3). In both healthy and infected samples, the main components were serpentine in stems and roots, and vindoline in leaves and explants. In infected plants, however, these alkaloids increased in all plant organs. Vinblastine has been found in large amounts in infected roots, while it was not detectable in healthy ones. Since vindoline is one of two precursors of vinblastine, it seems that phytoplasma infection could shift the reaction from vindoline, present in healthy roots, to vinblastine, abundant in infected ones.

Vindoline increased in the infected leaves and in the infected stems but decreased in infected roots and in micropropagated explants. Ajmalicine increased in the infected micropropagated explants (doubled) and, at the same time, a fourfold increase in serpentine has been found. A notable increase in the production of vinblastine in phytoplasma-infected roots has also been observed (Favali *et al.*, 2004).

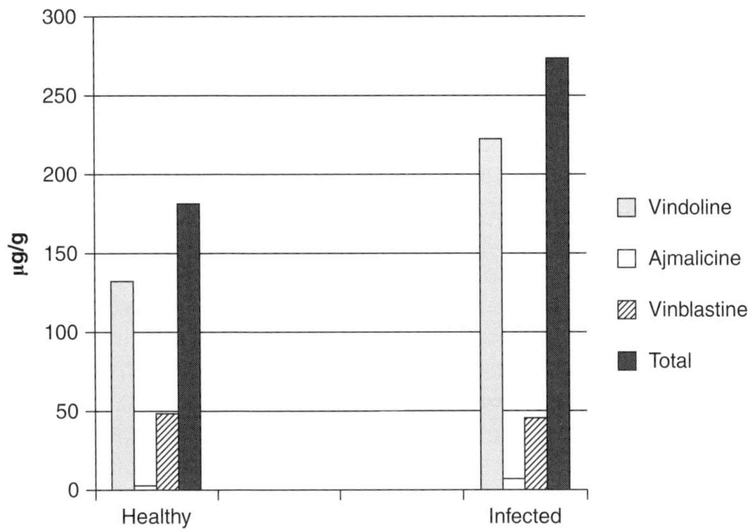

Fig. 8.3. Alkaloid composition in healthy and phytoplasma-infected *Catharanthus roseus* leaves.

Using ^{1}H-NMR spectroscopy, Choi *et al.* (2004) confirmed the fact that *C. roseus* leaves infected with phytoplasmas contained higher vindoline levels (two- to fourfold increase) than healthy ones. From these observations, the authors suggested that indole alkaloid metabolites are major discriminating factors for characterizing phytoplasma-infected *C. roseus* leaves. A higher alkaloid content has also been reported in *Spartium junceum* plants affected by spartium witches'-broom (Mancini *et al.*, 2008).

In *Hypericum perforatum* infected with ash yellows phytoplasma essential oils were analysed by means of gas chromatography (Bruni *et al.*, 2005). The affected plants exhibited a drastic decrease in the amount of essential oils and the abundance of sesquiterpenes. A correlation between these biochemical changes and the presence of phloem necrosis associated with phytoplasma infection has been hypothesized.

As a whole, the published data suggest changes in secondary metabolites in phytoplasma-affected plants. They mainly focus on biochemical differences among healthy and phytoplasma-infected plants, but the precise role of secondary metabolites in defence or recovery mechanisms is still unknown.

Biochemical Changes and Recovery from Phytoplasma Diseases

Recovery in diseased plants is a spontaneous remission of symptoms, which has been reported in grapevine, apple and apricot plants affected by phytoplasmas (Musetti *et al.*, 2007). The physiological basis for this phenomenon is not yet understood. On the basis of the behaviour of pathogens closely related to phytoplasmas (Gram-positive bacteria) (Martini, 2004), we can correlate recovery to various events, including the activity of particular substances or plant secondary metabolites, or the induction of systemic acquired resistance (SAR).

Recently, it has been observed that, in apple (Musetti *et al.*, 2004), apricot (Musetti *et al.*, 2005) and grapevine (Musetti *et al.*, 2007), recovery from phytoplasma-associated diseases was accompanied by an overproduction of hydrogen peroxide (H_2O_2), localized in the phloem tissues. No such H_2O_2 accumulation was detected in infected individuals with symptoms and in healthy control plants. Overproduction of H_2O_2 requires the intervention of antioxidant systems, which include metabolites such as ascorbate (AsA) and reduced glutathione (GSH), and scavenging enzymes, such as superoxide dismutase (SOD), catalase (CAT), ascorbate peroxidase (APX) and hydrogen donor-aspecific peroxidases, e.g. guaiacol peroxidase (GPX). In recovered plants, the activities of the two enzymes primarily involved in H_2O_2 scavenging, namely CAT and APX, significantly decreased when compared with healthy or diseased plants. Therefore it has been hypothesized that decreased scavenging, rather than enhanced synthesis, might be the likely cause of the increased H_2O_2 levels in plants recovering from phytoplasma infection. This led us to hypothesize an active role of H_2O_2, and possibly other reactive oxygen species (ROS), in counteracting pathogen virulence and contributing to the establishment of recovery.

It has been hypothesized that Ca^{2+}-dependent signalling activities, in particular those connected with plant resistance, increase in recovered plants. Indeed, it has been shown recently that Ca^{2+} concentration in the cytosol is remarkably increased in recovered apple plants as compared with the healthy or infected plants (Musetti *et al.*, 2008).

These observations, together with the fact that recovered plants can be reinfected in nature to a lesser extent than non-infected plants, indicate that a type of SAR could be involved in the induction of recovery. Current investigations aim at studying expression of defence-related genes and determining the genetic bases of recovery.

Summary and Looking Forward

In this chapter, studies on changes of the most important metabolic pathways of plants following phytoplasma infection have been reviewed. Phytoplasma diseases, in particular those affecting crops, fruit trees and grapevines, are economically significant and very dangerous for that reason. No curative methods are available against these pathogens; therefore management of phytoplasma-infected plants has mainly focused on controlling the insect vectors and on roguing infected plants from crops and weeds. Resistant cultivars are rare (Thomas and Mink, 1998; Bisognin *et al.*, 2008). Phytoplasma pathogenicity has been poorly understood. Recent advances in biotechnology have permitted the sequencing of full-length phytoplasma genomes, offering new insights into plant/phytoplasma interactions.

It has been demonstrated that phytoplasma genomes contain clusters of repeated gene sequences, named putative mobile units (MPUs), involved in major phytoplasma genome rearrangements and size changes. The finding that MPUs contain genes for membrane target proteins suggests that they may mediate phytoplasma interaction with hosts. The fact that phytoplasmas could adapt to different hosts by varying the number and types of MPUs has been hypothesized (Hogenhout *et al.*, 2008). Further characterization of MPU sequences is in progress, with the expectation that more will be revealed about relationships between phytoplasmas, plants and insect vectors.

Other genes that regulate host/phytoplasma interactions are those encoding various membrane target and secreted proteins (effector proteins). It has been hypothesized that some of these effectors interact with proteins that are conserved amongst plants and animals, explaining why phytoplasmas have broad host ranges.

Individuation of plant genes that are differentially expressed after infection represents another important contribution to clarifying plant/phytoplasma relationships. Biochemical modifications in host plants and changes of plant gene expression induced by infection are the basis for understanding how phytoplasmas cause diseases and consequently how plants react to phytoplasma challenge, activating resistance mechanisms that in turn lead to recovery. In the long term, these studies should provide clues for developing new control strategies against phytoplasma diseases.

Acknowledgements

The author is grateful to Dr Laurence Cantrill for revisions to the English for this chapter.

References

Agrios, G.N. (1997) *Plant Pathology*, 4th edn. Academic Press, San Diego California.

André, A., Maucourt, M., Moing, A., Rolin, D. and Renaudin, J. (2005) Sugar import and phytopathogenicity of *Spiroplasma citri*: glucose and fructose play distinct roles. *Molecular Plant–Microbe Interactions* 18, 33–42.

Bagni, N. and Torrigiani, P. (1992) Polyamines: a new class of growth substances. In: Karssen, C.M., van Loon, L.C. and Vreugdenhil, D. (eds) *Progress in Plant Growth Regulation*. Kluwer Academic Publishers, Dordrecht, the Netherlands, pp. 264–275.

Bertamini, M., Grando, M.S., Muthuchelian, K. and Nedunchezhian, N. (2002a) Effect of phytoplasmal infection on photosystem II efficiency and thylakoid membrane protein changes in field grown apple (*Malus pumila*) leaves. *Physiological and Molecular Plant Pathology* 61, 349–356.

Bertamini, M., Nedunchezhian, N., Tomasi, F. and Grando, S. (2002b) Phytoplasma (Stolbur subgroup (Bois Noir-BN)) infection inhibits photosynthetic pigments, ribulose-1,5-biphosphate carboxylase and photosynthetic activities in field grown grapevine (*Vitis vinifera* L. cv. Chardonnay) leaves. *Physiological and Molecular Plant Pathology* 61, 357–366.

Bisognin, C., Schneider, B., Salm, H., Grando, M.S., Jaraush, W., Moll, E. and Seemüller, E. (2008) Apple proliferation resistance in apomictic rootstocks and its relationship to phytoplasma concentration and simple sequence repeat genotypes. *Phytopathology* 98, 153–158.

Braun, E.J. and Sinclair, W.A. (1976) Translocation in phloem necrosis-diseased American elm seedlings. *Phytopathology* 68, 1733–1737.

Bruni, R., Pellati, F., Bellardi, M.G., Benvenuti, S., Paltrinieri, S., Bertaccini, A. and Bianchi, A. (2005) Herbal drug quality and phytochemical composition of *Hypericum perforatum* L. affected by ash yellows phytoplasma infection. *Journal of Agricultural and Food Chemistry* 53, 964–968.

Carginale, V., Capasso, M.G., Ionata, E., La Cara, F., Pastore, M., Bartaccini, A. and Capasso, A. (2004) Identification of genes expressed in response to phytoplasma infection in leaves of *Prunus armeniaca* by messenger RNA different display. *Gene* 332, 29–34.

Chang, C.-J. (1998) Pathogenicity of aster yellows phytoplasma and *Spiroplasma citri* on periwinkle. *Phytopathology* 88, 1347–1350.

Choi, Y.H., Tapias, E.C., Kim, H.K., Lefeber, A.W.M., Erkelens, C., Verhoeven, J.T.J., Brzin, J., Zel, J. and Verpoorte, R. (2004) Metabolic discrimination of *Catharanthus roseus* leaves infected by phytoplasma using HNMR spectroscopy and multivariate data analysis. *Plant Physiology* 135, 2398–2410.

Davey, J.E., Van Staden, J. and De Leeuw, G.T.N. (1981) Endogenous cytokinin levels and development of flower virescence in *Catharanthus roseus* infected with mycoplasmas. *Physiological Plant Pathology* 19, 193–200.

Favali, M.A., Sanità di Toppi, L., Vestena, C., Fossati, F. and Musetti, R. (2001) Phytoplasmas associated with tomato stolbur disease. *Acta Horticulturae* 551, 93–99.

Favali, M.A., Musetti, R., Benvenuti, S., Bianchi, A. and Pressacco, L. (2004) *Catharanthus roseus* L. plants and explants infected with phytoplasmas: alkaloid production and structural observations. *Protoplasma* 223, 45–51.

He, F.-T., Lee, I.-M., Hammond, R.W. and Zhao, Y. (1998) Preliminary investigation of the role of protein kinase gene expression in the induction of phyllody symptoms on plants infected with phytoplasmas. *Program Abstract, 12th Conference International Organization for Mycoplasmology*. Sydney, Australia, p. 184.

Hogenhout, S.A., Oshima, K., Ammar, E., Kakizawa, S., Kingdom, E.N. and Namba. S. (2008) Phytoplasmas: bacteria that manipulate plants and insects. *Molecular Plant Pathology* 9, 403–423.

Jagoueix-Eveillard, S., Tarendeau, F., Guolter, K., Danet, J.-L., Bové, J.M. and Garnier, M. (2001) *Catharanthus roseus* genes regulated differentially by mollicute infections. *Molecular Plant–Microbe Interactions* 14, 225–233.

Junqueira, A., Bedendo, I. and Pascholati, S. (2004) Biochemical changes in corn plants infected by the maize bushy stunt phytoplasma. *Physiological and Molecular Plant Pathology* 65, 181–185.

Kartte, S. and Seemüller, E. (1991) Histopathology of apple proliferation in *Malus* taxa and hybrids of different susceptibility. *Journal of Phytopathology* 131, 149–160.

Kesumawati, E., Kimata, T., Uemachi, T., Hosokawa, M. and Yazawa, S. (2006) Correlation of phytoplasma concentration in *Hydrangea macrophylla* with green-flowering stability. *Scientia Horticulturae* 108(1), 74–78.

Knoblauch, M., Peters, W.S., Ehlers, K. and Van Bel, A.J.E. (2001) Reversible calcium-regulated stopcocks in legume sieve tubes. *Plant Cell* 13, 1221–1230.

Köhle, H., Jeblick, W., Poten, F., Blascheck, W. and Kauss, H. (1985) Chitosan-elicited callose synthesis in soybean cells as a Ca^{2+}-dependent process. *Plant Physiology* 77, 544–551.

Krapp, A., Hofmann, B., Schafer, C. and Stitt, M. (1993) Regulation of the expression of rbcS and other photosynthetic genes by carbohydrates: a mechanism for 'sink-regulation' of photosynthesis. *Plant Journal* 3, 817–828.

Lee, I.-M., Davis, R.E. and Gundersen-Rindal, D.E. (2000) Phytoplasma: phytopathogenic mollicutes. *Annual Review of Microbiology* 54, 221–255.

Leon, R., Santamaria, J.M., Alpizar, L., Escamilla, J.A. and Oropeza, C. (1996) Physiological and biochemical changes in shoots of coconut palms affected by lethal yellowing. *New Phytologist* 134(2), 227–234.

Lepka, P., Stitt, M., Moll, E. and Seemüller, E. (1999) Effect of phytoplasmal infection on concentration and translocation of carbohydrates and amino acids in periwinkle and tobacco. *Physiological and Molecular Plant Pathology* 55, 59–68.

Machenaud, J., Henri, R., Dieuaide-Noubhani, M., Pracros, P., Renaudin, J. and Eveillard, S. (2007) Gene expression and enzymatic activity of invertase and sucrose synthase in *Spiroplasma citri* or stolbur phytoplasma infected plants. *Bulletin of Insectology* 60, 219–220.

Mancini, E., Marcone, C. and De Feo, V. (2008) Indagini sul metabolismo secondario di piante di *Spartium junceum* affette dalla malattia degli scopazzi della ginestra. *Petria* 18, 348–351.

Martini, M. (2004) Ribosomal protein gene-based phylogeny: a basis for phytoplasma classification. PhD thesis, University of Udine, Italy.

Martin-Tanguy, J. (1997) Conjugated polyamines and reproductive development biochemical, molecular and physiological approaches. *Physiologia Plantarum* 100, 675–688.

Maust, B.E., Espadas, F., Talavera, C., Aguilar, M., Santamaria, J.M. and Oropeza, C. (2003) Changes in carbohydrate metabolism in coconut palms infected with the lethal yellowing phytoplasma. *Phytopathology* 93, 976–981.

Musetti, R. (2006) Patogeni e piante di interesse agronomico: un approccio morfologico. In: Quaglino, D., Falcieri, E., Catalano, M., Diaspro, A., Montone, A., Mengucci, P. and Pellicciari, C. (eds) *1956–2006: 50 anni di Microscopia in Italia tra storia, progresso ed evoluzione.* PI.ME Editrice, Pavia, Italy, pp. 325–334.

Musetti, R. and Favali, M.A. (2003) Calcium localization and X-ray microanalysis in *Catharanthus roseus* L. infected with phytoplasmas. *Micron* 34, 387–393.

Musetti, R., Scaramagli, S., Vighi, C., Pressacco, L., Torrigiani, P. and Favali, M.A. (1999) The involvement of polyamines in phytoplasma-infected periwinkle (*Catharanthus roseus* L.) plants. *Plant Biosystems* 133, 37–45.

Musetti, R., Favali, M.A. and Pressacco, L. (2000) Histopathology and polyphenol content in plants infected by phytoplasmas. *Cytobios* 102, 133–147.

Musetti, R., Sanità di Toppi, L., Ermacora, P. and Favali, M.A. (2004) Recovery in apple trees infected with the apple proliferation phytoplasma: an ultrastructural and biochemical study. *Phytopathology* 94, 203–208.

Musetti, R., Sanità di Toppi, L., Martini, M., Ferrini, F., Loschi, A., Favali, M.A. and Osler, R. (2005) Hydrogen peroxide localisation and antioxidant status in the recovery of apricot plants from European stone fruit yellows. *European Journal of Plant Pathology* 112, 53–61.

Musetti, R., Marabottini, R., Badiani, M., Martini, M., Sanità di Toppi, L., Borselli, S., Borgo, M. and Osler, R. (2007) On the role of H_2O_2 in the recovery of grapevine (*Vitis vinifera*, cv. Prosecco) from Flavescence dorée disease. *Functional Plant Biology* 34, 750–758.

Musetti, R., Tubaro, F., Polizzotto, R., Ermacora, P. and Osler, R. (2008) Il 'Recovery' da apple proliferation in melo è associato all'aumento della concentrazione dello ione calcio nel floema. *Petria* 18(2), 380–383.

Nicolaisen, M. and Horvath, D.P. (2008) A branch-inducing phytoplasma in *Euphorbia pulcherrima* is associated with changes in expression of host genes. *Journal of Phytopathology* 156, 403–407.

Perica, M. (2008) Auxin-treatment induces recovery of phytoplasma-infected periwinkle. *Journal of Applied Microbiology* 105(6), 1826–1834.

Pertot, I., Musetti, R., Pressacco, L. and Osler, R. (1998) Changes in indole-3-acetic acid level in micropropagated tissues of *Catharanthus roseus* L. infected by the agent of the clover phyllody and effect of exogenous auxins on phytoplasma morphology. *Cytobios* 95, 13–23.

Pracros, P., Renaudin, J., Eveillard, S., Mouras, A. and Hernould, M. (2006) Tomato flower abnormalities induced by stolbur phytoplasma infection are associated with changes of expression of floral development genes. *Molecular Plant–Microbe Interactions* 19, 62–68.

Ricci, A., Branca, C., Osler, R. and Favali, M.A. (1988) Ultrastructural and hormonal alterations in oat leaves infected with barley yellow dwarf virus. *Giornale Botanico Italiano* 122, 62–63.

Romanazzi, G., Prota, V.A., Casati, P., Murolo, S., Silletti, M.R., Di Giovanni, R., Landi, L., Zorloni, A., D'Ascenzo, D., Virgili, S., Garau, R., Savino, V. and Bianco, P.A. (2007) Incidenza del recovery in viti infette da fitoplasmi in diverse condizioni climatiche e varietali italiane e tentativi di comprensione ed induzione del fenomeno. *Atti Convegno nazionale 'Nuove possibilità di lotta contro le fitoplasmosi della vite e dei fruttiferi basate su recovery, resistenze indotte e antagonisti'*, Ancona, Italy, pp. 9–11.

Roustan, J.P., Latchè, A. and Fallot, J. (1992) Influence of ethylene on the incorporation of 3,4-C14methionine into polyamines in *Daucus carota* cells during somatic embryogenesis. *Plant Physiology and Biochemistry* 30, 201–205.

Scaramagli, S., Bueno, M., Torrigiani, P., Altamura, M.M., Capitani, F. and Bagni, N. (1995) Morphogenesis in cultured thin layers and pith explants of tobacco. II. Early hormone-modulated polyamine biosynthesis. *Journal of Plant Physiology* 147, 113–117.

Smart, C.D. and Kirkpatrick, B.C. (1996) Identification of host plant genes whose expression is altered upon aster yellows phytoplasma infections. *IOM Letters* 4, 274.

Tan, P.Y. and Whitlow, T. (2001) Physiological response of *Catharanthus roseus* (periwinkle) to ash yellows phytoplasmal infection. *New Phytologist* 150, 757–769.

Thomas, P.E. and Mink, G.L. (1998) Tomato hybrids with nonspecific. immunity to viral and mycoplasma pathogens of potato and. tomato. *Hortscience* 33, 764–765.

Torrigiani, P., Rabiti, A.L., Bortolotti, C., Betti, L., Marani, F., Canova, A. and Bagni, N. (1997) Polyamine synthesis and accumulation in the hypersensitive response to TMV in *Nicotiana tabacum*. *New Phytologist* 135, 467–473.

Van Der Heijden, R., Verpoorte, R. and Ten Hoopen, H.J.G. (1989) Cell and tissue cultures of *Catharanthus roseus* (L.) G. Don: a literature survey. *Plant Cell Tissue Organ Culture* 18, 231–280.

Weisburg, W.G., Tully, J.G., Rose, D.L., Petzel, J.P., Oyazu, H., Yang, D., Mandelco, L., Sechrest, J., Lawrence, T.G., Van Etten, J., Maniloff, J. and Woese, C.R. (1989) A phylogenetic analysis of the mycoplasmas: basis for their classification. *Journal of Bacteriology* 171, 6455–6467.

Zhong, B.-X. and Shen, Y.-W. (2004) Accumulation of pathogenesis-related type-5-like proteins in phytoplasma infected garland chrysanthemum *Chrysanthemum coronarium*. *Acta Biochimica et Biophysica Sinica* 36(11), 773–779.

1

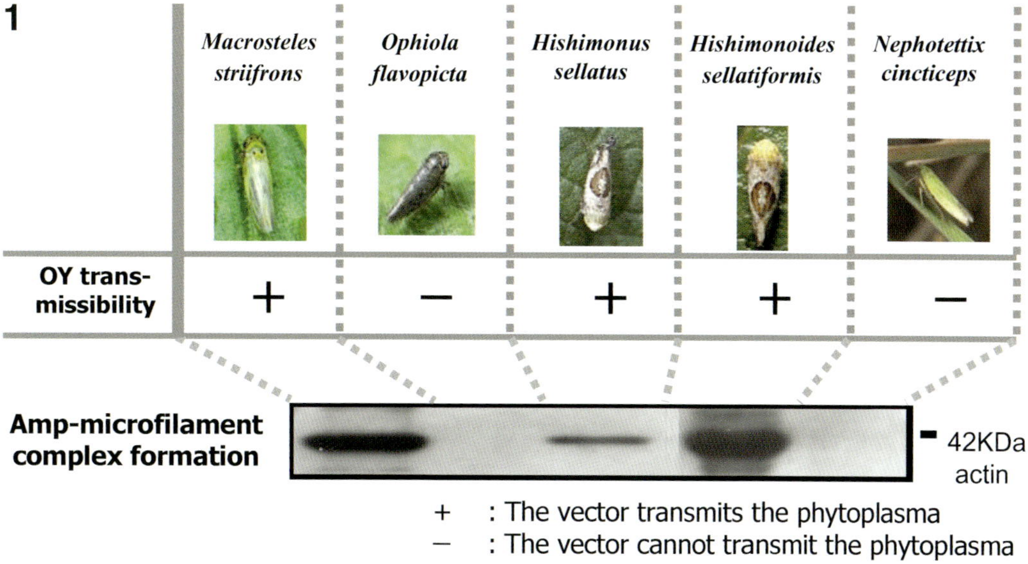

2

Plate 1. Correlation between Amp–microfilament complex formation and insect transmission. In nature, phytoplasmas are transmitted by specific leafhopper vectors. The OY phytoplasma is transmitted by *Macrosteles striifrons, Hishimonus sellatus* and *Hishimonoides sellatiformis* but not by *Ophiola flavopicta* or *Nephotettix cincticeps.* The Amp–microfilament complexes were detected by the Amp affinity column assay and Western blot analysis with an anti-actin antibody. The Amp–microfilament complexes were detected in the samples from the phytoplasma-transmitting insect species but not in those from the non-transmitting species.
Plate 2. Witches'-brooms, specific phytoplasma symptoms on *Spartium junceum* (Spanish broom) affected by the spartium witches'-broom disease.

3

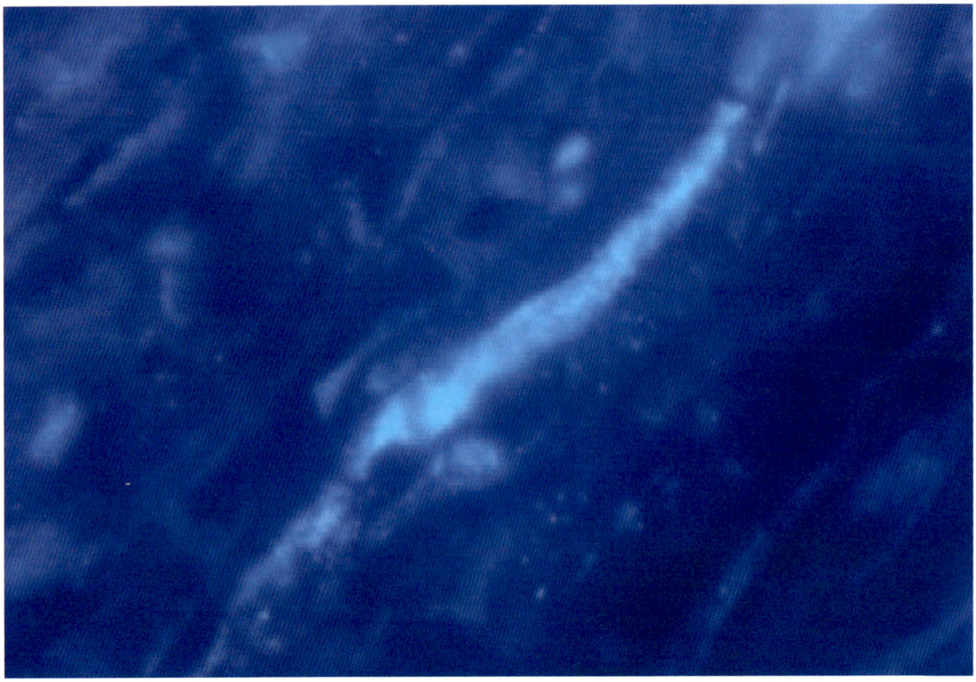

4

Plate 3. Phytoplasma colonization in petioles of alder yellows-affected *Alnus glutinosa* (alder), as viewed by fluorescence microscopy using the DNA dye 4'-6-diamidino-2-phenylindole (DAPI fluorescence test). Several sieve tubes show phytoplasmal infections as single particles, while others show them as small aggregates or larger fluorescent areas.

Plate 4. (a) *Cacopsylla picta*, (b) *Cacopsylla melanoneura*, (c) *Cacopsylla pruni*, (d) *Cacopsylla pyri* female and (e) *Cacopsylla pyri* male.

9 Plant Resistance

Erich Seemüller[1] and Hugh Harries[2]

[1]Julius Kühn Institute, Germany; [2]Centro de Investigación Científica de Yucatán, Mexico

Introduction

In nature, phytoplasmas are mainly spread by phloem-feeding insects, reside in phloem sieve tubes and persistently colonize their hosts. The persistent infection implies that economic losses are often most severe in perennial plants, particularly in trees and shrubs. The life cycle of phytoplasmas also accounts for the fact that the diseases they cause are difficult to control. Standard recommendations include phytosanitary and other preventive measures and the control of insect vectors. However, none of these approaches is fully satisfactory. The same applies for treatments with tetracycline antibiotics, which have been discontinued in fruit growing and forestry for several reasons.

A more promising approach to controlling phytoplasmoses is the use of resistant plants. Intra- and interspecific variation in susceptibility to these diseases has been reported for several decades. However, only some of the work provides firm data on the response of the host, on host–pathogen interaction and on the anatomical, physiological and molecular basis of resistance. Also, relatively few breeding projects have been carried out that resulted in information on inheritance of resistance. For these reasons, this chapter will focus on a few more extensively studied examples of resistance to phytoplasmal diseases: coconut lethal yellowing in the tropics and three major diseases of temperate fruit trees, namely apple proliferation (AP), pear decline (PD) and European stone fruit yellows (ESFY). AP and PD are examples showing that diseases of grafted trees can be suppressed by the use of resistant rootstocks. Other aspects, such as plant resistance to insect vectors and significant findings on plant resistance to other phytoplasmoses, will be treated as well.

Coconut Lethal Yellowing

Currently, the value of lethal yellowing (LY)-resistant coconut (*Cocos nucifera* L.) germplasm has been called into question, in part by the failure to identify highly resistant varieties in Ghana (Dery *et al.*, 2008) and Tanzania (Schuiling *et al.*, 1992) but, critically, by an even more virulent epidemic in Jamaica, which, over the last decade, has destroyed not only the palms planted under the rehabilitation programme (Baudouin *et al.*, 2008; Lebrun *et al.*, 2008) but also the older survivors from the earlier epidemic. Yet in Jamaica and in Florida there was a period of four decades between 1955 and 1995 when plant breeders were able to select LY-resistant coconut varieties and produce resistant hybrids that farmers planted with confidence. During that time, the pathogen was recognized as a phytoplasma of the coconut lethal yellowing or 16Sr IV group; symptom remission was achieved by antibiotic application (oxytetracycline); and the vector (in Florida) was identified as the planthopper *Myndus crudus*. Despite these results, the existing diseased areas expanded (most notably in Ghana, Haiti and Mozambique but, exceptionally, not in the Dominican Republic) while jump-spread (or careless quarantine) started new outbreaks, regardless of national boundaries (Mexico, 1978; Belize, 1992; Honduras, 1996; Turkey, 1999; Guatemala, 2001; Nevis, 2005). In addition to *C. nucifera*, LY susceptibility has been reported in over 30 other (mainly) palm species, and a group of diverse but closely related phytoplasmas are associated with palm diseases elsewhere in the Americas, Africa and Asia.

Concurrently, since the early 1960s, the coconut palm ceased to be the premier source of vegetable oil in world trade. Recovery of LY resistance and its market position are both jeopardized by its recalcitrance to manipulation: a single species, propagated only from seed nuts, the annual multiplication rate is low and the generation time is long; it cannot be budded or grafted; on the rare appearance of suckers, branches or inflorescence bulbils, these are difficult to propagate vegetatively or do not revert to flower normally; and sophisticated biotechnology has failed to routinely produce clonal plantlets by tissue culture. Embryo rescue is all that has been achieved, initially for the 'Makapuno' cultivar group (which has unusual, jelly-like endosperm), then recommended for the safe transfer of disease-free planting material and long-term *in vitro* conservation. Dividing the embryo or the young plumule does offer the potential for cloning such material, but it is too soon to speculate when, or whether, transgenic or genetically modified coconut palms can be developed that are resistant, tolerant or possibly immune to phytoplasma diseases.

Yet the pattern, now being experienced in Jamaica and elsewhere, of an uncontrollable epidemic, introduction of exotic germplasm and subsequent exposure of survivors to further epidemic cycles has already occurred in Cuba, where the involvement of an insect vector was postulated over 65 years ago (Bruner and Boucle, 1943) and when it was observed that 'certain plants. . . remain alive and . . . could represent cases of partial innate resistance . . . plant breeding . . . seems the only recourse'.

Introducing resistant germplasm

Yellowing and dying coconut palms were reported in the Caribbean region in the 19th century (Cayman Islands, 1843; Cuba, 1870; Jamaica, 1872; Haiti, 1880), just as coir, copra and coconut oil (CNO) became commercially important to European and North American traders. The first serious LY epidemic, in Baracoa, Cuba from 1905 to 1910 (Bruner and Boucle, 1943), decimated production to such an extent that entrepreneurs from the USA chose to plant and process coconuts in the Philippines rather than import copra from closer sources in Latin America and the Caribbean. Coming just as the demand for CNO (a raw material for candles, soap, margarine and high explosives) was stimulated by the 1914–1918 war, this made the Philippines the dominant coconut-producing country, an indirect and hitherto unrecognized effect of LY.

The rising demand for CNO encouraged colonial and commercial investment in plantations and caused a worldwide shortage of planting material. For that reason, shipments of seed nuts were sent from the Federated Malay States to St Lucia, Trinidad and other British Caribbean islands in the early 1920s. Around the same time, and probably from the same sources, coconut seed nuts were introduced by American plantation companies to Panama, Honduras, Cuba, possibly to Puerto Rico and some even went to Florida. Ironically, in the light of subsequent events, no Malayan seed nuts went directly to Jamaica. Not until the late 1930s did estates in Jamaica obtain planting material, when, to speed recovery after hurricanes in 1944 and 1947, seed nuts of early-germinating, -flowering and -fruiting 'Malayan Dwarf' were imported from St Lucia.

Disease resistance was not a consideration because 'West End Bud Rot', as LY was known when it began to cause considerable damage between Lucea and Montego Bay, did not occur in the main copra-producing eastern parishes of Jamaica. Cutting out and burning did not control spread, but at Roundhill three red dwarf palms that did survive were multiplied until the whole estate and neighbouring properties were planted with thousands of red-fruited dwarf palms. Originating from a 1921 Malayan shipment to Trinidad and sent from there to Jamaica in 1940, it was the eye-filling novelty of their red colour en masse that made their survival so 'spectacular' to scientists, who were not aware of the priority of research in Cuba or of the 'Indio' coconuts when renaming the disease Lethal Yellowing (Nutman and Roberts, 1955).

Resistance screening and ranking

When it was confirmed that the lethal yellowing resistance of the green and yellow dwarf colour forms in the Jamaican government immunity trials was equal to that of the red dwarf (Whitehead, 1966a), more dwarf seed nuts were imported from St Lucia in 1968 specifically to supply the lethal yellowing rehabilitation programme. Only after surveying local material was exotic coconut germplasm imported from many Asian and Pacific countries for LY screening (Whitehead, 1966b, 1968). Field exposure trials were a major part of the breeding programme, and, without benefit of aerial surveys or

Table 9.1. Varieties ranked on a five-point scale of resistance (based on Harries, 1995).

Rank	Resistance	Varieties
1	Highest	'Malayan Dwarf' (yellow, red, green); 'Sri Lanka Green Dwarf' (yellow, green); 'King'; 'Indian Dwarf' (green)
2	High	'Maypan' ('Malayan Dwarf' × 'Panama Tall'; 'Malayan Dwarf' × 'Niu Leka'; any rank 1 × rank 3).
3	Intermediate	'Panama Tall' (local); tall varieties from Malaysia, Peru and Thailand; 'Fiji Dwarf' ('Niu Leka'); 'Mayjam' ('Malayan Dwarf' × 'Jamaica Tall'; any rank 1 × rank 5).
4	Low	Varieties from some South-east Asian and Pacific islands.
5	Least	Tall varieties from the Caribbean and Atlantic coast of America, West and East Africa, India and Sri Lanka.

laboratory-based confirmation, the efficiency of disease assessment depended on regular monthly visits to a dozen or so trials, at scattered locations, by trained field assistants under supervision of field officers, data entry by office or laboratory staff, frequent 3- to 6-monthly follow-up visits by the scientist in charge, and the publication of results in readily available annual research reports. The attention to detail by repeated individual examination of palms of every age for visual symptoms (premature nut fall, black inflorescences, leaf yellowing, spear leaf collapse and death within 4–6 months) took into account losses from other causes and even identified rare instances of natural symptom remission (Harries, 1974).

Such was the confidence in 11 years of trial data (1968–1978) that the varieties in the Jamaican germplasm collection were ranked on a five-point scale of resistance (Table 9.1). The three local varieties and two hybrids were each assigned to different ranks and, with additional information from fruit component analysis (FCA) data that had been recorded at the time the germplasm collection was made (Whitehead, 1966b, 1968), the resistance rankings of other varieties could be predicted, even if they were not available for testing in Jamaica.

For instance, the local tall varieties in the Caribbean region and the Atlantic coasts of America and Africa could be grouped with 'Jamaica Tall' (rank 5). In contrast, countries like Mexico, with both Caribbean and Pacific coastlines, would have rank 5 and rank 3 tall varieties, respectively. The actual level of resistance or susceptibility is assumed to be influenced by growing conditions, by vector activity and phytoplasma type or strain, and by exposure to earlier LY outbreaks if any occurred.

Evolution, prediction and heritability of resistance

The obvious geographical grouping in both FCA data and LY resistance rankings, taken together with other significant differences, such as speed of germination, leaf and stem dimensions, flowering pattern and windstorm tolerance, led to the creation of a theory of coconut evolution. Briefly stated, the primordial wild coconut, evolved by floating between uninhabited oceanic islands, beyond the reach of insect vectors, would not need disease resistance. On the continental lowlands of South-east Asia, where an insect-vectored disease could reach epidemic proportions, domestic types, selected by exposure and survival, would have durable, horizontal resistance (Robinson, 1977). Introgressive hybridization between wild and domestic types would generate local populations with different levels of resistance. The LY-like diseases, if present (Maramorosch, 1964), would be endemic, but epidemics would be possible where wild-type characteristics chanced to predominate in the local coconut populations (Allorerung *et al.*, 1999).

From the results in Jamaica it was possible to consider how resistance might be inherited (Harries, 1995; Ashburner and Been, 1997). For instance, if resistance was due to a single gene, the 'Malayan Dwarf' might be homozygous dominant, the 'Jamaica Tall' homozygous recessive and their F_1 hybrid, 'Mayjam', heterozygous. Varieties such as the 'Panama Tall', which show an intermediate degree of resistance, might have the genetic pattern of an F_2 generation, while 'Maypan', produced by crossing 'Malayan Dwarf' and 'Panama Tall', would be the equivalent of a backcross and have a better degree of resistance. Although it is technically possible now to acquire breeders' seed in any quantity (the F_2 by open pollination and backcross, and later generations by modified mass-controlled pollination), the space and time needed for field exposure trials, and their cost, make resistance screening impractical. The requirements of an efficient selection method cannot be met because it is not possible to subject enough test plants to a high infection pressure at a developmental stage in which symptom expression can be rapidly and confidently determined. Unavoidably, coconut planting density is low (80–140/ha), palms cannot be artificially infected and infection pressure depends on disease incidence in the immediate neighbourhood of the trial, over which there is no control. The incubation time to symptom appearance may be 3–6 months after infection in young palms and 7–15 months in mature palms, and a full syndrome of symptoms shows only in bearing palms. The suspected presence of the pathogen in the palm tissue can be tested in the field by tetracycline-mediated symptom remission and in the laboratory by fluorescence microscopy, electron microscopy or by PCR techniques. Positive laboratory results are confirmed (often in advance) by palm death within 4–6 months.

It should be borne in mind that, because of the perennial nature of a tree crop, when susceptible palms die of disease the remaining resistant palms, whether homozygous or heterozygous, will cross- and self-pollinate to give progeny which, in its turn, will be exposed to the disease. This is how resistant varieties must have originally developed over many generations. Those varieties which showed most promise in Jamaica can trace an origin to South-east Asia. This finding had some limited support from field-exposure trials in Tanzania (Harries, 1995) and Ghana (Dery *et al.*, 2008).

The mode of action of any resistance gene(s) also needs to be investigated. It is thought to be unlikely that coconut palms possess genetically based, post-infection resistance to phytoplasma diseases. If the primary reason palms are not infected is due to vector feeding preference, this might explain why interplanting with immune crops may be effective.

Rehabilitation, cyclical epidemics and loss of resistance

Over a period of 15 years in Jamaica, differences in resistance became clear, and an acceptable hybrid, 'Maypan', was mass produced by emasculating 'Malayan Dwarf' seed parents in an isolated seed garden and applying pollen collected from selected survivors in LY areas. Although a range of different hybrids was possible by this means, only small quantities were produced for performance testing. As none appeared to be any better than 'Maypan' they were not used in the lethal yellowing rehabilitation scheme, and for a further 18 years the 'Maypan' hybrid and its 'Malayan Dwarf' parent continued to give commercial yields in Jamaica and good resistance, both there and in Florida.

The first LY symptom, premature nut fall, allows susceptible germplasm to survive, so the possibility that lethal yellowing might have cyclic recurrence in Jamaica was always admitted. By the late 1980s, unusually high lethal yellowing disease incidence began to be noticed amongst 'Malayan Dwarf' coconut palms at localized sites in Jamaica and Florida (Howard *et al.*, 1987). The sites tended to be golf courses or tourist resorts rather than farms, but by the turn of the century an island-wide epidemic was evident – 'Large-scale replanting with Maypan . . . took place in the 1970s . . . These palms, now up to 30 years old, are now succumbing to LY. This means a massive breakdown in resistance' (Mark Schuiling, personal communication, http://tech.groups.yahoo.com/group/CICLY/message/555; 28/1/2001). Yet, almost concurrently over the same period, lethal yellowing in the Dominican Republic had not developed any epidemic, any jump-spread, any great loss of existing coconut palms and any need for a coordinated lethal yellowing rehabilitation programme. As a result, earlier cut-and-burn quarantine measures were discontinued and germplasm introduced for screening was ignored. However, the recent identification of phytoplasma at a new location (Martinez *et al.*, 2008) leaves no room for complacency.

With the benefit of hindsight, it is possible to see what caused the loss of resistance in Jamaica. In the early 1980s the LY aid projects, having achieved positive results, were terminated, the expatriate research staff left and, with the withdrawal of supplementary funding, the local scientists turned to other pressing problems, in the knowledge that resistant varieties were being planted. But, also in the early 1980s, policy decisions taken to resuscitate the ailing banana industry in Jamaica changed the traditional mixed cropping system, which did not give the unblemished fruit quality demanded of that important export crop, and bananas were no longer planted between coconut palms. A barrier of an immune intercrop, such as bananas, is a possible contributing factor protecting both susceptible and resistant coconut varieties from LY. The presence of royal palms (*Roystonia* spp.) demonstrates this, both amongst the ordinary tall coconuts in the Dominican Republic and also in Cuba, where the 'Dorado Cubano' (a selection from progenies of the Cuban 'Criollo' and the introduced 'Indio' coconuts) is reported to perform well in mixed plantings where LY is present.

Current activities

Following a meeting of 'LY experts' in 2002, a tripartite project, 'Sustainable Coconut Production through Control of Coconut Lethal Yellowing Disease', was set up between Jamaica, Mexico and Honduras. The introduction and screening of germplasm is ongoing, but research results have implicated alternative vectors and host weed species that were not previously suspected (Brown *et al.*, 2006, 2007).

Elsewhere the use of potentially highly productive hybrid coconuts has met with mixed success commercially and proved disappointing against lethal yellowing-related diseases in areas of East and West Africa. In Tanzania, the national coconut development programme initially ignored local varieties and imported germplasm to make F_1 hybrids developed in the Ivory Coast, on the assumption that adequate resistance would be inherited from the 'Malayan Dwarf' seed parent. Resistance (and drought tolerance) proved inadequate and lethal disease rehabilitation received a setback. Now, resistance is being tested in trials involving 29 subpopulations of the local 'East African Tall', from areas in Kenya and Tanzania where the disease has existed for a long time. Other recent strategies include the use of molecular techniques for genetic improvement and early elimination of infected palms on farmers' plantations (Kullaya and Mpunami, 2008). In Ghana, disease screening trials of varieties and hybrids from the Ivory Coast succumb under intense disease pressure. There are significantly different, but not fully LY-resistant, varieties (Quaicoe *et al.*, 2008), and a new coconut breeding project (M. Dickinson, Nottingham, 2008, personal communication) will make genetic improvement a component of an integrated control strategy, because coconut–food crop intercropping systems had negligible levels of major diseases and pests (Ennin *et al.*, 2008).

In the near future a lethal yellowing rehabilitation project is to start in Mozambique and it is anticipated that both local and introduced varieties and hybrids will be tested and intercropping will be encouraged.

Apple Proliferation

Apple proliferation (AP) is caused by '*Candidatus* (*Ca.*) Phytoplasma mali' and is one of the most damaging phytoplasmal diseases in Europe. Diseased trees of the cultivated apple (*Malus* × *domestica*) are characterized by witches'-broom formation, growth suppression and undersized, poor-tasting fruit.

An indication of an approach to control the disease was obtained in work on the colonization behaviour of the causal agent. These studies have shown that the pathogen, which depends on intact sieve tubes, is usually eliminated in the stem during winter due to the degeneration of the phloem. Overwintering occurs in the roots, where functional sieve tubes are present throughout the year. The stem may be recolonized from the roots in spring when new phloem is being formed (Schaper and Seemüller, 1982; Seemüller *et al.*, 1984). This fluctuation in colonization pattern has led to the presumption that growing the mostly susceptible scion cultivars on resistant rootstocks can prevent the disease or reduce its impact.

Identification of resistance and factors involved

Screening of many established and experimental rootstocks, which were mainly based on *M.* × *domestica*, has shown that there is no satisfactory resistance in this group. Trees on these stocks frequently developed symptoms, remained permanently infected and showed a high phytoplasma titre (Seemüller *et al.*, 1992). Furthermore, examination of wild and ornamental, mostly exotic, *Malus* taxa revealed that many of them, as evidenced by a high mortality rate, are even more susceptible than *M.* × *domestica* genotypes, probably because in evolution they were never exposed to the pathogen. Satisfactory resistance was only observed in progenies of open-pollinated experimental apomictic rootstock selections consisting of interspecific hybrids between the apomictic *M. sieboldii* and genotypes of the non-apomictic taxa *M.* × *domestica* and *M.* × *purpurea* 'Eleyi'. Trees grown on such stocks never developed AP symptoms or only temporary mild ones. In these rootstocks the pathogen was either not detected or difficult to detect by fluorescence microscopy, indicating a low phytoplasma titre (Kartte and Seemüller, 1991b; Seemüller *et al.*, 1992).

AP resistance of *M. sieboldii*-derived rootstocks was examined in more detail in several long-term field trials in which trees on seedlings of open-pollinated *M. sieboldii* and 15 *M. sieboldii* hybrids were compared with trees on *M.* × *domestica*-based rootstocks. Also, progenies of the apomictic species *M. sargentii* and *M. sargentii* × *M.* × *domestica* F_1 hybrids were included in this work, which was carried out following experimental inoculation or under natural infection conditions. Similar results were obtained in all trials. Satisfactory resistance was only shown by trees on progenies of *M. sieboldii* and *M. sieboldii* hybrids 3432, 4551, 4556, 4608, 4637, C1907, D1131, D2118, D2212, Gi477/4, H0801 and H0909. The resistance level of this group was usually significantly higher than that of trees on *M.* × *domestica*-based stocks and, particularly, on *M. sargentii*-derived roots. Also, *M. sieboldii*-based hybrids 20186 and H0901, and selection C0725, which has *M. sieboldii* and *M. sargentii* as parents, were significantly more susceptible than the resistant *M. sieboldii*-derived stocks. This indicates that the *M. sieboldii*-based resistance is a segregating trait and that *M. sargentii* has a negative effect on resistance (Seemüller *et al.*, 2008b). Examples of the response to infection of trees on various rootstocks are depicted in Figs 9.1 and 9.2.

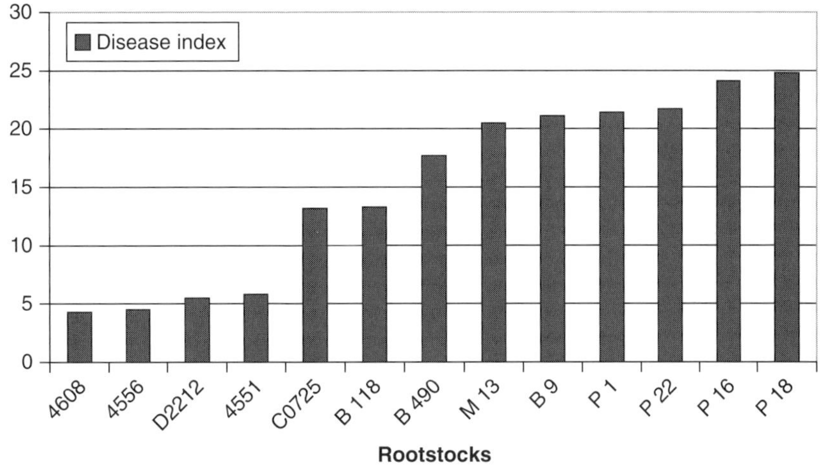

Fig. 9.1. Susceptibility of Golden Delicious apple trees on various rootstocks to apple proliferation, as expressed by disease indices accumulated over 14 years. The trees were experimentally inoculated at the beginning of the trial. 4608, D2212, 4551 and 4556 are progenies of resistant *Malus sieboldii* selections, and C0725 is a susceptible (*M. sieboldii* × *Malus* × *domestica*) × *Malus sargentii* hybrid progeny. The other stocks are from the Polish P, the Russian B and the British M series.

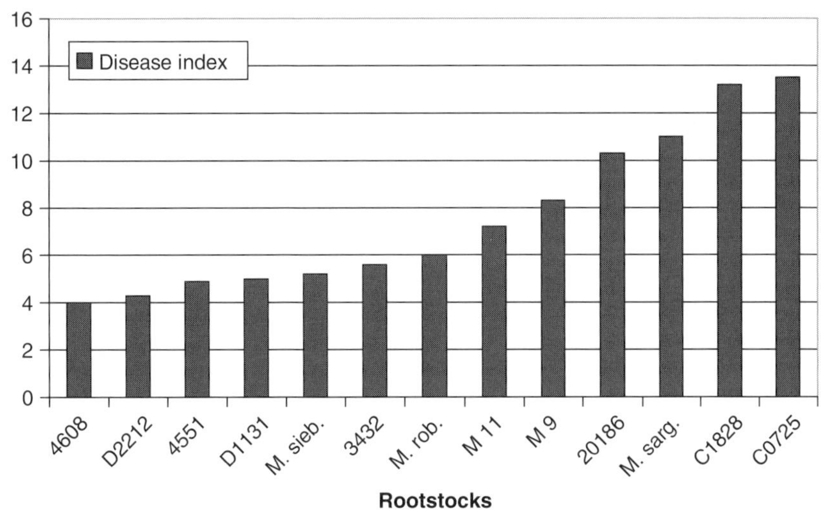

Fig. 9.2. Susceptibility of Golden Delicious apple trees on various rootstocks to apple proliferation under natural infection conditions, as expressed by disease indices accumulated over 12 years. 4608, D2212, 4551 and D1131 are progenies of resistant *Malus sieboldii* selections. 3432 and C0725 are (*M. sieboldii* × *Malus* × *domestica*) × *Malus sargentii* hybrid progenies, and C1828 is a *M.* × *domestica* × *M. sargentii* progeny. M. sieb., M. rob., M. sarg., progenies of *M. sieboldii*, *Malus robusta* and *M. sargentii*, respectively.

Great differences in resistance were observed within progenies, similar to the variability between progenies. Severely affected trees occurred in all of them. However, their number was low in the progenies of resistant seed parents and high in progenies of susceptible seed parents, such as those of *M. sargentii*-derived stocks (Seemüller *et al.*, 2008b). This phenomenon is mainly due to the fact that apomixis, the asexual formation of seeds that are genetically identical to the mother, is not obligate in the genus *Malus*. Thus, progenies contain a variable percentage of hybrids from reduced and unreduced maternal gametes (see below for details). Their resistance is influenced by recombination events and the genetic background of the unknown pollen parent. The level of resistance also depends on the screening procedure. The differences between resistant and susceptible rootstocks were smaller under natural infection conditions than following experimental inoculation, because naturally infected trees become diseased considerably later than graft-inoculated trees (Figs 9.1 and 9.2).

Phytoplasma concentration in the rootstocks examined, as measured by quantitative real-time PCR, varied greatly. While the titre in the susceptible *M.* × *domestica* rootstocks M 9 and M 11 was approximately 1×10^9 cells/g root phloem, the concentration in resistant apomictic rootstocks was much lower, ranging from 2×10^5 to 3×10^6 cells/g phloem. However, the concentration in the roots of the susceptible *M. sargentii*-derived apomicts was in the same low range (Bisognin *et al.*, 2008). This indicates that host suitability per se, as expressed in phytoplasma titre, is obviously not the only factor for resistance. Instead it appears that the response of the phloem to infection plays an important role. It has been shown that the phloem-residing phytoplasmas cause extended sieve tube necrosis and depletion of starch in the roots of trees grown on *M. sargentii*-derived apomicts, whereas starch was accumulated in the aerial parts, in particular in the leaves. In contrast, infected trees on resistant selection 4551 showed little phloem necrosis in the roots and, like healthy trees, high starch contents in the roots and little starch accumulation in the leaves (Kartte and Seemüller, 1991a). This indicates that the response of the phloem to infection is an important factor in resistance and that this response is not markedly influenced by phytoplasma concentration.

Recolonization of the stem from the roots seems to be another factor in AP resistance. In the study on phytoplasma titres in trees on resistant and susceptible rootstocks it became evident that only a minority of the trees on resistant rootstocks were infected in the top, whereas most trees on susceptible stocks were colonized. Also, the titre in the stem of trees on susceptible rootstocks was usually higher than that of trees on resistant stocks (Bisognin *et al.*, 2008). It thus appears that the low infection rate and the low titre in the stem of trees on resistant apomictic stocks result from low phytoplasma concentrations in the roots. The low starting concentration in the roots and poor host suitability of resistant apomictic genotypes may have a negative effect on the spread of the pathogen from the roots into the scion in spring. Thus, the low titre in the roots is likely to contribute to the resistance of *M. sieboldii*-derived stocks. It is well established that severe symptoms such as witches'-brooms and undersized fruits only develop when the phytoplasma concentration in the stem is high.

Resistance breeding

Pomological evaluation revealed that trees grown on most *M. sieboldii*-derived resistant apomicts are more vigorous than trees on M 9, the most important stock for commercial apple growing in Europe. Also, productivity is mostly lower, mainly due to alternate bearing (Seemüller *et al.*, 2008b). For this reason, a breeding programme has been initiated, with the aim of reducing vigour and improving yield capacity. Between 2001 and 2006, 18 major crosses were made, in which *M. sieboldii* (4n), two *M. sieboldii* F_1 hybrids (3n, genotypes 4551 and 4608) and six F_2 hybrids that were obtained by backcrossing of F_1 hybrids with M 9 (4n, genotypes H0901, H0909) or by open pollination of F_1 hybrids (4n, genotypes C1907, D2118, D2212, H0801) were used as seed parents. These accessions were usually pollinated with M 9, but in a few cases with other *M.* × *domestica* genotypes (Bisognin *et al.*, 2009).

More than 3000 offspring from these crosses were genetically examined using simple sequence repeat (SSR) analysis supported by flow cytometry. With this approach, it was possible to distinguish sexually derived seedlings from apomicts, to determine the ploidy level of parents and offspring and to explore the segregation mode of polyploid genomes. This work showed that an average of 58% (37–83%) of the seedlings exhibited the same set of SSR alleles as the apomictic seed parents and were thus grouped as 'motherlike'. Seedlings displaying pollen-parent-specific alleles were assigned to two groups named 'hybrid I' and 'hybrid II'. Hybrid I showed the whole marker profile of the seed parent at each locus plus one paternally derived allele. In contrast, hybrid II exhibited half of the specific alleles of both the seed and the pollen parents. Crosses with triploid seed parents resulted only in recombinants of the hybrid I class. However, three groups of recombinants were obtained from tetraploid seed parents, namely hybrids I and II and offspring derived from autopollination. The percentage of the three classes varied considerably with the crosses and was, on average, 16, 20 and 6%, respectively (Bisognin *et al.*, 2009). Similar numbers were observed in progenies of open-pollinated apomicts (Bisognin *et al.*, 2008). The low number of fully recombinant offspring of apomictic seed parents considerably impedes the selection of genotypes showing the expected properties.

AP resistance was assessed by observation of graft-inoculated offspring in the nursery for 2 years, accumulating the annual disease rating values to obtain cumulative indices (CDI). Compared with an average CDI of 4.1 across all progenies, offspring of the crosses 4608 × M 9 and D2212 × M 9 showed a remarkable inheritance of resistance, as expressed by CDI values of 1.1 and 1.3, respectively. Two-thirds of these offspring were classified as resistant because they never developed symptoms or only mild symptoms in the first year post inoculation. In other progenies, the number of resistant offspring ranged from 8 to 42%. Plants classified as resistant and transplanted for further observation under commercial growing conditions confirmed the good performance of the selected offspring. Among the various SSR categories there was usually no difference in resistance between trees on motherlike and hybrid I roots. This is probably due to the fact that both rootstock classes have the full genetic endowment of the resistant mother. However, resistance of trees on hybrid II roots was mostly significantly lower than that of trees on motherlike and hybrid I rootstocks. Also, the average phytoplasma titre in H0909 × M 9 offspring was significantly higher in hybrid II roots than in motherlike and hybrid I roots (Seemüller *et al.*, 2008a).

Pear Decline

Pear decline (PD) is a serious threat to the cultivated European or French pear *Pyrus communis*. The disease is widespread in Europe and North America and is induced by '*Ca.* Phytoplasma pyri', which is phylogenetically closely related to '*Ca.* Phytoplasma mali'. The disease is associated with a range of non-specific symptoms, such as foliar reddening, leaf curl, premature leaf drop, poor fruit development, growth suppression, slow decline and quick decline. Since the outbreak of PD in western North America in the 1940s and 1950s it has been observed that the severity and course of the disease varied greatly with the rootstock. As described above for apple, the importance of the rootstock can be explained by the annual fluctuation of the phytoplasma colonization in the stem and the overwintering of the pathogen in the roots (Schaper and Seemüller, 1982; Seemüller *et al.*, 1984). For this reason the usually susceptible scion cultivars can successfully be grown on resistant stocks.

Identification of resistance and factors involved

During the PD epiphytotic in western North America, trees on the oriental stocks *Py. pyrifolia* (syn. *Py. serotina*) and *Py. ussuriensis* were suffering most from the disease, while trees on seedlings of *Py. calleryana* and *P. communis* 'Bartlett' were moderately and slightly affected, respectively. Severe histopathological symptoms in the secondary phloem of susceptible rootstocks were observed to consist of deposition of pathological callose on sieve areas, sieve tube necrosis and formation of replacement phloem. In addition, accumulation of starch was identified above the bud union and depletion of starch below this line. These symptoms were closely related to the severity of aerial symptoms and were thus mild in the rootstocks of little affected trees or absent in those not affected. From these findings, it was concluded that translocation of carbohydrates from the stem to the roots is impaired by the disease, leading to decline of trees on susceptible stocks (Batjer and Schneider, 1960; Blodgett *et al.*, 1962).

In subsequent work, additional *Pyrus* taxa were evaluated under natural infection conditions in the Pacific North-west. Seedlings derived from open pollination of *Py. betulifolia*, *Py. elaeagrifolia*, *Py. nivalis*, *Py. pashia*, *Py. syriaca* and *Py. communis* 'Kirchensaller', clonal Quince A and C (*Cydonia oblonga*), and own-rooted *Py. communis* cvs 'Old Home' (OH), 'Anjou', 'Bartlett' and 'Winter Nelis' were classified as highly resistant or resistant, whereas seedlings of *Py. amygdaliformis*, *Py. caucasica*, *Py. cordata*, *Py. fauriei* and unspecified *Py.*

communis seedlings from France were reported to be susceptible. A range of OH × F clonal rootstocks deriving from the cross of the *Py. communis* cvs 'Old Home' × 'Farmingdale', which were developed for fire blight resistance, were assessed as highly resistant (Westwood and Lombard, 1982).

Only some of these evaluations were confirmed in Europe. Work in Germany showed that several OH × F selections are susceptible to PD (Seemüller *et al.*, 1998a). In Italy, own-rooted trees of *Py. communis* cvs 'Bartlett', 'Conference' and 'Abate Fetel' were little affected, while own-rooted 'Comice' trees developed severe symptoms under the same conditions. Of the quince rootstocks tested, trees on Quince A and Quince C suffered more from disease than trees on Quince BA29 and Quince CTS212 (Giunchedi *et al.*, 1995). In Germany, trees on Quince A were much less affected than trees on 'Kirchensaller', which proved rather susceptible (Seemüller *et al.*, 1986). This difference was explained by unsuitable host properties of quince for '*Ca.* Phytoplasma pyri', which result in low phytoplasma titres and poor survival of the pathogen. The poor host suitability of quince has been confirmed by others (Poggi Pollini *et al.*, 1995). As described above for AP, there is an indication that the low phytoplasma titre in the roots impairs recolonization of the stem in spring, resulting in no symptoms or a milder form of the disease. However, trees on quince are often severely affected in Italy. These differences may be explained by a heavier psyllid problem in Italy. It is supposed that under these circumstances the trees are repeatedly reinfected from the beginning of the growing season. Therefore, the top of the tree is infected earlier and perhaps more heavily than is case when recolonization takes place from the roots. This results in a greater severity of disease (Poggi Pollini *et al.*, 2001). In Germany, psyllid infestation is lower and so is the severity of disease in trees on quince.

Resistance screening by experimental inoculation

Due to the continuing PD problem, the unsatisfactory resistance of at least some *Py. communis* stocks and the susceptibility of quince rootstocks under certain conditions in Europe, a screening project was established in Germany, in which progenies of 39 open-pollinated genotypes belonging to 26 *Pyrus* taxa were examined. Nearly 1200 seedlings were graft-inoculated and observed for at least 18 years (Seemüller *et al.*, 1998a, 2009). The *Pyrus* progenies differed significantly in PD resistance and there was also a great variability within the progenies. Unaffected and little, moderately and severely affected trees were observed in all of them. However, there were great quantitative differences in the occurrence of these resistance categories. In the progenies of about one-third of the genotypes, the majority of the seedlings mediated a substantial level of resistance to grafted trees, as expressed by low CDI values (4.6–9.9), more than 50% of unaffected or slightly affected trees and low mortality rates. Trees on progenies of particular *Py. communis*, *Py. calleryana* 'Bradford' and *P. betulifolia* accessions were most resistant. Significantly different from this group was another third of the progenies, on which most grafted trees were susceptible, as evidenced by CDI values between 13.5 and 20.1, a low percentage of little affected or unaffected trees and generally high mortality rates. The most susceptible trees were on progenies of *Py. kunariana*, *Py. lecontii* and a particular *P. ussuriensis* accession. Between these two groups, there was another third of progenies, which was not statistically different from the resistant or the susceptible group. These progenies were defined as moderately resistant.

The variation in resistance within progenies corresponds to results obtained by Westwood (1976) in analysing offspring of crosses of *Pyrus* genotypes of different susceptibility. In his work, crosses of resistant parents, for instance of *Py. betulifolia* genotypes, resulted in a high proportion of resistant offspring and few plants that were susceptible. In contrast, offspring of susceptible parents such as *Py. pyrifolia* genotypes were predominantly susceptible, while resistant seedlings were rare. Crosses of resistant and susceptible genotypes yielded approximately 50% resistant and 50% susceptible offspring.

Considerable variation in susceptibility to PD was also observed between progenies of different accessions of the same taxon. Striking examples of such differences were *Py. communis* and *Py. betulifolia* accessions, of which the progenies were amongst the most resistant, moderately resistant or very susceptible. For example, a Muscovite accession of *P. communis* was the most resistant one, while the French *P. communis* stock Feudière proved highly susceptible. Similar instances were observed in *P. calleryana* 'Bradford', *Py. pyraster* and other taxa. Even *P. ussuriensis*, reported to be very susceptible, had two progenies that were moderately resistant, while a third accession was the most susceptible (Seemüller *et al.*, 2009). These results show that resistance is a segregating trait and cannot be assigned to a certain taxon. Because resistant individuals segregated in all progenies examined, it appears that resistance genes were present in at least one of the parents.

Resistance to psyllid vectors

In nature '*Ca.* Phytoplasma pyri' is vectored by the psyllids *Cacopsylla pyri* and *C. pyricola*, which are difficult to control. Most or all *P. communis* cultivars are highly suitable reproduction hosts for both vectors. In contrast, *Py. ussuriensis* and *Py. pyrifolia* are unsuitable for *Cacopsylla* feeding. Thus, as a different approach to controlling PD, crosses and backcrosses of *Py. communis* cultivars and *Py. ussuriensis* genotypes have been made since 1920, attempting to develop scion cultivars that resist the vectors. Such crosses resulted in 60% psyllid-resistant offspring. More recently, *Py. pyrifolia* is also being used as donor of resistance. The *Py. ussuriensis*-derived resistance is based on antixenosis, as expressed by poor settling of the adults, reduced oviposition and nymphal antibiosis. The physiological and biochemical mechanisms behind these traits are poorly understood (see Pasqualini *et al.*, 2006, for references).

Negative effects on reproduction biology and feeding behaviour of pear psylla, *C. pyricola*, were also observed on a transgenic clone of *P. communis*

'Bartlett' expressing the synthetic antimicrobial peptide *D5C1* and the selection marker *npt*II. Short-term studies (up to 7 days) indicated that pear psylla adults preferred to settle and oviposit, and nymphs fed more and developed slightly faster on transgenic pear than on non-transgenic pear. In contrast, a 32-day study on psylla colony development showed that considerably fewer eggs, nymphs and adults were produced on transgenic pear. These results suggest that chronic exposure of psylla populations to transformed pear plants that express the lytic peptide and *npt*II marker had detrimental effects on pear psylla reproductive biology (Puterka *et al.*, 2002).

European Stone Fruit Yellows

European stone fruit yellows (ESFY) is caused by '*Ca.* Phytoplasma prunorum', a close relative of '*Ca.* Phytoplasma mali' and '*Ca.* Phytoplasma pyri', and affects most cultivated stone fruit species. It is known to induce apricot (*Pr. armenica*) chlorotic leaf roll, leptonecrosis of Japanese plum (*Pr. salicina*), and yellows and decline diseases of peach (*Pr. persica*), almond (*Pr. dulcis*), European plum (*Pr. domestica*), flowering cherry (*Pr. serrulata*) and several rootstocks for stone fruits. There is an indication that the commonly used sweet cherry (*Pr. avium*) rootstock/scion combinations do not suffer or suffer little from ESFY, even though *P. avium* genotypes are hosts of the pathogen (Giunchedi *et al.*, 1982; Jarausch *et al.*, 1999; Kison and Seemüller, 2001). The same seems to be true for sour cherry (*Pr. cerasus*). The strategy to control stone fruit phytoplasmoses by the use of resistant plants differs from those of pome fruits because the pathogen persists in the top throughout the year (Seemüller *et al.*, 1998b). This was unexpected because the stem phloem of cherry and peach is reported to degenerate at the end of the growing season. However, in contrast to the degeneration in apple and pear, there remains a thin layer of small sieve tubes adjacent to the cambial zone in cherry, the so-called winter phloem, in which it is conceivable that phytoplasmas are able to persist. It is also possible that the pathogen may survive in the functional sieve tubes that are present over winter in the pathological replacement phloem found in phytoplasma-affected peach (Schneider, 1945). Due to these stone-fruit-specific colonization conditions, successful disease control is not possible with resistant rootstocks alone but also requires a resistant scion cultivar. However, the response of the rootstock to disease plays a significant role in resistance because the performance of infected trees depends on the efficiency of the root system.

Screening for resistance and factors involved

Twenty-three clonal or seedling rootstocks for stone fruits belonging to several major rootstock groups were examined following experimental inoculation by Kison and Seemüller (2001). This germplasm differed considerably in its response to infection. Trees on *Pr. domestica* stocks Ackermann's,

Brompton and P 1275 and on *Pr. cerasifera* clonal stock Myrabi were unaffected or little affected, while trees on GF 677 (*Pr. dulcis* × *Pr. persica*), GF 8-1 (*Pr. cerasifera* × *Pr. munsoniana*) and the *Pr. insititia* stocks St Julien A and St Julien GF 655/2 were slightly more damaged. Rootstocks Ishtara ((*Pr. cerasifera* × *Pr. persica*) × *Pr. salicina*), Myrobalan (*Pr. cerasifera*) seedling and peach seedlings Higama and GF 305 mediated moderate susceptibility to grafted trees. Trees on peach rootstocks Montclar, unspecified peach seedling, Rutgers Red Leaf, and Rubira on apricot seedling and St Julien 2 (*Pr. insititia*) showed high susceptibility. Mortality in the latter group was up to 100%, depending on susceptibility of the scion and virulence of the inoculum. Flowering cherry was least affected on Gisela 3 (*Pr. fruticosa* × *Pr. avium*) and, in increasing order, considerably more on F 12/1 (*Pr. avium*), Gisela 1 (*Pr. cerasus* × *Pr. canescence*), Weihroot 158 (*Pr. cerasus*) and Gisela 5 (*Pr. fruticosa* × *Pr. cerasus*). The low susceptibility of trees on *Pr. domestica* stocks, GF 8/1 and certain *Pr. insititia* stocks, the moderate susceptibility on Myrobalan seedling and the high susceptibility of trees on peach and apricot rootstocks are in agreement with results reported from other work, mostly upon observation of naturally infected trees (Morvan, 1977; Devignes and Cornaggia, 1982; Dosba *et al.*, 1991).

Significant differences in susceptibility to ESFY were also identified in scion cultivars of stone fruits. Like the resistant *Pr. domestica* rootstocks, the majority of European plum cultivars develop few or no symptoms upon infection. Examples are 'Reine Claude' ('Greengage') and related cultivars, 'Ruth Gerstetter', 'Bluefree', 'President' and 'Stanley' (Carraro *et al.*, 1998; Jarausch *et al.*, 2000). Exceptions reported to be markedly affected are Prune d'Ente (Prune d'Agen)-related cultivars, including 'Primacotes', 'Tardicotes', 'Lorida' and 'Spurdente' (Jarausch *et al.*, 2000). A considerable variability was observed even in apricot and Japanese plum, the most susceptible stone fruit species. However, the main differences were that less susceptible cultivars such as apricot 'Hungarian Best' declined more slowly than highly susceptible cultivars such as 'Canino' (Morvan, 1977; Duval *et al.*, 1999). Differences between resistant *Pr. domestica* cultivars and highly susceptible *Pr. salicina* cultivars were also observed in a histopathological study. While in the leaf phloem of Japanese plum, large areas with collapsed, thick-walled sieve tubes with dense, opaque inclusions and other alterations were observed, the symptoms were mild or lacking in European plum (Musetti and Favali, 1999).

In the screening work by Kison and Seemüller (2001), phytoplasmas were detected by 4'-6-diamidino-2-phenylindole (DAPI) fluorescence microscopy or PCR in all rootstocks and scion cultivars tested. By PCR, '*Ca.* Phytoplasma prunorum' was detected in nearly all samples. However, detection frequency varied considerably when the less sensitive DAPI staining was employed. With this method, between 75 and 100% of stem samples from peach genotypes were phytoplasma positive. Phytoplasmas were detectable in about 50% of stem samples from apricot and almond genotypes, most *Pr. insititia*-based stocks, GF 8/1 and the cherry stocks F 12/1 and Gisela 1. They were not detected, or rarely, in stems of *Pr. domestica* stocks, Myrobalan seedling, Myrabi and the cherry stocks Weihroot 158, Gisela 3 and Gisela 5. These data reflect the host suitability of the genotypes and indicate a correlation of presence and concentration of phytoplasmas in rootstocks for peach, apricot, almond and plums and their resistance. Such a correlation does not exist in cherry rootstocks. The differences observed in the stem of apricot, peach, plum and almond genotypes did not occur in the roots, where phytoplasmas were detected by fluorescence microscopy in about the same concentration in nearly all samples.

Cross-protection

There is an indication that cross-protection is another approach to control ESFY. In France several decades ago, a few severely affected apricot trees that recovered from disease and then performed as well as healthy trees, or nearly so, were observed among many declining trees. Since 1973 it was shown in several trials that trees pre-immunized by grafting scions from recovered trees were little affected or unaffected when grown at heavily infested sites where non-pre-immunized trees declined. From this result, it was concluded that the recovered trees were harbouring a graft-transmissible agent conferring a pre-immunizing effect. In recovered and pre-immunized trees, a phytoplasma that was indistinguishable from severe ESFY phytoplasma strains was identified by RFLP analysis of PCR-amplified 16S rDNA. It thus appears that avirulent or mild strains of the ESFY agent are responsible for the cross-protection effect. Pre-immunization had greater success when the trees were grown on more tolerant stocks such as Myrobalan and Brompton than with trees on the more susceptible peach and apricot rootstocks. It was also shown that strains for cross-protection have to be carefully selected for avirulence and their pre-immunizing effect (Castelain *et al.*, 2007).

The mechanism of cross-protection is unknown. There are several possibilities to explain this phenomenon. Pre-immunizing agents may suppress severe strains by competition for nutrients or attachment sites or a better adaptation to the sieve tube environment. Also, production of inhibitory substances and horizontal gene transfer from avirulent or mild strains to severe strains are conceivable. In *Spiroplasma citri*-infected periwinkle (*Catharanthus roseus*) showing unusually mild symptoms, a virus was observed in the spiroplasma cells. Transmission of spiroplasmas containing the virus to plants already infected with a virulent *S. citri* strain resulted in suppression of symptoms and in a reduction in the number of viable spiroplasmas (Alivizatos *et al.*, 1982). Cross-protection is not used much in plant protection due to potential risks. It cannot be ruled out that mild strains may become virulent by mutation or by acquiring virulence genes. They also may be more aggressive in combination with other strains, other pathogens or in other plants than the original host. However, cross-protection has been widely applied in Brazil, where, over many years, millions of sweet orange and grapefruit trees have been treated against citrus tristeza virus disease (Lee and Rocha-Pena, 1992).

Resistance to Other Phytoplasmoses

Intra- and interspecific differences in the response of plants to many other phytoplasma diseases have been observed. More detailed information is available from elm yellows, ash yellows, mulberry dwarf, paulownia witches'-broom, jujube witches'-broom, brinjal (aubergine) little leaf, rice yellow dwarf and sesame phyllody. Data on the resistance to some of these diseases will be treated briefly because they are confirming or extending the findings described above for palms and temperate fruit trees. One example is elm yellows, caused by '*Ca.* Phytoplasma ulmi'. Elm yellows is lethal to *Ulmus* spp. native to North America. Eurasian species such as *Ul. minor* and *Ul. parvifolia* are mostly more tolerant, but only *Ul. glabra*, *Ul. pumila* and *U. pumila* hybrids are highly resistant. Like diseased trees of apple and pear, diseased trees of the highly susceptible American elm (*Ul. americana*) show severe histopathological symptoms in the secondary phloem, whereas the phloem of more tolerant *U. parvifolia* is little affected. Similar to susceptible *M. sargentii*-derived genotypes, which do not develop specific symptoms but show growth suppression and decline, and the more tolerant *M.* × *domestica* genotypes, which are characterized by witches'-broom formation, the phytoplasma titre in the stem is low in rapidly declining *U. americana* and high in the more tolerant, witches'-broom-developing *U. minor* (Braun and Sinclair, 1976, 1979). A slightly different relationship was identified in paulownia witches'-broom-affected clones and hybrids of *Paulownia* spp., where the titre was high in genotypes that develop phyllody and witches'-broom symptoms. However, in slightly affected trees the titre was low (Tian *et al.*, 1994). Also, in witches'-broom-diseased jujube (*Ziziphus jujuba*) trees, resistance was associated with low phytoplasma numbers (Liu *et al.*, 2004). Both findings correspond to the low phytoplasma titre in resistant apple genotypes that have *M. sieboldii* as a parent.

Natural resistance to mulberry dwarf was reported to be associated with the phytoalexin concentration in the cortex of *Morus alba* trees. The amount of a small group of compounds isolated by thin-layer chromatography was fourfold higher in resistant cultivars than in susceptible ones (Kuai *et al.*, 1999). Transgenic resistance was obtained by expressing an antibacterial peptide encoded by the *shiva-1* gene in *Paulownia* hybrids. Both symptom development and phytoplasma titre were significantly reduced in transgenic plants (Du *et al.*, 2005). In other work, single-chain variable fragment (scFv) antibodies directed against the major membrane protein of the stolbur (STOL) phytoplasma were expressed in tobacco. The results obtained with the transgenic plants were inconsistent. In one experiment, resistance was markedly increased in transgenic plants, while in another the positive effect was weak and temporary (Le Gall *et al.*, 1998; Malembic-Maher *et al.*, 2005). In sesame (*Sesamum indicum*), resistance to the causative sesame phyllody phytoplasma and to the leafhopper vector *Orosius albicinctus* has been identified. Disease resistance in cultivated sesame varieties is governed by a single recessive gene, while the wild species *S. alatum* and *S. mulayanum* possess a single dominant gene conferring resistance to sesame phyllody. *Sesamum alatum* also seems be a donor of vector resistance (Parani *et al.*, 1996; Singh *et al.*, 2007). In rice, too, resistance to the rice yellow dwarf phytoplasma '*Ca.* Phytoplasma oryzae' and to its vectors (*Nephotettix* spp.) has been recognized. As in wild sesame species, disease resistance is reported to be controlled by a single dominant gene (Muniyappa and Raychaudhuri, 1988).

Conclusions

There are many reports on the variable response of plants to phytoplasma infection. Although, especially in earlier reports, the data are often based on observations of naturally infected plants at a time when detection and characterization of the pathogens were difficult or impossible, there is no doubt that the development of resistant plants is a promising approach to controlling phytoplasmoses. Due to biological constraints and, perhaps, to increased vector activity or phytoplasma virulence, the resistance approach in palms is more difficult to realize than in other plants. Control strategies include selection or breeding of plants resistant to the diseases and/or the insect vectors, as well as engineering of transgenic plants. However, any effort to develop resistant plants is considerably hampered by our unsatisfactory knowledge in phytoplasmology, mainly due to the uncultivability of the pathogens under axenic conditions. Despite the progress made by the use of molecular technologies, firm data on phytoplasma pathogenicity, phytoplasma–host interactions and the molecular basis of resistance are sparse. However, it is well established that phytoplasmas severely affect phloem function in susceptible plants, impairing the transport of soluble organic material, particularly to the roots. These symptoms are lacking or mild in resistant plants. Another factor of resistance is phytoplasma concentration. In several plant species it could be shown that resistance is associated with low phytoplasma titre. However, low concentration alone is not the only basis of resistance, because in highly susceptible plants that do not develop specific symptoms phytoplasma numbers are also low. More information on how phytoplasmas induce plant diseases can be expected from nucleic acid sequence analysis of entire phytoplasma genomes. There are several candidate genes that are suspected to be involved in pathogenicity, such as those encoding immunodominant membrane proteins or other proteins that are secreted or have transmembrane domains. Effector proteins have been identified that accumulate in the nuclei of tobacco plants (Hogenhout *et al.*, 2008). Progress in elucidating phytoplasma–plant interactions and fostering resistance research can also be expected from analysing host plant genomes or mapping of loci linked to resistance. The availability of markers linked to resistance would greatly facilitate selection and breeding projects.

References

Alivizatos, A.S., Townsend, R. and Markham, P.G. (1982) Effects of infection with a spiroplasma virus on the symptoms produced by *Spiroplasma citri*. *Annals of Applied Biology* 101, 85–92.

Allorerung, D., Harries, H.C., Jones, P. and Warokka, S. (eds) (1999) *Proceedings of the Workshop on Lethal Diseases of Coconut Caused by Phytoplasma and their Importance in Southeast Asia.* Asian Pacific Coconut Community, Jakarta, Indonesia.

Ashburner, G.R. and Been, B.O. (1997) Characterization of resistance to lethal yellowing in *Cocos nucifera* and implications for genetic improvement of this species in the Caribbean region. In: Eden-Green S.J. and Ofori, F. (eds) *Proceedings of an International Workshop on Lethal Yellowing-like Diseases of Coconut.* Elmina, Ghana, 1995. Natural Resources Institute, Chatham, UK, pp. 173–183.

Batjer, L.P. and Schneider, H. (1960) Relation of pear decline to rootstocks and sieve-tube necrosis. *Proceedings of the American Society for Horticultural Science* 76, 85–97.

Baudouin, L., Lebrun, P., Berger, A., Myrie, W., Been, B. and Dollet, M. (2008) The Panama Tall and the Maypan hybrid coconut in Jamaica: did genetic contamination cause a loss of resistance to lethal yellowing? *Euphytica* 161, 353–360.

Bisognin, C., Schneider, B., Salm, H., Grando, M.S., Jarausch, W., Moll, E. and Seemüller, E. (2008) Apple proliferation resistance in apomictic rootstocks and its relationship to phytoplasma concentration and simple sequence repeat genotypes. *Phytopathology* 98, 153–158.

Bisognin, C., Seemüller, E., Citterio, S., Velasco, R., Grando, M.S. and Jarausch, W. (2009) Use of SSR markers to assess sexual vs apomictic origin and ploidy level of breeding progenies derived from crosses of apple proliferation-resistant *Malus sieboldii* and its hybrids with *Malus × domestica* cultivars. *Plant Breeding* DOI: 10.111/;.1439-05232008.01614.x.

Blodgett, E.C., Schneider, H. and Aichele, M.D. (1962) Behavior of pear decline disease on different stock–scion combinations. *Phytopathology* 52, 679–684.

Braun, E.J. and Sinclair, W.A. (1976) Histopathology of phloem necrosis in *Ulmus americana*. *Phytopathology* 66, 598–607.

Braun, E.J. and Sinclair, W.A. (1979) Phloem necrosis of elms: symptoms and histopathological observations in tolerant hosts. *Phytopathology* 69, 354–358.

Brown, S.E., Been, B.O. and McLaughlin, W.A. (2006) Detection and variability of the lethal yellowing group (16Sr IV) phytoplasmas in the *Cedusa* sp. (Hemiptera: Auchenorrhyncha: Derbidae) in Jamaica. *Annals of Applied Biology* 149, 53–62.

Brown, S.E., Been, B.O. and McLaughlin, W.A. (2007) First report of the presence of the lethal yellowing group (16Sr IV) of phytoplasmas in the weeds *Emelia* (*Emilia*) *fosbergii* and *Synedrella nodiflora* in Jamaica. *British Society for Plant Pathology New Disease Reports* 16, Aug. 2007–Jan. 2008 <http://www.bspp.org.uk/ndr/volume16.asp>.

Bruner, S.C. and Boucle, L. (1943) La enfermendad conocida por 'pudricionn de cogollo en Cuba'. *Revista Cubana de Ciencia Agricola* 26, 132–141.

Carraro, L., Loi, N., Ermacora, P. and Osler, R. (1998) High tolerance of European plum varieties to plum leptonecrosis. *European Journal of Plant Pathology* 104, 141–145.

Castelain, C., Jullian, J.-P., Lemaire, J.-M. and Morvan, G. (2007) Lutte biologique contre l'ECA: la prémunition. *L'Arboriculture* No. 611, 33–38.

Dery, S.K., Philippe, R., Baudouin, L., Quaicoe, R.N., Nkansah-Poku, J., Owusu-Nipah, J., Arthur, R., Dare, D., Yankey, N. and Dollet, M. (2008) Genetic diversity among coconut varieties for susceptibility to Cape St Paul Wilt Disease. *Euphytica* 164, 1–11.

Devignes, J.C. and Cornaggia, D. (1982) Observations on apricot chlorotic leaf roll (ACLR): sensitiviness of different *Prunus* species, detection, spread in plum orchards. *Acta Horticulturae* 130, 249–256.

Dosba, F., Lansac, M., Mazy, K., Garnier, M. and Eyquard, J.P. (1991) Incidence of different diseases associated with mycoplasma-like organisms in different species of *Prunus*. *Acta Horticulturae* 283, 311–320.

Du, T., Wang, Y., Hu, Q.X., Chen, J., Liu, S., Huang, W.J. and Lin, M.L. (2005) Transgenic *Paulownia* expressing *shiva-1* gene has increased resistance to Paulownia witches' broom disease. *Journal of Integrative Plant Biology* 47, 1500–1506.

Duval, H., Jullian, J. and Lemaire, J. (1999) Enroulement chlorotic de l'abricotier: évaluation de la sensibilité de principales variétés de prunier japonais. *Phytoma* No. 516, 38–40.

Ennin, S., Banful, A., Andoh-Mensah, E., Issaka, R.N., Lambtey, J.N., Dzomeku, B.M., Manu, J.A., Bolfrey-Arku, G. and Dery, S.K. (2008) Food crop intercropping alternative for replanting tolerant LYD coconut hybrid in Ghana. *Proceedings of an International Workshop on Lethal Yellowing Disease of Coconut*, 3–5 June 2008, Accra, Ghana. CIRAD (Centre de coopération internationale en recherche agronomique pour le développement), France <http://www.cirad.bf/fr/anx/conf_lyd00.php>.

Giunchedi, L., Poggi Pollini, C. and Credi, R. (1982) Suspectibility of stone fruit trees to the Japanese plum-tree decline causal agent. *Acta Horticulturae* 130, 285–290.

Giunchedi, L., Poggi Pollini, C., Bissani, R., Babini, A.R. and Vicchi, V. (1995) Etiology of a pear decline disease in Italy and susceptibility of pear variety and rootstock to phytoplasma-associated pear decline. *Acta Horticulturae* 386, 489–495.

Harries, H.C. (1974) Natural symptom remission of lethal yellowing disease of coconut. *Tropical Agriculture* 51, 575–576.

Harries, H.C. (1995) The genetics of durable resistance to lethal yellowing disease. In: Orepeza, C., Howard, F.W. and Ashburner, G.R. (eds) *Lethal Yellowing: Research and Practical Aspects*. Kluwer Academic Publishers, Dordrecht, the Netherlands, pp. 149–171.

Hogenhout, S.A., Oshima, K., Ammar, E.D., Kakizawa, S., Kingdom, H. and Namba, S. (2008) Phytoplasmas: bacteria that manipulate plants and insects. *Molecular Plant Pathology* 9, 403–423.

Howard, F.W., Atilano, R., Barrant, C.I., Harrison, N.A. and Theobold, W.F.W. (1987) Unusually lethal yellowing disease incidence in Malayan dwarf coconut palms on localized sites in Jamaica and Florida. *Journal of Plantation Crops* 15, 86–100.

Jarausch, W., Eyquard, J., Mazy, K., Lansac, M. and Dosba, F. (1999) High level of resistance of sweet cherry (*Prunus avium* L.) towards European stone fruit yellows phytoplasmas. *Advances in Horticultural Science* 13, 108–112.

Jarausch, W., Eyquard, J.P., Lansac, M., Mohns, M. and Dosba, F. (2000) Susceptibility and tolerance of new French *Prunus domestica* cultivars to European stone fruit yellows phytoplasmas. *Journal of Phytopathology* 148, 489–493.

Kartte, S. and Seemüller, E. (1991a) Histopathology of apple proliferation in *Malus* taxa and hybrids of different susceptibility. *Journal of Phytopathology* 131, 149–160.

Kartte, S. and Seemüller, E. (1991b) Susceptibility of grafted *Malus* taxa and hybrids to apple proliferation disease. *Journal of Phytopathology* 131, 137–148.

Kison, H. and Seemüller, E. (2001) Differences in strain virulence of the European stone fruit yellows phytoplasma and susceptibility of stone fruit trees on various rootstocks to this pathogen. *Journal of Phytopathology* 149, 533–541.

Kuai, Y.Z., Zhu, F.P., Xia, Z.S. and Chen, P.G. (1999) Studies on the mechanism of mulberry varietal resistance to mulberry yellow dwarf disease. *International Journal of Tropical Plant Diseases* 17, 103–114.

Kullaya, A. and Mpunami, A. (2008) Breeding for resistance to coconut lethal disease in Tanzania. *Proceedings of an International Workshop on Lethal Yellowing Disease of Coconut*, 3–5 June 2008, Accra, Ghana. CIRAD (Centre de coopération internationale en recherche agronomique pour le développement), France <http://www.cirad.bf/fr/anx/conf_lyd00.php>.

Lebrun, P., Baudouin, L., Myrie, W., Berger, A. and Dollet, M. (2008) Recent lethal yellowing outbreak: why is the Malayan Yellow Dwarf coconut no longer resistant in Jamaica? *Tree Genetics and Genomes* 4, 125–131.

Lee, R.F. and Rocha-Pena, M.A. (1992) Citrus tristeza virus. In: Mukhopadhyay, A.N., Chaube, H.S., Kumar, J. and Singh, U.S. (eds) *Plant Diseases of International Importance.* Prentice Hall, Englewood Cliffs, New Jersey, pp. 226–249.

Le Gall, F., Bové, J.M. and Garnier, M. (1998) Engineering of a single-chain variable-fragment (scFv) antibody specific for the stolbur phytoplasma (Mollicute) and its expression in *Escherichia coli* and tobacco plants. *Applied and Environmental Microbiology* 64, 4566–4572.

Liu, M.J., Zhou, J.Y. and Zhao, J. (2004) Screening of Chinese jujube germplasm with high resistance to witches' broom disease. *Acta Horticulturae* 663, 575–579.

Malembic-Maher, S., Le Gall, F., Danet, J.L., de Borne, F.D., Bové, J.M. and Garnier-Semancik, M. (2005) Transformation of tobacco plants for single-chain antibody expression via apoplastic and symplasmic routes, and analysis of their susceptibility to stolbur phytoplasma infection. *Plant Science* 168, 349–358.

Maramorosch, K. (1964) *A Survey of Coconut Diseases of Unknown Etiology.* Food and Agricultural Organization, Rome, Italy.

Martinez, R.T., Narvaez, M., Fabre, S., Harrison, N., Oropeza, C., Dollet, M. and Hichez, E. (2008) Coconut lethal yellowing on the southern coast of the Dominican Republic is associated with a new 16SrIV group phytoplasma. *Plant Pathology* 57, 366.

Morvan, G. (1977) Apricot chlorotic leaf roll. *EPPO Bulletin* 7, 37–55.

Muniyappa, V. and Raychaudhuri, S.P. (1988) Rice yellow dwarf disease. In: Maramorosch, K. and Raychaudhuri, S.P. (eds) *Mycoplasma Diseases of Crops.* Springer-Verlag, New York, pp. 233–284.

Musetti, R. and Favali, M.A. (1999) Histological and ultrastructural comparative study between *Prunus* varieties of different susceptibility to plum leptonecrosis. *Cytobios* 99, 73–82.

Nutman, F.J. and Roberts, F.M. (1955) Lethal yellowing: the 'unknown' disease of coconut palms in Jamaica. *Empire Journal of Experimental Agriculture* 23, 257–267.

Parani, M., Singh, K.N., Rangasamy, S.R.S. and Ramalingam, R.S. (1996) A study on mechanism of phyllody disease resistance in *Sesamum alatum* Thonn. *Current Science* 70, 86–89.

Pasqualini, E., Civolani, S., Musacchi, S., Ancarini, V., Dondini, L., Robert, P. and Baronio, P. (2006) *Cacopsylla pyri* behaviour on new pear selections for host resistance programs. *Bulletin of Insectology* 59, 27–37.

Poggi Pollini, C., Bissani, R. and Giunchedi, L. (1995) Overwintering of pear decline agent in some quince rootstocks. *Acta Horticulturae* 386, 496–499.

Poggi Pollini, C., Bissani, R., Guerrini, S., Firrao, G. and Giunchedi, L. (2001) Epidemiological studies on pear decline (PD) in Italy. *Acta Horticulturae* 550, 361–364.

Puterka, G.J., Bocchetti, C., Dang, P., Bell, R.L. and Scorza, R. (2002) Pear transformed with a lytic peptide gene for disease control affects nontarget organism, pear psylla (Homoptera : Psyllidae). *Journal of Economic Entomology* 95, 797–802.

Quaicoe, R., Dery, S.K., Philippe, R., Baudouin, L., Nipah, O., Nkansah-Poku, J., Arthur, R., Dare, D., Yankey, E.N., Pilet, F. and Dollet, M. (2008) Resistance screening trials on coconut varieties to Cape Saint Paul wilt in Ghana. *Proceedings of an International Workshop on Lethal Yellowing Disease of Coconut*, 3–5 June 2008, Accra, Ghana. CIRAD (Centre de coopération internationale en recherche agronomique pour le développement), France <http://www.cirad.bf/fr/anx/conf_lyd00.php>.

Robinson, R.A. (1977) The international program on horizontal resistance. *FAO Plant Protection Bulletin* 25, 197–200.

Schaper, U. and Seemüller, E. (1982) Condition of the phloem and the persistence of mycoplasmalike organisms associated with apple proliferation and pear decline. *Phytopathology* 72, 736–742.

Schneider, H. (1945) Anatomy of buckskin-diseased peach and cherry. *Phytopathology* 35, 610–635.

Schuiling, M., Mpunami, A., Kaiza, D.A. and Harries, H.C. (1992) Lethal disease of coconut palm in Tanzania. 3. Low resistance of imported germplasm. *Oléagineux* 47, 693–698.

Seemüller, E., Schaper, U. and Zimbelmann, F. (1984) Seasonal variation in the colonization patterns of mycoplasmalike organisms associated with apple proliferation and pear decline. *Zeitschrift für Pflanzenkrankheiten und Pflanzenschutz* 91, 371–382.

Seemüller, E., Schaper, U. and Kunze, L. (1986) Effect of pear decline on pear trees on 'Quince A' and *Pyrus communis* seedling rootstocks. *Zeitschrift für Pflanzenkrankheiten und Pflanzenschutz* 93, 44–50.

Seemüller, E., Kartte, S. and Kunze, L. (1992) Resistance in established and experimental apple rootstocks to apple proliferation disease. *Acta Horticulturae* 309, 245–251.

Seemüller, E., Lorenz, K.-H. and Lauer, U. (1998a) Pear decline resistance in *Pyrus communis* rootstocks and progenies of wild and ornamental *Pyrus* taxa. *Acta Horticulturae* 472, 681–691.

Seemüller, E., Stolz, H. and Kison, H. (1998b) Persistence of the European stone fruit yellows phytoplasma in aerial parts of *Prunus* taxa during the dormant season. *Journal of Phytopathology* 146, 407–410.

Seemüller, E., Bisognin, C., Schneider, B., Hergenhahn, F., Velasco, R., Jarausch, W. and Grando, M.S. (2008a) Programma di breeding per lo sviluppo di portinnesti resistenti ad apple proliferation. In: Ioriatti, C. and Jarausch, W. (eds) *Scopazzi del Melo – Apple Proliferation*. Fondazione Edmund Mach, San Michele all'Adige, Italy, pp. 173–185.

Seemüller, E., Moll, E. and Schneider, B. (2008b) Apple proliferation resistance of *Malus sieboldii*-based rootstocks in comparison to rootstocks derived from other *Malus* species. *European Journal of Plant Pathology* 121, 109–119.

Seemüller, E., Moll, E. and Schneider, B. (2009) Pear decline resistance in progenies of *Pyrus* taxa used as rootstocks. *European Journal of Plant Pathology* 123, 217–223.

Singh, P.K., Akram, M., Vajpeyi, M., Srivastava, R.L., Kumar, K. and Naresh, R. (2007) Screening and development of resistant sesame varieties against phytoplasma. *Bulletin of Insectology* 60, 303–304.

Tian, G.Z., Xiong, Y., Wang, Y., Zhao, D., Li, F. and Xu, G. (1994) Resistances of different clones of *Paulownia* to witches' broom agent mycoplasma-like organisms (MLO). *Forest Research* 2, 155–161.

Westwood, M.N. (1976) Inheritance of pear decline resistance. *Fruit Varieties Journal* 30, 63–64.

Westwood, M.N. and Lombard, P.B. (1982) Rootstocks for pear. *Proceedings of Oregon Horticultural Society* 73, 64–79.

Whitehead, R.A. (1966a) Some notes on dwarf coconut palms in Jamaica. *Tropical Agriculture* 43, 277–294.

Whitehead, R.A. (1966b) *Sample Survey and Collection of Coconut Germplasm in the Pacific Islands*. Ministry of Overseas Development, London.

Whitehead, R.A. (1968) *Collection of Coconut Germplasm from the Indian/Malaysian Region, Peru and the Seychelles Islands and Testing for Resistance to Lethal Yellowing Disease*. Food and Agriculture Organization, Rome, Italy.

10 Phytoplasma Diseases of the Gramineae

Yaima Arocha Rosete and Phil Jones

Rothamsted Research, UK

Introduction

Grasses belong to the family Gramineae (= Poaceae) and encompass some 600 genera and between 9000 and 10,000 species, contributing to food, forage, industry and ornamental production systems. About 70% of all crops are included in this family. Major cereal crops in various continents include: rice, in southern and eastern Asia; wheat and barley, in Europe, northern Asia and the Americas; and maize (corn), in Central and South America. Minor crops are: oats; most millets; fescues; ryegrasses, the most widely used forage grass in the world; and sugarcane, which besides sugar is also used in biofuel production. Natural grassland communities such as the steppes (Asia) and pampas (South America) are estimated to contribute 20% of the earth's vegetation cover.

Phytoplasmas are uncultured, obligate, intracellular prokaryotic plant pathogens of the class *Mollicutes*, causing diseases of crops, ornamentals and weeds from temperate to tropical regions worldwide, leading to phytosanitary constraints and considerable losses for world economies (Lee *et al.*, 2000). Molecular methods are the best approach for their study in both plant hosts and insect vectors, and particularly the knowledge of their complete sequence will be the tool to control their associated diseases in nature.

Phytoplasmas are naturally transmitted by phloem-feeding insects of the Hemiptera order (Weintraub and Beanland, 2006). In natural agroecosystems, both insect vectors and alternative hosts that serve as reservoirs of infection play a fundamental role in phytoplasma host-range and disease spread, since it is a result of the three-way vector–phytoplasma–plant interactions. The Gramineae have the largest number of species associated with phytoplasma diseases worldwide, and are also the one plant family where the majority of phytoplasma vector species (Delphacidae) have been found. A comprehensive relationship among selected representative '*Ca*. Phytoplasma' strains infecting Gramineae in different countries worldwide is shown in Table 10.1 and the phylogenetic tree of Fig. 10.1.

Phytoplasma Diseases of Rice

Rice is a staple food for more than half the world's population. Domesticated rice comprises two species in the *Oryza* genus: *O. sativa* (Asian rice), native to tropical and subtropical southern Asia, and *O. glaberrima* (African rice), native to West Africa. The major rice-growing areas of the world are Southeast Asia, Indo-Asia, the Americas and Africa.

Rice yellow dwarf (RYD)

RYD, a serious problem for rice farmers, has only been detected to date in Asia, where it is recorded from most rice-growing countries (Nakashima *et al.*, 1993). Infected rice turns pale yellow and gradually starts to decay and produce numerous tillers. Sometimes only a faint mottling occurs. Plants infected early normally do not die but produce abnormal heads or no heads, and ultimately show stunted growth and fail to produce grain (Jung *et al.*, 2003). For many years, the agent associated with RYD was believed to be a virus, until it was identified as a phytoplasma, designated as a '*Candidatus* Phytoplasma oryzae' (Jung *et al.*, 2003).

Rice and the grass *Alopecurus aequalis* appear to be the only natural hosts for the phytoplasma (Nakashima *et al.*, 1993). The closest relatives to '*Ca.* Phytoplasma oryzae' are those associated with sugarcane white leaf (SCWL), sugarcane grassy shoot (SCGS), annual blue grass white leaf (ABGWL), Bermuda grass white leaf (BGWL) and Brachiaria grass white leaf (BraWL) phytoplasmas (Tran-Nguyen *et al.*, 2000).

'*Ca.* Phytoplasma oryzae' is transmitted by three species of leafhoppers that are found only in Asia (Nakashima *et al.*, 1993): *Nephotettix cincticeps* (Uhler), *N. virescens* (Distant) and *N. nigropictus* (Stål). The phytoplasma can overwinter in leafhoppers and the wild grass *Alopecurus aequalis*, disseminated primarily by the leafhopper. Leafhoppers acquire the phytoplasma by feeding on infected plants for 1–3 h, and, after a latent period of 20–39 days, the phytoplasma passes from the gut to the salivary gland of the insect. Leafhoppers are then capable of inoculating healthy plants in usually less than 1 h of feeding. The latent period in rice is about 1 month in 'warm' weather and 3 months in 'cool' weather. There is limited evidence for the spread of RYD phytoplasma from rice to other members of the Gramineae, although it is believed to occur through root grafts and occasionally by leafhopper transmission (Jung *et al.*, 2003). The host range of '*Ca.* Phytoplasma oryzae', in nature, may be determined by its vector feeding preferences, which are controlled by biophysical and biochemical mechanisms and ultimately by genetic factors. The plant host specificity may also be due to resistance of a particular plant, since '*Ca.* Phytoplasma oryzae' has not been transmitted to periwinkle or other plants by *Cuscuta* spp.

Table 10.1. Gramineae-infecting phytoplasmas selected for phylogenetic analysis.

GenBank acc. no.	Phytoplasma	Plant host	Country	16Sr group
D12581	Rice yellow dwarf	Rice	Japan	16SrXI, '*Ca.* Phytoplasma oryzae'
AB052873	Rice yellow dwarf	Rice	Thailand	16SrXI, '*Ca.* Phytoplasma oryzae'
AB052870	Rice orange leaf	Rice	Philippines	16SrI, '*Ca.* Phytoplasma asteris'
AF487779	Maize bushy stunt	Maize	Mexico	16SrI, '*Ca.* Phytoplasma asteris'
AY265208	Maize bushy stunt	Maize	Mexico	16SrI, '*Ca* Phytoplasma asteris'
DQ222972	Corn reddening	Maize	Serbia	16SrI, '*Ca.* Phytoplasma asteris'
AY734453	Barley deformation	Barley	Lithuania	16SrI, '*Ca.* Phytoplasma asteris'
AF453416	Oat proliferation	Oat	Lithuania	16SrI, '*Ca.* Phytoplasma asteris'
DQ078304	Wheat blue dwarf	Wheat	China	16SrI, '*Ca.* Phytoplasma asteris'
AY635145	Bamboo witches'-broom	Bamboo	China	16SrI, '*Ca.* Phytoplasma asteris'
EF012650	Napier grass stunt	Napier grass	Uganda	16SrXI, '*Ca.* Phytoplasma oryzae'
AY377876	Napier grass stunt	Napier grass	Kenya	16SrXI, '*Ca.* Phytoplasma oryzae'
DQ305977	Napier grass stunt	Napier grass	Ethiopia	16SrIII, Western X-disease
AB052874	Sugarcane white leaf	Sugarcane	Thailand	16SrXI, '*Ca.* Phytoplasma oryzae'
EF614269	Sugarcane grassy shoot	Sugarcane	India	16SrXI, '*Ca.* Phytoplasma oryzae'
AF056095	Sugarcane yellows	Sugarcane	South Africa	16SrIII, Western X-disease
EU170474	Sugarcane yellows	Sugarcane	India	16SrXII, '*Ca.* Phytoplasma australiense'
AY725228	Sugarcane yellows	Sugarcane	Cuba	16SrXII, '*Ca.* Phytoplasma australiense'
AF509322	Sorghum bunchy shoot	Sorghum	Australia	16SrXI, '*Ca.* Phytoplasma oryzae'
AF509324	Sorghum grassy shoot	Sorghum	Australia	16SrXI, '*Ca.* Phytoplasma oryzae'
EU032485	Bermuda grass white leaf	Bermuda grass	India	16SrXIV, '*Ca.* Phytoplasma cynodontis'
EU377477	Bermuda grass white leaf	Bermuda grass	China	16SrXIV, '*Ca.* Phytoplasma cynodontis'
AJ550986	Bermuda grass white leaf	Bermuda grass	Italy	16SrXIV, '*Ca.* Phytoplasma cynodontis'
AJ550984	Bermuda grass white leaf	Bermuda grass	Italy	16SrXIV, '*Ca.* Phytoplasma cynodontis'
AB052871	Bermuda grass white leaf	Bermuda grass	Thailand	16SrXIV, '*Ca.* Phytoplasma cynodontis'
AF248961	Bermuda grass white leaf	Bermuda grass	Thailand	16SrXIV, '*Ca.* Phytoplasma cynodontis'
EU29411	Bermuda grass white leaf	Bermuda grass	Malaysia	16SrXIV, '*Ca.* Phytoplasma cynodontis'
EF444486	Bermuda grass white leaf	Bermuda grass	Iran	16SrXIV, '*Ca.* Phytoplasma cynodontis'
FJ348654	Bermuda grass white leaf	*Dichantium annulatum*	India	16SrXIV, '*Ca.* Phytoplasma cynodontis'
AB052872	Bermuda grass white leaf	*Brachiaria* sp.	Thailand	16SrXIV, '*Ca.* Phytoplasma cynodontis'

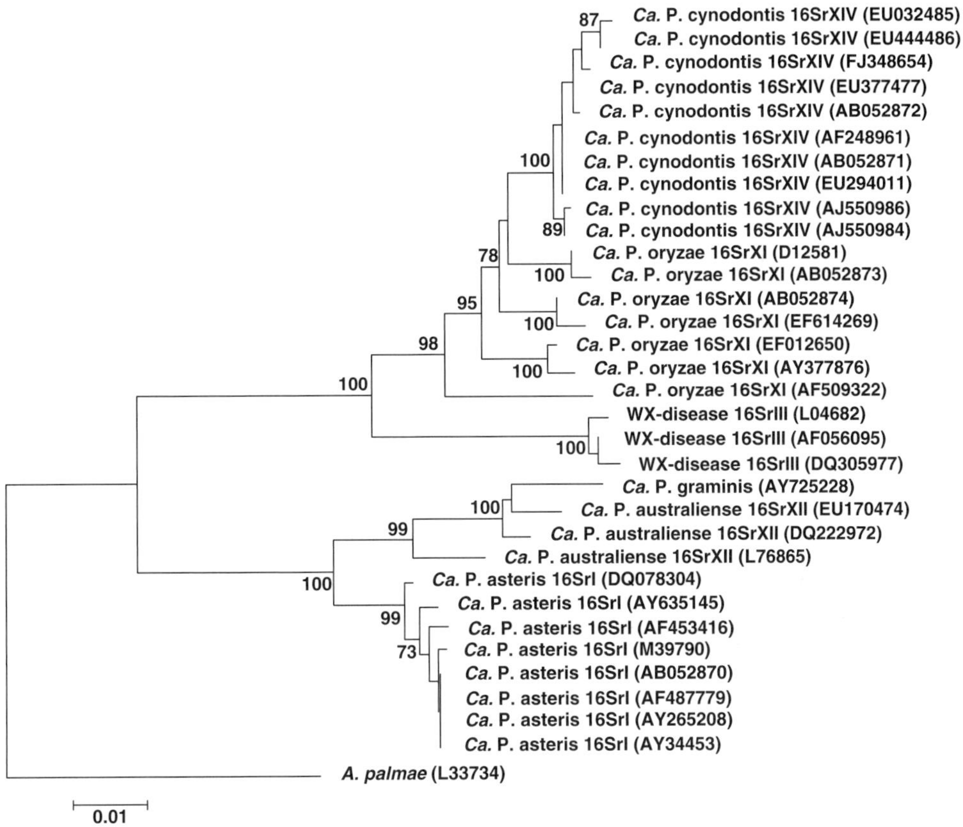

Fig. 10.1. Phylogenetic relationships of selected phytoplasmas affecting Gramineae. *Ca.* P. equivalent to '*Candidatus* Phytoplasma'.

There is little information on sequence data for '*Ca.* Phytoplasma oryzae'. Negative results were obtained when 16SrI isolates including RYD were tested for ribosomal proteins due to specific constraints on their genomes (Martini *et al.*, 2007). However, the role and gene organization of immunodominant membrane proteins have been predicted for several phytoplasmas, including '*Ca.* Phytoplasma oryzae' and 16SrXI-related strains (Kakizawa *et al.*, 2006). The future availability of the full genome sequence of '*Ca.* Phytoplasma oryzae' and identification of genes involved in vector interactions will help formulate improved strategies for RYD management.

Rice orange leaf (ROL)

ROL was widely distributed in South and South-east Asia (Hibino *et al.*, 1987), including Thailand, Malaysia, Indonesia, China, Sri Lanka and the Philippines. Symptoms are typified by orange-coloured leaves, which later roll inward and desiccate. Infected plants die 2–3 weeks after symptoms appear. The diseased plants are generally distributed sporadically in the field, and the disease does not cause serious yield loss. ROL was associated with a phytoplasma in Thailand, Malaysia, Indonesia and the Philippines (Hibino *et al.*, 1987) following electron microscopy evidence. ROL is transmitted by the leafhopper *Recilia dorsalis* Motchulsky, which also transmits rice dwarf virus and rice tungro virus. These diseases differentiate from ROL on symptoms and geographic distribution. In the Philippines, the ROL phytoplasma is transmitted by *R. dorsalis* in a persistent manner, with an incubation period of 15–33 days.

Phytoplasma and Spiroplasma Diseases of Maize and Sorghum

Maize (*Zea mays* L.) ranks third in production following wheat and rice, with an average of 380 million tons produced annually on 120 million ha by 53 countries. It is the world's most widely grown crop in almost all tropical areas of the world, including tropical highlands over 3000 m in altitude, to temperate regions. To date, two mollicute plant pathogens affect maize: the corn stunt spiroplasma (CS) and maize bushy stunt phytoplasma, causing stunting and maize redness (MR), corn reddening (CR) and reduced yield.

Maize bushy stunt (MBS) and corn stunt (CS)

These two diseases have been reported in maize (corn) in North, Central and South America (Bedendo *et al.*, 2000; Lee *et al.*, 2000; Gomes *et al.*, 2004). Symptoms are associated with an initial development of marginal yellow and orange colour of the older leaves, and subsequently are characterized by marginal chlorosis, tearing, shortening and twisting of young leaf tips. Numerous tillers develop at the base of the plant and at leaf axils. MBS can be distinguished from CS on the basis of symptom appearance on maize at elevated temperature, the range of insect vector species, plant host range and length of the latent period in the natural leafhopper vector *Dalbulus maidis* DeLong and Wolcott. Despite these distinctions, diagnosis of MBS may be confounded by mixed infections with CS or maize viruses, and symptom variations due to host genotype, pathogen strain and environmental conditions.

In nature, CS is transmitted by *D. maidis* and *Dalbulus elimatus* (Ball) in a persistent and propagative manner. *D. maidis* is the most efficient vector of CS, with an acquisition access period of 15 min and an incubation period of 7 days. MBS is transmitted by *D. maidis*. It is also transmitted by *D. elimatus*, *Baldulus tripsaci* Kramer and Whitcomb and *Graminella nigrifrons* (Fobes). *D. maidis* is the principal vector of maize rayado fino marafivirus, CS spiroplasma and MBS phytoplasma in tropical and subtropical areas of the western hemisphere. A non-specific vector–plant interaction has been hypothesized for a single *Dalbulus*–mollicute combination, so future work must be focused on both the range and the intraspecific variation across populations of potential vectors.

MBS is associated with a phytoplasma that is member of the 16SrI '*Ca.* Phytoplasma asteris' group, subgroup 16SrI-B. CS is a disease caused by *Spiroplasma kunkelii*, a cultivable mollicute with helical morphology. Both plant pathogens are strikingly similar in their infection patterns of insects and plants (Bai *et al.*, 2004). They are restricted to plant phloem tissues, from where they are acquired by phloem-feeding insects, and subsequently invade and replicate in the cells of the insect gut and other tissues. The fact that phytoplasmas overlap plant hosts and vectors gives ample opportunities for these two mollicutes to interact and exchange genetic information. MBS phytoplasma and CS spiroplasma share plant hosts such as maize and the vector *D. maidis*.

The MBS phytoplasma genome annotation is in progress, and it will be the first full sequence available for a phytoplasma from a Gramineae host. Comparative genome studies between a '*Ca.* Phytoplasma asteris'-related strain, aster yellows witches'-broom phytoplasma (AY-WB) and *S. kunkelii* have shown that both insect-transmitting pathogens possess four proteins (PNPases for gene expression regulation; CBF for regulation of plasmid replication; cytosine deaminase involved in nucleotide metabolism; and YlxR, a RAN-binding protein) that are absent from mycoplasmas (Bai *et al.*, 2004). In contrast, *S. kunkelii* uses PTS systems for the import of fructose, glucose and trehalose, while phytoplasmas import maltose through ABC transporters. More valuable insights will derive once the MBS phytoplasma full genome sequence is available to understand pathogenicity and evolutionary relationships between phytoplasmas and spiroplasmas affecting similar plant hosts and transmitted by common insect hemipteran vectors.

Maize redness (MR) and corn reddening (CR)

CR was observed for the first time in 1957 in the middle-south Banat region of Serbia in Europe, while MR has been reported from Serbia, Romania and Bulgaria for 50 years (Jovic *et al.*, 2007). CR symptoms can be present in up to 90% of the plants during CR epidemics, and yield losses can be over 50% (Duduk and Bertaccini, 2006; Bekavac *et al.*, 2007).

Plants infected by CR have the same size and appearance as healthy ones, but the disease reduces grain filling and maize cob weight. The cobs of infected plants contain grains that are shrivelled, nutritionally poor and ripen and dry earlier than those of healthy maize. Symptoms always appear first around the borders of fields, suggesting that infection is coming from outside the field through infective vectors (Duduk and Bertaccini, 2006) and then are found randomly distributed in maize fields following no specific pattern. The first symptom consists of a red to reddish-violet colour that appears on the midrib of leaves immediately above the ear (Bekavac *et al.*, 2007). From the midribs, the discoloration extends to the basal part of the leaf margin, then to the tip of the blade, then spreads to the top leaves and finally to the leaves below the ear. Reddening then spreads to the veins of neighbouring leaves, and from these to leaf laminae and sheaths, ear husks and, after flowering, to all parts of the plant. Reddening intensity varies from lightly to heavily discoloured plants in the same field (Bekavac *et al.*, 2007) and occurs only on exposed parts, as covered parts, e.g. the internodes beneath leaf sheaths, remain green until the end of the season. When reddening involves the whole plant, it quickly wilts; the roots become necrotic and the plant dies and can be easily uprooted. Symptoms of reddening begin in the second half of July on the main leaf midrib, then they spread to the stalk and eventually affect the whole plant, reaching a peak by mid-August and the first half of September (Duduk and Bertaccini, 2006; Bekavac *et al.*, 2007). A maize population (NS 1-257 CRS) has been found as a possible source for CR resistance and breeding, showing adequate levels of available genetic variability and high heritability (Bekavac *et al.*, 2007).

Recent epiphytotics of MR reduced crop yields by 40–90% in southern Banat, Serbia (Jovic *et al.*, 2007). MR symptoms appear late in July and include reddening of the leaf midrib, followed by reddening of leaves and stalks. Some maize plants can also show red streaks along the midrib; ear development is abnormal but no dwarfing or phyllody is associated (Jovic *et al.*, 2007).

The CR phytoplasma was associated with stolbur (STOL), '*Ca.* Phytoplasma australiense', 16SrXII-A (Duduk and Bertaccini, 2006). STOL disease in Serbia, first noticed in 1949 on pepper, has its natural reservoir in *Convolvulus arvensis* (bindweed) and its most significant vector is *Hyalesthes obsoletus* Signoret, which overwinters on the roots of *C. arvensis*. Duduk and Bertaccini (2006) reported a 99% similarity in the 16S rDNA and 100% of the spacer regions between the CR phytoplasma and a group 16SrXII phytoplasma infecting pepper.

MR is also associated with the presence of a 16SrXII phytoplasma (Jovic *et al.*, 2007) for which *Reptalus panzeri* (Löw) has been identified as the insect vector (Jovic *et al.*, 2007). Current research in this area is directed at identification of the natural reservoir(s) of phytoplasma in Serbia and studying the relationship between host, phytoplasma and vector.

Sorghum bushy shoot (SBS) and grassy shoot (SGS)

Sorghum (*Sorghum* L.) is an annual grass native to tropical and subtropical regions of all continents, in addition to the south-west Pacific and Australasia. It is one of the five top cereal crops in the world, and Africa is still the largest producer. Numerous *Sorghum* species are used for food, fodder and the production of alcoholic beverages, as well as biofuels.

SGS and SBS have been associated with symptoms of yellow, white or creamy leaves, grassy shoots, bunchy shoots, witches'-brooms, abnormal tillers, stunting and floral deformation. SGS phytoplasma was found in *Sorghum stipoideum* and *Whiteochloa cymbiformis* in northern Australia (Schneider *et al.*, 1999) and identified as a member of group 16SrXI, '*Ca.* Phytoplasma oryzae'. Additional surveys of these and other grasses in northern Australia revealed a new phytoplasma within the group 16SrXI, associated with SBS (Tran-Nguyen *et al.*, 2000). Blanche *et al.* (2003) revealed that the location of the SGS phytoplasma varies in *S. stipoideum* in different parts of the plant, including leaves, stem, flowers and roots, and at different times of the year. These studies also confirmed the poor relationship between symptoms and phytoplasma presence; therefore, symptoms alone are not reliable indicators of phytoplasma presence or identity in grass hosts (Tran-Nguyen *et al.*, 2000). Additionally, two RFLP variants, resembling SGS, were identified in *Dactyloctenium aegyptium* and *Chloris inflate*, as confirmed by 16S rDNA sequence analysis, as SGS var. I and SGS var. II (Blanche *et al.*, 2003); however, they are not associated with a specific set of symptoms.

Phytoplasma Diseases of Small Grains: Oats, Barley and Wheat

Small grains comprise a large percentage of cropped land area. Oat (*Avena sativa* L.), barley (*Hordeum vulgare* L.), wheat (*Triticum aestivum* L.) and *Triticosecale* Wittm. ex A. Camus (*Triticum* L. × *Secale* L.), or triticale (polyploid resultant from cross between *Triticum* L. and rye, *Secale* L.), are economically important cereal grain crops, grown in many countries of Europe and North America. Oat and ryegrass are economically important grain crops in Lithuania, usually considered as secondary crops (Urbanaviciene *et al.*, 2007). Barley is used for beer production and livestock feed, and in some countries is the major human feed grain, as in Lithuania, where it is one of the primary food crops.

Symptoms associated with phytoplasmas are typically yellowing of leaves and spikes, general stunting, sterility and deformation of spikes, dwarfed spikes and twisted awns. Hollingsworth *et al.* (2008) described symptoms in wheat and barley plants from chlorotic leaf blotches to complete plant collapse. Many plants exhibited red to purplish blotches on leaves that turned necrotic with time. Symptoms were indistinguishable from those caused by barley yellow dwarf virus (BYDV).

Phytoplasmas of the '*Ca.* Phytoplasma asteris' 16SrI group have been the main ones associated with diseases in small grains in Lithuania and North America. Phytoplasmas have been identified in oat, common meadow-grass, barley, triticosecale, ryegrass, smooth broomgrass and tall fescues, and classified in different subgroups, including 16SrI-A, 16SrI-B, 16SrI-C and 16SrI-L (Urbanaviciene *et al.*, 2007; Hollingsworth *et al.*, 2008).

In Europe much of the lowland grassland is cultivated and occupied by a restricted number of species, mainly Gramineae plant species. In Lithuania, 16SrI phytoplasmas have been identified in *Bromopsis inermis*, *Lolium multiflorum*, *Poa pratensis* and *Festuca arundinaceae* (Urbanaviciene *et al.*, 2007), which impacts on phytoplasma disease epidemiology, since the ryegrasses, *Lolium perenne* and *L. multiflorum*, occupy about 70% of the agricultural areas together with the fescues and cocksfoot.

Although *Macrosteles laevis* (Rib.) was shown to transmit 16SrI-associated symptoms to and from graminaceous plants, no vectors have been identified for phytoplasmas affecting the Gramineae hosts previously mentioned.

However, recent studies on regional and field distributions of the 16SrI phytoplasma in Minnesota and North Dakota (Hollingsworth *et al.*, 2008) have revealed that its high incidence in small grain crops may be more dependent on the prevalence of specific crops rather than vector feeding preference. The spatial patterns of the 16SrI phytoplasma were not apparent within or between years, but suggest that vectors migrate into fields from perimeter areas, such as grass ditches and field margins.

Phytoplasma Diseases of Sugarcane

Sugarcane (*Saccharum* sp.) is a perennial grass native to warm temperate and tropical regions of the Old World. By 2005, the world's largest producer of sugarcane was Brazil, followed by India. Uses of sugarcane include the production of sugar, molasses, cane spirits such as rum and ethanol for biofuel; raw material for paper and cardboard; and eating utensils branded as 'environmentally friendly' as they are made from a by-product of sugar production.

A number of sugarcane diseases have been associated with phytoplasmas, some of which are region or country specific (sugarcane white leaf, Thailand; sugarcane Ramu stunt, Papua New Guinea).

Yellow leaf syndrome (YLS)

YLS was first reported in the 1960s from East Africa and referred to as 'yellow wilt'; later similar symptoms were reported from Hawaii. It has since been recorded from all continents, including Africa (South Africa, Mauritius and Reunion Island), North America and Australia (Rassaby *et al.*, 2004; Arocha and Jones, 2008); Asia (Thailand and India) (Viswanathan and Balamuralikrishnan, 2004; Lehrer *et al.*, 2008); the Caribbean and Central America (Cuba, Nicaragua, Guatemala, Barbados and the French West Indies) (Daugrois *et al.*, 1999); and South America (Brazil, Colombia, Peru, Ecuador, Venezuela) (Victoria-Kafure *et al.*, 1998; Alegría *et al.*, 2000; Comstock *et al.*, 2002; Izaguirre-Mayoral *et al.*, 2002; Arocha and Jones, 2008).

Symptoms generally appear in maturing canes but can also be triggered by stress. For many years, symptoms of yellow leaf were attributed to factors such as stress, insect damage, waterlogging, cool winters, nutrient deficiency, low soil fertility and restricted root growth resulting from soil compaction. Symptoms are characterized by an intense yellowing of the midrib and lamina on the abaxial surface of mature leaves; this discoloration often occurs while the lamina is still green (Arocha and Jones, 2008). The colour gradually extends to the leaf blade and is sometimes accompanied by shortening of the upper internodes, producing a fan-like appearance. In some sugarcane cultivars, leaves show a red coloration of the midrib on the adaxial surface and tissue necrosis may also eventually be observed.

YLS was first associated with a virus (sugarcane yellow leaf polerovirus, SCYLV), family Luteoviridae, which was widespread through many cane-growing countries (Arocha and Jones, 2008). The sugarcane yellows phytoplasma (SCY) has been consistently associated with YLS in South Africa, Mauritius, Cuba and recently India (Arocha and Jones, 2008; Gaur *et al.*, 2008). SCY has been identified as a member of the 16SrIII, 'X-disease' phytoplasma group in South Africa and Mauritius (Arocha and Jones, 2008), and the 16SrXII group in India (Gaur *et al.*, 2008). However, further studies are required to test these hypotheses. In Cuba, YLS was associated with group 16SrI '*Ca.* Phytoplasma asteris' (Arocha *et al.*, 1999) throughout the country, and later with '*Ca.* Phytoplasma graminis' in the western and central regions (Arocha *et al.*, 2005b).

Both SCYLV and SCY have been associated with latent infection under field conditions, and mixed infections SCYLV/SCY have been shown to occur naturally (Arocha and Jones, 2008), where the presence of both SCYLV and SCY has a synergistic effect. Studies in Australia revealed a poor correlation between phytoplasmas and YLS symptoms (Tran-Nguyen *et al.*, 2000); however, in fact, phytoplasmas have been more consistently detected in cane with symptoms of YLS (Arocha and Jones, 2008).

Only one vector, *Saccharosydne saccharivora* (Westwood), has been proven for SCY group 16SrI in Cuba (Arocha *et al.*, 2005a). In addition, a species of the genus *Cedusa* was found as a putative vector of '*Ca.* Phytoplasma graminis' (Arocha *et al.*, 2005b).

As phloem-restricted organisms in a vegetatively propagated crop, both SCYLV and SCY can be readily spread to new locations through infected stem cuttings if suitable precautions are not taken. For SCYLV, this includes hot-and-cold-water treatment. Both SCYLV and SCY were eliminated from infected cane by tissue culture of callus derived from leaf rolls (Arocha and Jones, 2008). However, SCYLV is more readily eliminated than SCY when meristematic tissue culture is used; hence rigorous indexing of progeny must be carried out.

There has been little research on the epidemiology of SCYL, but it is apparent that sugarcane germplasm exchange has been a major contributor to its spread. In Cuba, a number of other plant species have been found to act as reservoirs for SCYP, including *Cynodon dactylon*, *Conyza canadensis*, *Sorghum halepense* and *Macroptilium lathyroides* (Arocha *et al.*, 2005b).

Grassy shoot (SCGS) and white leaf (SCWL)

Sugarcane white leaf (SCWL) and sugarcane grassy shoot (SCGS) occur throughout Asia and cause significant losses. Together with sugarcane green grassy shoot (SCGGS), they are considered the most economically damaging of the sugarcane phytoplasmas reported in Asia.

SCGS disease has been reported to occur in India, Bangladesh, Malaysia, Nepal and Pakistan, whereas SCWL is predominant in Taiwan, Sri Lanka, Japan and Thailand (Nakashima *et al.*, 1994). These phytoplasmas have not been detected in Australian sugarcane but SCWL-related strains have been found in grasses in northern Australia (Tran-Nguyen *et al.*, 2000; Blanche *et al.*, 2003). Both SCWL and SCGS cause stunting of infected plants, profuse tillering and chlorotic stripes of the leaves. SCWL symptoms consist of abnormal tillering, side shoots on the upper part of infected stalks and, most characteristically, bleached white or striped leaf blades. The most common foliar symptoms associated with SCGS are narrowing and partially or almost totally chlorotic leaf lamina, excessive tillering and witches'-broom-type symptoms by producing a large number of tillers. Severely infected younger plants that appear as yellowish or whitish rosettes of grass may die eventually.

SCWL, SCGS and SCGGS have been associated with distinct phytoplasma strains within the16SrXI '*Ca.* Phytoplasma oryzae', subgroup 16SrXI-B (Marcone *et al.*, 2004; Ariyarathna *et al.*, 2007). Very little or no information on SCWL or SCGS ribosomal, *tuf* or *sec* genes is available, and that on 16S rDNA is very poor, which has hampered studies of their phylogenetic relationships and genetics. In Thailand, SCGS was thought to be an early developmental stage of SCWL and caused by the same phytoplasma; however, 16S rDNA sequence analyses revealed that SCWL and SCGS, although within the same phylogenetic group, are two distinct phytoplasmas. The Indian isolate of SCGS phytoplasma shares very high (>99%) 16S rDNA sequence identity with other 16SrXI phytoplasmas, including SCWL. However, in studies based on the 16S–23S rRNA spacer region sequences, two isolates sharing 79 and 84% of identity, respectively, with 16SrXI members have been reported as two new phytoplasmas associated with significant losses in sugarcane (Nasare *et al.*, 2007).

Matsumuratettix hiroglyphicus (Matsumura) and *Yamatotettix flavovittatus* Matsumura are the known vectors of SCWL (Hanboonsong *et al.*, 2002, 2006). In India, the leafhopper *Proutista moesta* (Westwood) was shown to transmit SCGS, and recently *Deltocephalus vulgaris* Dash and Viraktamah has been identified as a potential vector for SCGS (Srivastava *et al.*, 2006). The transmission efficiency of *M. hiroglyphicus* (55%) is higher than that of *Y. flavovittatus* (45%) (Hanboonsong *et al.*, 2006). Populations of these two species peak at different times of the year and therefore complement each other in the transmission of the SCWL phytoplasma. Transovarial transmission has been shown for *M. hiroglyphicus* (Hanboonsong *et al.*, 2002) and confirmed it as a reservoir for the SCWL phytoplasma. This has important implications for the control of the disease, because there are no known alternative host plants for SCWL, so management of the disease will necessarily require the control of both *Y. flavovittatus* and *M. hiroglyphicus*.

Tissue culture has been demonstrated as a feasible method for the maintenance and '*in vivo*' investigation of the SCWL phytoplasma (Wongkaew and Fletcher, 2004), as both symptoms and the phytoplasma persist for over 6 years. Disease control depends on the management of insect vectors, traditional methods of thermotherapy and elimination of the infection source. The identification of a *Wolbachia* strain in *M. hiroglyphicus* could offer a new approach for the symbiotic control of SCWL. There are still no consistent reports on alternative hosts for SCWL or SCGS. Weeds have been suspected as reservoirs for these two phytoplasmas, from which healthy cane plants could be reinfected (Nakashima *et al.*, 1994), but this has never been proven (Tran-Nguyen *et al.*, 2000; Blanche *et al.*, 2003).

In the cane-growing areas of Asia and Australia, a number of grasses with white leaf symptoms have been shown to be infected by phytoplasmas related but not identical to those causing SCWL and SCGS. These plants include Bermuda grass (*C. dactylon*), crowfoot grass (*D. aegyptium*), Brachiaria grass (*Brachiaria distachya*), native annual sorghum (*S. stipoideum*), creeping panic (*W. cymbiformis*), *Whiteochloa biciliata* and *Dactyloctenium radulans* (Nakashima *et al.*, 1994; Schneider *et al.*, 1999; Tran-Nguyen *et al.*, 2000; Blanche *et al.*, 2003; Jung *et al.*, 2003). None of these SCWL- or SCGS-related strains have been transmitted to sugarcane (Blanche *et al.*, 2003). Perhaps these phytoplasmas are not able to infect sugarcane, or insect vectors that could transmit these phytoplasmas to sugarcane are not found in these particular habitats.

Ramu stunt (SCRS)

Sugarcane Ramu stunt (SCRS) disease of sugarcane is only found in Papua New Guinea (PNG). It was responsible for large yield losses in commercial sugarcane varieties (interspecific hybrids of *Saccharum* spp.) in the Ramu Valley, north-east PNG, during the late 1980s, and poses a major disease threat to the neighbouring sugar industries in Australia and Indonesia.

Typical symptoms start as short, irregular streaks or flecks, pale to creamy green in colour (Suma and Jones, 2000). As symptoms develop, the streaks become yellowish-green in colour, continuous or interrupted, and interspersed by apparently healthy green tissue. Streaks can vary from several millimetres in length to run the full length of the leaf blade and range in width from 2 to 5 mm or more. Leaves are short, stiff and erect, and become senescent prematurely. Leaf symptoms differ considerably, depending on the variety affected. Diseased canes are thinner than healthy canes. Stools are severely stunted and there is progressive death of stalks. Diseased stools ratoon poorly. In the field, infection of a susceptible variety can lead to total ratoon failure. Root systems are severely reduced and stunted.

SCRS has been consistently associated with a phytoplasma of group 16SrXI, '*Ca.* Phytoplasma oryzae' (Suma and Jones, 2000), and a delphacid planthopper, *Eumetopina flavipes* Muir, has been found as the vector (Kuniata *et al.*, 1994). More information regarding the full length of the 16S rDNA is required for the complete identification and characterization of the SCRS phytoplasma.

Control of SCRS can be achieved through the use of resistant varieties, but where these are not available vector management and elimination of the infection source are still being applied. Symptoms similar to those of SCRS have also been seen on *Imperata cylindrica* L. (Suma and Jones, 2000); however, more evidence is required to identify this plant species definitively as a SCRS reservoir. Viral particles have been reported in SCRS-infected canes but their role, if any, in the pathology of SCRS is uncertain.

Phytoplasma Diseases of Forage Grasses

Napier or elephant grass (*Pennisetum purpureum*) is an important grass in smallholder production systems in East Africa, increasingly associated with intensive (zero grazing) and semi-intensive dairy cattle production systems to meet the increasing demand for milk.

Napier grass stunt (NGS)

One of the most important threats to the productivity of Napier grass is a stunt disease (NGS) associated with phytoplasmas. NGS was first reported in 1997 from the Bungoma district of Kenya bordering Uganda (Jones *et al.*, 2004), then Ethiopia (Jones *et al.*, 2007) and Uganda (Nielsen *et al.*, 2007). It spreads quickly and covers several districts of western Kenya, causing serious economic losses in the smallholder dairy industry.

NGS symptoms include foliar yellowing, little leaves, bushy appearance, yellow to purple streaking, proliferation of tillers and shortening of internodes, to the extent that clumps are severely stunted and have a low biomass yield. Affected shoots become pale yellow-green in colour and seriously dwarfed. Often the whole stool is affected, with complete loss in yield and eventual death. Many smallholders have lost up to 100% of their Napier crop and are forced to reduce their number of animals or purchase fodder from the local market.

Phytoplasmas of group 16SrXI, '*Ca.* Phytoplasma oryzae' have been associated with NGS in Kenya (Jones *et al.*, 2004) and Uganda (Nielsen *et al.*, 2007), and of group 16SrIII, Western-X-disease, particularly subgroup A, in Ethiopia (Jones *et al.*, 2007; Arocha *et al.*, 2009). Both 16SrXI and 16SrIII phytoplasmas are able to cause identical symptoms in Napier grass in different areas within the East African region, which indicates that geographic location and vector distribution is an important factor for NGS phytoplasmas and that these cannot be distinguished by symptoms alone.

Research into the epidemiology of NGS is in its early stages. *Leptodelphax dymas* Fennah and an *Exitianus* sp. were identified as potential vectors, and *Medicago sativa* and *C. dactylon* could act as alternative hosts for the NGS phytoplasma in Ethiopia (Arocha *et al.*, 2009). Transmission studies are needed to confirm the vector status of either *L. dymas* or the *Exitianus* sp. and their relationship with alternative hosts. Investigations on the identification of the hemipteran vector of the NGS phytoplasma in Kenya are still ongoing.

As Napier grass is vegetatively propagated, NGS presents a serious phytosanitary problem. Roguing of infected plants can help lower infection pressure by providing fewer NGS plants for vectors to feed on. An alternative control strategy would be to identify Napier grass genotypes that are resistant to the pathogen or less attractive to vectors; such projects are already under way in Africa.

Phytoplasma Diseases of Other Gramineae

Bermuda grass white leaf (BGWL)

Bermuda grass (*C. dactylon)* is a creeping, stoloniferous grass that is native to many subtropical and tropical regions but probably originated in India. BGWL was first reported in Taiwan and is known to occur in Sudan and several Asian countries (Rao *et al.*, 2008; Snehi *et al.*, 2008), Italy (Marcone *et al.*, 1997) and Cuba (Arocha *et al.*, 2005c). BGWL is a destructive disease of Bermuda grass; symptoms are characterized by an extensive chlorosis of the leaves, proliferation of auxiliary shoots, bushy growth habit, small leaves, shortened stolons and rhizomes, stunting and death of the plant (Marcone *et al.*, 2004). Similar diseases, cynodon white leaf (CWL) and carpet grass (*Axonopus compressus*) white leaf (CGWL), have been reported in Australia (Tran-Nguyen *et al.*, 2000; Blanche *et al.*, 2003).

The phytoplasma associated with BGWL belongs to the 16SrXIV '*Ca.* Phytoplasma cynodontis' group. A number of white leaf diseases of other grasses are associated with phytoplasmas that also fall within the group; these include *B. distachya* (Brachiaria grass), *Poa annua* (annual blue grass) and *D. aegyptum* (crowfoot grass) (Lee *et al.*, 2000; Blanche *et al.*, 2003). Phytoplasmas related to BGWL have been associated with diseases of date palm (*Phoenix dactylifera*) in North Africa, where they cause white tip dieback (WTD) and slow-decline (SD); however, they cannot be classified as the same '*Ca.* Phytoplasma' species as '*Ca.* Phytoplasma cynodontis', since insufficient sequence data are available (Firrao *et al.*, 2005).

The BGWL phytoplasma is estimated to have the smallest genome size (530 kbp) of all phytoplasmas and may have the smallest chromosome known for any living cell (Lee *et al.*, 2000). However, in spite of the very limited 16S or 16S–23S rDNA information for phytoplasmas of this group, evidence suggests that these phytoplasmas must have undergone a dramatic degenerative evolution (Lee *et al.*, 2000). The loss of genes has led to the selection of a phytoplasma that can survive successfully in a complex of grass and crop species within the Gramineae.

Recent work in Iran has shown that the leafhopper *Exitianus capicola* (Stål), one of the main species of the Bermuda grass fauna, is both a natural and experimental vector of the BGWL phytoplasma (Salehi *et al.*, 2009). This leafhopper was previously reported as a vector of a phytoplasma infecting *Limonium* hybrid crops in Israel (Weintraub *et al.*, 2004).

'*Ca.* Phytoplasma cynodontis' is distantly related to phytoplasmas associated with sugarcane and sorghum diseases, which belong to the 16SrXI group. BGWL-related phytoplasmas have been identified in two dicotyledons, *Cirsium arvensis* (Canada thistle) and *Galactia tenuifolia*, and in the leafhopper *Psammotettix cephalotes* (Herrich-Schaeffer) (Marcone *et al.*, 2004). BGWL isolates found in Italy are identical to *B. distachya* phytoplasma, differing in only four nucleotide positions in their 16S rDNA from BGWL Thailand isolates (Firrao *et al.*, 2005), and can be distinguished from members of the group 16SrXI '*Ca.* Phytoplasma oryzae' based on RFLP banding patterns.

Summary

The importance of grasses and cereals to global food production systems means that diseases such as those associated with phytoplasmas which are vectored by insects pose a particular threat to food security and sustainability of farming systems. Their ability to cause epidemics in crops such as rice and sugarcane and our difficulty in controlling them should be noted as a high priority for agricultural research. For sugarcane in particular, the influence of human activities has spread the crop around the world, mainly at a time when there was little or no appreciation of latent infection and of pathogens being distributed along with the plants.

Grasses support a rich insect fauna, such as leafhoppers, planthoppers and psyllids, some of which have the capacity to transmit phytoplasmas. There is a growing acceptance that climate changes resulting from global warming will drive the spread of vectors (see Foissac and Wilson, Chapter 17, this volume), along with phytoplasmas, to new plant hosts and countries. Putative vectors that are not currently able to overwinter on grasses may be able to survive during warmer winters, which will have implications for crop protection strategies.

The future will offer many research opportunities for those working with phytoplasma diseases of grasses. None of the genomes of phytoplasmas which infect grasses have been completely sequenced; the maize bushy stunt phytoplasma genome is expected to be the first full sequence available and that will contribute significantly to our attempts to devise sustainable disease management strategies.

References

Alegría, O.M., Chatenet, M. and Girard, J.C. (2000) First report of sugarcane yellow leaf virus in Peru. *Plant Disease* 84, 1342.

Ariyarathna, H., Everard, J. and Karunanayake, E.H. (2007) Diseased sugarcane in Sri Lanka is infected with sugarcane grassy shoot and/or sugarcane white leaf phytoplasma. *Australasian Plant Disease Notes* 2, 123–125.

Arocha, Y. and Jones, P. (2008) Phytoplasmas affecting sugarcane and papaya in Cuba. In: Harrison, N., Rao, G.P. and Marcone, C. (eds) *Characterization, Diagnosis and Management of Phytoplasmas*. Studium Press LLC, Texas, pp. 249–273.

Arocha, Y., Peralta, E.L., González, L. and Jones, P. (1999) First report of a virus and a phytoplasma pathogen associated with sugarcane yellow leaf syndrome in Cuba. *Plant Disease* 83, 1171.

Arocha, Y., López, M., Piñol, B., Fernández, M., Picornell, B., Almeida, R., Palenzuela, I., Wilson, M.R. and Jones, P. (2005a) Transmission of sugarcane yellow leaf phytoplasma by the delphacid leafhopper *Sacharosydne saccharivora*, a new vector of sugarcane yellow leaf disease. *Plant Pathology* 54, 634–642.

Arocha, Y., López, M., Piñol, B., Fernández, M., Picornell, B., Almeida, R., Palenzuela, I., Wilson, M.R. and Jones, P. (2005b) '*Candidatus* Phytoplasma graminis' and '*Candidatus* Phytoplasma caricae', two novel phytoplasmas associated with diseases of sugarcane, weeds and papaya in Cuba. *International Journal of Systematic and Evolutionary Microbiology* 55, 2451–2463.

Arocha, Y., Piñol, B., Palenzuela, I., Almeida, R. and Jones, P. (2005c) First report of a phytoplasma associated with Bermuda-grass white leaf disease in Cuba. *Plant Pathology* 54, 233.

Arocha, Y., Zerfy, T., Abebe, G., Proud, J., Hanson, J., Wilson, M. and Jones, P. (2009) Identification of potential vectors and alternative plant hosts for the phytoplasma associated with Napier Grass Stunt disease in Ethiopia. *Journal of Phytopathology* 157, 126–132.

Bai, X., Zhang, J., Holford, I. and Hogenhout. S. (2004) Comparative genomics identifies genes shared by distantly related insect-transmitted plant pathogenic mollicutes. *FEMS Microbiology Letters* 235, 249–258.

Bedendo, I.P., Davis, R.E. and Dally, E.L. (2000) Detection and identification of the maize bushy stunt phytoplasma in corn plants in Brazil using PCR and RFLP. *International Journal of Pest Management* 46, 73–76.

Bekavac, G., Purar, B. and Jockovic, D. (2007) Corn reddening: the disease and breeding for resistance. *Journal of Plant Pathology* 89, 397–404.

Blanche, K.R., Tran-Nguyen, L. and Gibb, K.S. (2003) Detection, identification and significance of phytoplasmas in grasses in northern Australia. *Plant Pathology* 52, 505–512.

Comstock, J.C., Pena, M., Vega, J., Fors, A. and Lockhart, B. (2002) Report of sugarcane yellow leaf virus in Ecuador, Guatemala and Nicaragua. *Plant Disease* 86, 74.

Daugrois, J.H., Jean-Baptiste, I., Lockhart, B.E.L., Irey, M.S., Chatenet, M., Rott, P. (1999) First report of Sugarcane Yellow Leaf Virus in the French West Indies. *Plant Disease* 83, 588.

Duduk, B. and Bertaccini, A. (2006) Corn with symptoms of reddening: new host of stolbur phytoplasma. *Plant Disease* 90, 13–19.

Firrao, G., Gibb, K. and Streten, C. (2005) Short taxonomic guide to the genus '*Candidatus* Phytoplasma'. *Journal of Plant Pathology* 87, 249–263.

Gaur, R.K., Raizada, R. and Rao, G.P. (2008) Sugarcane yellow leaf phytoplasma associated for the first time with sugarcane yellow leaf syndrome in India. *Plant Pathology* 57, 772.

Gomes, A., Jardim, S., Teixeira-Guimarães, C., de Souza, E. and de Oliveira, E. (2004) Genetic variability of Brazilian phytoplasma and spiroplasma isolated from maize plants. *Pesquisa Agropecuaria Brasiliera* 39, 61–65.

Hanboonsong, Y., Panyim, S. and Damak, S. (2002) Transovarial transmission of sugarcane white leaf phytoplasma in the insect vector *Matsumuratettix hiroglyphicus* (Matsumura). *Insect Molecular Biology* 11, 97–103.

Hanboonsong, Y., Ritthison, W., Choosai, C. and Sirithorn, P. (2006) Transmission of sugarcane white leaf phytoplasma by *Yamatotettix flavovittatus*, a new leafhopper vector. *Journal of Economic Entomology* 99, 1531–1537.

Hibino, H., Jonson, G.B. and Sta Cruz, F.C. (1987) Association of mycoplasma-like organisms with rice orange leaf in the Philippines. *Plant Disease* 71, 792–794.

Hollingsworth, C.R., Atkinson, L.M., Samac, D.A., Larsen, J.E., Motteberg, C.D., Abrahamson, M.D., Glogoza, P. and MacRae, I.V. (2008) Region and field level distributions of aster yellows phytoplasma in small grain crops. *Plant Disease* 92, 623–630.

Izaguirre-Mayoral, M.L., Carballo, O., Alceste, C., Romano, M. and Nass, H.A. (2002) Physiological performance of asymptomatic and yellow leaf syndrome-affected sugarcanes in Venezuela. *Journal of Phytopathology* 150, 13–19.

Jones, P., Devonshire, J., Holman, T. and Ajanga, S. (2004) Napier grass stunt: a new disease associated with a 16SrXI group phytoplasma in Kenya. *Plant Pathology* 53, 519.

Jones, P., Arocha, Y., Zerfy, T., Proud, J., Abebe, G. and Hanson, J. (2007) A stunting syndrome of Napier grass in Ethiopia is associated with a 16SrIII group phytoplasma. *Plant Pathology* 56, 345.

Jovic, J., Cvrkovic, T., Mitrovic, M., Kranjajic, S., Redinbaugh, M., Pratt, R.C., Gingery, R.E., Hogenhout, S. and Tosevski, I. (2007) Role of stolbur phytoplasma and *R. panzeri* in the epidemiology of maize redness in Serbia. *European Journal of Plant Pathology* 118, 85–89.

Jung, H.Y., Sawayanagi, T., Wongkaew, P., Kakizawa, S., Nishigawa, H., Wei, W., Oshima, K., Miyata, S., Ugaki, M., Hibi, T. and Namba, S. (2003) '*Candidatus* Phytoplasma oryzae', a novel phytoplasma taxon associated with rice yellow dwarf disease. *International Journal of Systematic and Evolutionary Microbiology* 53, 1925–1929.

Kakizawa, S., Oshima, K. and Namba, S. (2006) Diversity and functional importance of phytoplasma membrane proteins. *Trends in Microbiology* 14, 254–256.

Kuniata, L.S., Young, G.R., Pais, E., Jones, P. and Nagaraja, H. (1994) Preliminary observations on *Eumetopina* sp. (Hemiptera: Delphacidae) as a vector of Ramu stunt disease of sugarcane in Papua New Guinea. *Journal of Australian Entomological Society* 33, 185–186.

Lee, I.M., Davis, R.E. and Gundersen-Rindal, D.E. (2000) Phytoplasma: phytopathogenic mollicutes. *Annual Review of Microbiology* 54, 221–255.

Lehrer, A.T., Kusalwong, A. and Komor, E. (2008) High incidence of sugarcane yellow leaf virus (SCYLV) in sugar plantations and germplasm collections in Thailand. *Australasian Plant Disease Notes* 3, 89–92.

Marcone, C., Ragozzino, A. and Seemüller, E. (1997) Detection of Bermuda grass white leaf disease in Italy and genetic characterization of the associated phytoplasma by RFLP analysis. *Plant Disease* 81, 862–866.

Marcone, C., Schneider, B. and Seemüller, E. (2004) '*Candidatus* Phytoplasma cynodontis', the phytoplasma associated with Bermuda grass white leaf disease. *International Journal of Systematic and Evolutionary Microbiology* 54, 1077–1082.

Martini, M., Lee, I.M., Bottner, D., Zhao, Y., Botti, S., Bertaccini, A., Harrison, N., Carraro, L., Marcone, C., Khan, A. and Osler, R. (2007) Ribosomal protein gene-based phylogeny for finer differentiation and classification of phytoplasmas. *International Journal of Systematic and Evolutionary Microbiology* 57, 2037–2051.

Nakashima, K., Kato, S., Iwanami, S. and Murata, N. (1993) DNA probes reveal relatedness of rice yellow dwarf mycoplasmalike organisms (MLOs) and distinguish them from other MLOs. *Applied and Environmental Microbiology* 59, 1206–1212.

Nakashima, K., Chaleeprom, W., Wongkaew, P. and Sirithorn, P. (1994) Detection of mycoplasma-like organisms associated with white leaf disease of sugarcane in Thailand using DNA probes. *JIRCAS* 1, 57–67.

Nasare, K., Yadav, A., Singh, A., Shivasharanappa, K., Nerkar, Y. and Reddy, V. (2007) Molecular and symptoms analysis reveal the presence of new phytoplasmas associated with sugarcane grassy shoot disease in India. *Plant Disease* 91, 1413–1418.

Nielsen, S.L., Ebong, C., Kabirizi, J. and Nicolaisen, M. (2007) First report of a 16SrXI stunt disease in Uganda. *Plant Pathology* 56, 1039.

Rao, G.P., Srivastava, S., Gupta, P.S., Sharma, S.R., Singh, A., Singh, S., Singh, M. and Marcone, C. (2008) Detection of sugarcane grassy shoot phytoplasma infecting sugarcane in India and its phylogenetic relationships to closely related phytoplasmas. *Sugarcane Technology* 10, 74–80.

Rassaby, L., Girard, J.C., Lemaire, O., Costet, L., Irey, M.S., Kodja, H., Lockhart, B. and Rott, P. (2004) Spread of sugarcane yellow leaf virus in sugarcane plants and fields on the island of Réunion. *Plant Pathology* 53, 117–125.

Salehi, M., Izadpanah, K., Siampour, M. and Taghizadeh, M. (2009) Molecular characterization and transmission of Bermuda grass white leaf phytoplasma in Iran. *Journal of Plant Pathology* (in press).

Schneider, B., Padovan, A., De La Rue, S., Eichner, R., Davis, R., Bernuetz, A. and Gibb, K. (1999) Detection and differentiation of phytoplasmas in Australia: an update. *Australian Journal of Agricultural Research* 50, 333–342.

Snehi, S.K., Khan, M.S., Raj, S.K., Mall, S., Singh, M. and Rao, G.P. (2008) Molecular identification of '*Candidatus* Phytoplasma cynodontis' associated with Bermuda grass white leaf disease in India. *Plant Pathology* 57, 770–771.

Srivastava, S., Singh, V., Gupta, P.S., Sinha, O.K. and Baitha, A. (2006) Nested PCR assay for the detection of sugarcane grassy shoot phytoplasma in the vector *Deltocephalus vulgaris*: a first report. *Plant Pathology* 55, 25–28.

Suma, S. and Jones, P. (2000) Ramu stunt. In: Rott, P., Bailey, R.A., Comstock, J.C., Croft, B.J. and Saumtally, A.S (eds) *A Guide to Sugarcane Diseases*. CIRAD and ISSCT, France, pp. 226–230.

Tran-Nguyen, L., Blanche, K.R., Egan, B. and Gibb, K.S. (2000) Diversity of phytoplasmas in northern Australian sugarcane and other grasses. *Plant Pathology* 49, 666–679.

Urbanaviciene, L., Jomantiene, R., Valiunas, D. and Davis, R.E. (2007) Molecular identification of 16SrI-A, 16SrI-B, 16SrI-C, and 16SrI-L subgroups of phytoplasmas in gramineous plants in Lithuania. *Bulletin of Insectology* 60(2), 127–128.

Victoria-Kafure, J.I., Garcés-Obando, F.F., Guzmán-Romero, M.L. and Ángel, F. (1998) Síndrome de la hoja amarilla en Colombia ScYLV (sugarcane yellow leaf virus). Carta Trimestral. *Cenicaña* 20(2–3), 3–7.

Viswanathan, R. and Balamuralikrishnan, M. (2004) Detection of sugarcane yellow leaf virus, the causal agent of yellow leaf syndrome in sugarcane by DAS-ELISA. *Archives of Phytopathology Plant Protection* 37, 169–176.

Weintraub, P.G. and Beanland, L.A. (2006) Insect vectors of phytoplasmas. *Annual Review of Entomology* 51, 91–111.

Weintraub, P.G., Pivonia, S., Rosner, A. and Gera, A. (2004) A new disease in *Limonium* hybrids. II. Insect vectors. *Horticultural Science* 39, 1060–1061.

Wongkaew, P. and Fletcher, J. (2004) Sugarcane white leaf phytoplasma in tissue culture: long-term maintenance, transmission, and oxytetracycline remission. *Plant Cell Report* 23, 426–434.

11 Phytoplasma Epidemiology: Grapevines as a Model

Fiona E. Constable

Department of Primary Industries, Australia

Introduction

Epidemiology within plant pathology was defined by Kranz (1990) as 'the science of populations of pathogens in populations of host plants, and the diseases resulting therefrom under the influence of the environment and human interferences'. The study of epidemiology aims to clearly describe the disease triangle representing the interaction between the host, the pathogen and the environment and use this information to develop control strategies. Control of phytoplasma-associated diseases relies on prevention rather than cure. Consequently, the epidemiology of many phytoplasma-associated diseases has been well studied, particularly in high-value crops such as pome and stone fruit and grapevines.

Identifying and characterizing the pathogen is often the first step in describing the epidemiology of a phytoplasma-associated disease, particularly as specific species or strains may have a specific biology that is different from other similar or closely related phytoplasmas. These differences may be associated with disease expression and development, host range and vectors. Studying the relationships between specific species or strains of phytoplasmas can assist in identifying their origin and means of dispersal.

The spatial and temporal patterns of phytoplasma-associated diseases have been used to improve understanding of their ecological and biological properties. Many statistical tools exist to quantify disease in space and time, and an excellent guide is provided by Madden *et al.* (2007). Specific patterns can indicate from where the disease has originated and assist in identifying the location of alternative hosts and vectors of phytoplasma. Spatial and temporal analyses can also assist in identifying some of the behavioural characteristics of a vector, such as movement and feeding in the affected crop.

The epidemiology of phytoplasma-associated diseases is intrinsically linked to the biology of their insect vectors. Identification of specific vectors, their behaviours and their preferred host plants is vital in understanding the epidemiology of phytoplasma-associated diseases. Linking the spatial and temporal patterns of diseases with the spatial and temporal patterns of a vector can assist in identifying times of high and low risk for movement to or within the crop and subsequent infection events. This information can be used to develop management strategies aimed at controlling the vector, alternative or preferred hosts for the vector and/or alternative hosts for the phytoplasma, and subsequent spread of disease.

Grapevine yellows disease epidemiology

Extensive epidemiological studies have been carried out for four of the grapevine yellows (GY) diseases, including flavescence dorée (FD), bois noir (BN), Australian grapevine yellows (AGY) and North American grapevine yellows (NAGY). Although the symptomatology of the GY diseases is nearly identical regardless of the location, the epidemiology associated with each phytoplasma species, and even amongst strains of the same phytoplasma, can vary. Consequently, these diseases provide a unique opportunity to compare and highlight biological attributes that are important to the epidemiology of phytoplasma-associated disease.

A summary of some of the important epidemiological information available for each of the GY diseases reported worldwide is presented in Table 11.1 (after Boudon-Padieu, 2003, 2005). At least ten phytoplasmas have been associated with diseases of grapevines in many viticultural regions worldwide. All GY diseases exhibit most of the following symptoms:

- irregular yellowing in white varieties or reddening in red varieties;
- the yellow leaf tissue may become necrotic;
- backward curling of the leaves;
- overlapping of leaves on affected shoots;
- rows of black pustules can develop on the green bark of affected shoots;
- tips of affected shoots may die and shoots may die back;
- affected shoots fail to harden off and remain rubbery;
- flowers on affected shoots may abort; and
- berries may shrivel and whole clusters of berries can be subject to early drying.

Grapevine varieties differ in severity of expression of GY diseases. Chardonnay and Riesling are more severely affected than many other varieties, whilst some rootstocks may be infected by phytoplasmas but do not show disease (Boudon-Padieu, 1999). Significant reductions in yields have been reported for some GY diseases (Caudwell, 1964; Magarey and Wachtel, 1986b). Remission and recovery from disease are reported for FD, BN and AGY, but grapevines affected by NAGY often decline and die.

Table 11.1. Current status of molecular characterization, biology and vectors of phytoplasmas associated with grapevine yellows diseases (after Boudon-Padieu, 2003, 2005).

Grapevine yellows disease	Phytoplasma name	Ribosomal group (subgroup)	Known insect vector to grapevine	Preferred host plants of vector	Alternative hosts of the phytoplasma	Occurrence
Flavescence dorée	Flavesence dorée (FD; '*Candidatus* Phytoplasma vitis'*)	16SrV (-C, -D) or EY	*Scaphoideus titanus* Ball	*Vitis* sp.	*Clematis alba*	France, Italy, Spain, Serbia, Slovenia, Switzerland
Palatinate grapevine yellows	Palatinate grapevine yellows (PGY)	16SrV or EY	*Oncopsis alni* Schrank	*Alnus glutinosa*		Germany
Bois noir, Legno nero, Vergilbungskrankheit Schwarzholzkrankheit	Stolbur (STOL, '*Candidatus* Phytoplasma solani'*)	16SrXII-A or stolbur	*Hyalesthes obsoletus* Signoret	*Convolvulus arvensis*, *Urtica dioica*, *Ranunculus* spp., *Solanum* spp., *Lavandula* spp.	*C. arvensis*, *U. dioica*, *Ranunculus* spp., *Solanum* spp., *Lavandula* spp.	Europe, Israel, Lebanon
Australian grapevine yellows	'*Candidatus* Phytoplasma australiense'	16SrXII-B	ND†	ND	*Maireana brevifolia*	Australia
Australian grapevine yellows	Tomato big bud (TBB)	16SrII -D	ND	ND		Australia

Buckland Valley grapevine yellows	Buckland Valley grapevine yellows (BVGY)	16SrI-related or AY-	ND	ND		Australia
Grapevine yellows	Aster yellows	16SrI (-B, -C) or AY	ND	ND		Italy, Chile, Tunisia
North American grapevine yellows (NAGY)	Virginia grapevine yellows I (NAGY I)	16SrI-A or AY	ND	ND	*Vitis* spp. Various herbaceous hosts	Virginia (USA)
	Western X Virginia grapevine yellows III (NAGYIII)	16SrIII-I or WX	ND	ND	*Vitis* spp. *Prunus* spp.	New York (USA) Virginia (USA)
Grapevine yellows	'*Candidatus* Phytoplasma fraxini'	16SrVII	ND	ND	ND	Chile
Grapevine yellows	X-disease	16SrIII	ND	ND	ND	Italy, Israel

*Suggested *Candidatus* phytoplasma names; however, the species are still to be described.
†ND = not determined.

Case Study 1: Flavescence Dorée – a Closed Epidemiological System

Background

Flavescence dorée was first described in 1957 from the Gascoyne region of France (Caudwell, 1957). The disease has subsequently been observed in various regions of France, including Corsica, and other European countries, including Italy, Portugal, Spain, Serbia, Slovenia and Switzerland (Boudon-Padieu, 2005). The association between phytoplasmas and FD disease was demonstrated when phytoplasma bodies were observed in grapevines, bean and chrysanthemum vascular tissue and in the salivary glands of the vector *Scaphoideus titanus*, which were experimentally infected after feeding on FD-affected grapevines (Caudwell *et al.*, 1971).

The FD is a member of the 16SrV taxonomic group. Three strain clusters of FD phytoplasma have been recognized, based on sequence analysis of the 16S rRNA, *secY*, *map* and *uvrB-degV* genes (Arnaud *et al.*, 2007). These strain clusters include FD-1, which comprises isolate FD70 (FD2000) and has low genetic variability; FD-2, which comprises isolates FD92 and FD-D and has no genetic variability; and FD-3, which comprises isolate FD-C and has more genetic variability (Arnaud *et al.*, 2007).

Each of the strains differs in its geographic range and possible region of origin and perhaps in its means of dispersal (Arnaud *et al.*, 2007). Strain cluster FD-1 is restricted to France, with higher incidence in south-western regions, and is also present in Piemonte and Lombardia regions of Italy. Strain cluster FD-1 is not as frequently detected as FD-2. FD-2 is the most commonly detected strain cluster in France and is also detected in Italy and Spain. Strain cluster FD-3 has been detected in Italy, particularly in the north-eastern regions, Serbia and Slovenia, and its occurrence coincides with the presence of *Clematis vitalba*, which is an alternative host for this FD strain (Filippin *et al.*, 2007). Interestingly FD-3-infected *C. vitalba* plants were found in regions where FD has not yet been detected in grapevines. So far, *S. titanus* has not been shown to transmit the phytoplasma from *C. vitalba* to grapevine and it is possible that another vector exists that occasionally feeds on and infects grapes and the phytoplasma only spreads in vineyards once *S. titanus* appears.

Phytoplasmas associated with Palatinate GY (PGY) in Germany and alder yellows (AldY) also belong to the 16SrV group and have a high sequence similarity with strains of FD (Arnaud *et al.*, 2007). It has been proposed that FD, PGY and AldY have a common origin in Europe.

Transmission of the flavescence dorée phytoplasma

FD is transmitted persistently by the leafhopper *S. titanus* (Schvester *et al.*, 1969). FD principally spreads in a 'closed system', i.e. after the introduction of FD into vineyards, either through infected planting material or from outside sources through infectious vectors, the phytoplasma is spread grapevine to grapevine by *S. titanus*. *S. titanus* lives and feeds exclusively on grapevine in Europe and can transmit all three FD strain clusters from grapevine to grapevine, often resulting in epidemics of the disease (Boudon-Padieu, 2005; Bressan *et al.*, 2006).

Understanding the biology of *S. titanus* is crucial to determining epidemiology and subsequent control measures. *S. titanus* is univoltine and the eggs overwinter on grapevine and hatch in the following year (Bressan *et al.*, 2006). Females deposit clusters of 10–12 eggs in cracks of the bark in late summer. In Europe, the eggs hatch from the middle of May through to July. The nymphs have five instar stages, each lasting a week. The adults appear in late July until early August and disappear after egg-laying in September.

It seems likely that *S. titanus* adults are more significant in the epidemiology of FD, as they have a better ability to acquire and transmit FD than earlier feeding stages (Bressan *et al.*, 2006). *S. titanus* can acquire and transmit FD at all feeding growth stages, from nymph to adult (Bressan *et al.*, 2006). However, acquisition is thought to be low early in the season, due to the high proportion of first- and second-instar nymphs at this stage, which do not acquire the phytoplasma as efficiently as the older feeding stages (Bressan *et al.*, 2006). The lack of acquisition efficiency of the younger insects is possibly because they probe and feed less than older insects, thus reducing their access period. *S. titanus* requires a minimum acquisition access period (AAP) of 1 week for FD acquisition, and the probability of pathogen uptake increases with longer access periods (Schvester *et al.*, 1969). In addition, the younger nymphs may have less access to FD due to a lower FD titre in the host whilst feeding early in the growing season of the host. Also, the latent period between acquisition and transmission of FD by *S. titanus* is 32–40 days. Thus it is more likely that adults are infectious than earlier-instar stages. Transovarial transmission of FD does not occur, thus transmission can only occur once the eggs have hatched and the nymphs have fed on infected grapevines (Bressan *et al.*, 2006).

Scaphoideus titanus was less able to acquire FD from grapevine varieties that were less susceptible to FD (Bressan *et al.*, 2005). Lower acquisition and transmission rates may affect the numbers of infected leafhoppers occurring in vineyards and will reduce rate of spread.

The nymphs do not move far from where they are hatched and adults have greater flight ability; consequently the adult stage is more important in dispersing FD within and beyond the vineyard. The density of adults is positively correlated with grapevine density. That is, the larger and more closely planted the vines are within a vineyard, the larger the populations of *S. titanus* (Lessio and Alma, 2004). Consequently, more densely planted and diseased vineyards might expect greater FD pressure from a larger number of infectious adults.

This information about the life cycle and acquisition and transmission ability of the various developmental stages of *S. titanus* indicates that the rate FD spreads in vineyards can be lowered by reducing the population of first-

and second-instar nymphs. To reduce the spread of FD, it is recommended that insecticides are applied when nymphs first appear (Bressan *et al.*, 2006).

The geographic distribution of *S. titanus* is also expanding, with new reports in recent years in Hungary (Dér *et al.*, 2007) and Bosnia and Herzegovina (Delić *et al.*, 2007). It is hypothesized that *S. titanus* was introduced to Europe from the USA, where it is a native species, early in the 20th century, through infested planting material (Caudwell, 1983). *S. titanus* is then suspected to have been moved throughout Europe also on infested planting material, because the insect does not travel long distances from the vineyard (Bertin *et al.*, 2007). Evidence for this is provided through the widespread distribution of the FD-2 strain cluster, which has low genetic variability, in France and Italy (Arnaud *et al.*, 2007). Movement of both the vector and FD through planting material into previously unaffected regions represents a risk of the further spread of FD.

FD can be disseminated through planting material. However, not all cuttings from an infected grapevine or an infected cane produce infected plants. This is possibly due to the uneven distribution of phytoplasma in the plant (Boudon-Padieu, 1999). Transmission can occur through infected scion wood and through some rootstocks. Rootstocks represent a high risk of transmission as they are often symptomless hosts, and the transmission frequency of FD in rootstock cuttings is between 6 and 80% (Boudon-Padieu, 1999; Caudwell *et al.*, 1994). Likewise scion wood can also be a symptomless carrier of FD, particularly when propagation material is taken in the same year as an infection event, since infected vines may not show symptoms until the following year after an infection event (Morone *et al.*, 2007).

Host range of the flavescence dorée phytoplasma

At first glance it would appear that the FD epidemiological system is closed: i.e. FD is spread from grapevine to grapevine by its vector *S. titanus* without the involvement of alternative hosts. This is true once the phytoplasma has been introduced to the vineyard. But the introduction of FD into European vineyards is less clear, particularly as *S. titanus* is monophagous in Europe.

Initially it was hypothesized that FD was introduced from the USA with *S. titanus*, but the presence of FD in the USA has not been confirmed. Recent studies suggest that FD strains and PGY might, in fact, have a common European origin from AldY (Arnaud *et al.*, 2007). FD, PGY and AldY belong to the same phylogenetic subclade of phytoplasmas and AldY has not been reported in the USA (Arnaud *et al.*, 2007). The most recent hypothesis suggests that some strains of FD-related phytoplasmas may have been transmitted to grapevine by occasional grapevine-feeding vectors from AldY-infected alders (Arnaud *et al.*, 2007). Both PGY and AldY are transmitted by the alder leafhopper *Oncopsis alni* (Maixner *et al.*, 2000).

Wild *C. vitalba* is a host for the FD-3 strain cluster (Angelini *et al.*, 2004). It may be an original host of the FD-3 strain cluster, especially as this FD strain cluster has been detected in *C. vitalba* in regions in Europe where FD disease has not been observed in vineyards. However, *S. titanus* survive for only a short time and cannot acquire FD-3 strains from, and transmit to, *C. vitalba* (Filippin *et al.*, 2007) or from *C. vitalba* to grapevine (Arnaud *et al.*, 2007), suggesting the involvement of another vector. PCR techniques detected FD-3 in the leafhopper *Dictyophara europaea* L., which may represent a vector for FD-3 on *C. vitalba* and from *C. vitalba* to grapevine (Filippin *et al.*, 2007); however, transmission studies are required to confirm the vector status of *D. europaea*.

Alternatively, recent studies showed that FD-infectious nymphs and adults of *S. titanus* could be found on symptomless wild *Vitis riparia* in European woodlands (Lessio *et al.*, 2007). Although not strictly an alternative host as they are *Vitis* species, it is possible that these 'wild' vines might act as source of inoculum for nearby vineyards.

Flavescence dorée disease progression – recurrence, remission and recovery

Temporal analyses of FD disease incidence showed that grapevines can recover from FD disease (Caudwell, 1961; Osler *et al.*, 2003; Morone *et al.*, 2007). FDp was not detected in recovered grapevines (Osler *et al.*, 2003; Morone *et al.*, 2007). However, the ability to recover is dependent on variety (Bellomo *et al.*, 2007). For example, Merlot, which is less susceptible to FD infection, is less likely to express the disease in subsequent years, whilst Chardonnay, which is highly susceptible to FD, may continue to show symptoms and eventually die. The recovery response has been linked to continual accumulation of H_2O_2 in leaves, resulting in low FD titres or the prevention of further infection by FD (Musetti *et al.*, 2007). Recovery in subsequent years may not be strain dependent, as the phenomenon has been observed for both FD-2 and FD-3 phytoplasmas (Belli *et al.*, 1973; Caudwell *et al.*, 1994; Angelini *et al.*, 2006). Phytoplasmas could not be detected in the canopy of recovered vines and FD detection was strongly linked with symptoms. This result suggests that FD expression is related to phytoplasma location and/or titre.

The recovery rate of FD-affected grapevines is inversely correlated with the number of vectors present in vineyards. As the number of vectors decreased, the number of recovered plants increased (Morone *et al.*, 2007). This observation indicates that the control of the vector in the vineyard reduces the incidence of FD disease. Interestingly, the number of vines displaying FD symptoms was correlated with the number of vectors present in the previous year but not in the current year, indicating a 1-year lag in disease expression (Morone *et al.*, 2007).

Some vines, including less susceptible varieties, can show FD symptoms in subsequent years, and this phenomenon is often caused by reinfection events associated with the presence of infectious vectors (Morone *et al.*, 2007). It has also been shown that, in some instances, recurrent symptom expression is also associated with persistent infection, as symptoms were observed on vines enclosed in a screen house and in the absence of vectors (Osler *et al.*, 1993). Persistent infection of phytoplasmas is not unexpected as the sieve elements of grapevine phloem function for 2 or more years (Esau, 1948).

Flavescence dorée disease incidence in space

Most of the distribution studies for FD have described the detection, location and number of affected vineyards and incidence of disease within vineyards in the regions where FD has been found. These reports show that FD can be 'epidemic', i.e. the disease affected a large number of grapevines within a vineyard or a region at the same time, and disease incidence in vineyards may reach 95% affected vines (Bressan *et al.*, 2006).

The patterns of distribution of FD affected grapevines in vineyards were usually non-random and may occur in runs along rows (Credi and Callegari, 1988; Arno *et al.*, 1993). Disease in runs indicates the presence of vectors moving from grapevine to grapevine and potentially spreading phytoplasmas from grapevine to grapevine. An edge effect was also observed by Credi and Callegari (1988) and may indicate that a source of inoculum for FD occurred outside the vineyards that were being observed. A random distribution of FD disease was observed in some years in some vineyards where clustering had been observed in other years (Credi and Callegari, 1988), and the changing distribution may have been influenced by recovery of some vines, recurrent disease expression in others and the occurrence of newly affected vines.

In reviewing the literature, it can be seen that FD is spreading in Europe. It appears that in some regions, particularly in Serbia and Slovenia and perhaps in north-western Italy, the epidemic occurrence of FD followed the introduction of *S. titanus*. Whether FD-infectious *S. titanus* is distributed to vineyards through planting material or if the disease and phytoplasma are in fact present at low levels before the vector appears is not known. The occurrence of FD in these regions seems to be associated with the presence of the FD-3 phytoplasma strain. It is possible that this FD strain has a slightly different epidemiology from that of the FD-2 strain as it has an alternative host, *C. vitalba*, and possibly another vector (Fillipin *et al.*, 2007), However, once FD-3 enters a vineyard it is probably spread grapevine to grapevine like FD-2.

Case Study 2: Bois Noir – Grapevines as the Casualty of an Open Epidemiological System

Background

BN, also known as Vergilbungskrankheit and legno nero, occurs in many European countries and is also present in Israel and Lebanon (Boudon-Padieu, 2003, 2005). When BN was first described it was considered a 'form' of FD, with a possible common aetiology (Caudwell, 1961). Ten years later a different aetiology for BN was inferred when it was shown that the phytoplasma could not be transmitted by *S. titanus*, the vector of FD (Caudwell *et al.*, 1971). BN is now known to be associated with the stolbur phytoplasma (STOL), which is a member of the 16SrXII-A group, and the name '*Candidatus* Phytoplasma solani' has been proposed because it is known to infect various solonaceous plants (Firrao *et al.*, 2005). STOL is transmitted to grapevine by the planthopper *Hyalesthes obsoletus* Signoret (Sforza *et al.*, 1998).

Three STOL types, tuf type I, tuf type II and tuf type III, have been identified based on RFLP analysis of the *tuf* gene using the restriction enzyme *Hpa*II (Langer and Maixner, 2004). Tuf type I and tuf type II are most commonly detected in grapevine. They have different alternative hosts. Tuf type I is most frequently associated with *Urtica dioica* (nettle) and tuf type II is found in *Convolvulus arvensis* (bindweed). Further variation based on sequence analysis of other genes is observed amongst isolates of tuf types I and II, but the epidemiological significance is unknown (Pacifico *et al.*, 2007).

Primary and incidental hosts of the stolbur phytoplasma

STOL has a broad host range, including various herbaceous and woody host plants (Maixner, 2006), some of which are also hosts of *H. obsoletus*. In particular, bindweed and nettle play a significant role in the epidemiology of BN in Europe. Both plants are often found within or near BN-affected vineyards, and their location and density play an important role in the spatial and temporal incidence of the disease, as they affect the density and movement of the vector into and within the vineyard (Maixner, 2007). Whilst bindweed may display symptoms of proliferation, nettle is symptomless (Johannesen *et al.*, 2008).

Because grapevine is an incidental feeding host for *H. obsoletus* and a dead-end host for STOL phytoplasma (Lee *et al.*, 1998), it does not play a significant role in STOL epidemiology, rather it is a casualty of the true host–phytoplasma–vector system.

Transmission of the stolbur phytoplasma

In Europe, the epidemiology of STOL in grapevine is directly linked to the biology and life cycle of its primary vector *H. obsoletus* and its preferred host plants. Alternative vectors for STOL to grapevine are suspected, as BN occurs in some regions in Europe, Israel and Lebanon where *H. obsoletus* is scarce (Maixner, 2006). Various vectors of STOL are reported for other crops, but their role in BN epidemiology is unknown (Maixner, 2006). *Macrosteles quadripunctulatus* Forbes has been shown to transmit STOL to grapevine under experimental conditions (Batlle *et al.*, 2008). *Anaceratagallia ribauti* Ossiannilsson, which was collected from various hosts in Austrian vineyards, can transmit STOL tuf type II to the experimental host *Vicia faba* and acquire the phytoplasma from infected bindweed and may also contribute to the epidemiology of BN (Reidle-Bauer *et al.*, 2008).

In Europe, *H. obsoletus* has one generation per year (Bressan *et al.*, 2007). *H. obsoletus* lay their eggs in the soil at the base of their preferred hosts. Once hatched, the nymphal instar stages develop and feed on the roots, where they can acquire STOL. In Europe, overwintering occurs during the third-instar stage and adults fly from June until the end of August, depending on the host plant and climate. In Israel, two generations of *H. obsoletus* per year occur and adults are observed in spring and autumn (Sharon *et al.*, 2005).

STOL is acquired during the nymphal stages from the roots of infected bindweed or nettle. *H. obsoletus* transmits the phytoplasma to grapevine and other hosts as adults. Adults are less efficient in acquiring STOL from the aerial parts of the plant (Lessio *et al.*, 2007). *H. obsoletus* transmits STOL to herbaceous plants more efficiently than to grapevine (Maixner *et al.*, 2001). *H. obsoletus* activity, including walking on nettle and flights, is greatest during mid-afternoon to mid-evening (15:00–21:00 h), and grapevine is most likely to be inoculated with STOL at this time (Bressan *et al.*, 2007). The length of time that a vector feeds on a grapevine influences the rate of transmission. Transmission can occur after an inoculation access period of 3–6 h; the greatest efficiency was observed after 12 h (Bressan *et al.*, 2007). Interestingly, the survival of *H. obsoletus* on grapevine decreased after 12 h, indicating that grapevine is a poor host of the vector.

H. obsoletus are more likely to feed on grapevine by chance rather than actively seeking grapevine as a food source, and they only feed on grapevine as adults (Bressan *et al.*, 2007). Therefore the proximity of infected STOL host plants, which are also host to the vector, affects the incidence of BN in vineyards.

H. obsoletus reproduces and develops on a variety of hosts, some of which can be found in and around vineyards, but not on grapevine. Different preferred plant species for the vector are observed in different geographic regions in Europe and Israel. For example, in Germany bindweed has traditionally been the preferred host, although nettle is becoming more important; *Vitex agnus-castus* is an important host in Israel and in some parts of Italy nettle is the primary host (Johannesen *et al.*, 2008). Interestingly, *H. obsoletus* populations occurring on the two main hosts, nettle and bindweed, are distinct; those developing on nettle emerge earlier than those developing on bindweed (Johannesen *et al.*, 2008).

In Israel, *V. agnus-castus*, a shrub in the Verbenaceae, is a preferred host for *H. obsoletus* and is also resistant to STOL infection (Zahavi *et al.*, 2007). Studies are being done to determine if this host of the vector might be used in a 'push–pull strategy' to reduce the incidence of *H. obsoletus* in vineyards, and therefore disease.

In Germany, more severe outbreaks of BN have recently occurred (Johannesen *et al.*, 2008). These outbreaks were correlated with the recent colonization of nettle by the vector, increasing populations of the vector and a greater incidence of the STOL type I in grapevine (Johannesen *et al.*, 2008). The reason for the change in host preference from bindweed to nettle by the vector and expanded populations is unknown. One hypothesis suggests that a lengthening of the growth period of nettle is allowing *H. obsoletus* nymphs to feed and develop for extended periods and subsequently increase their population growth rate (Boudon-Padieu and Maixner, 2007). Another study, analysing the haplotypes of *H. obsoletus*, suggests that the increased prevalence of BN population coincides with a recent migration of 'nettle-adapted' *H.*

obsoletus northwards into German and northern French wine-growing regions (Johannessen *et al.*, 2008).

STOL can be transmitted through propagation material. The rate of transmission is thought to be low (Osler *et al.*, 1997) but propagation material does represent a pathway for spread of the phytoplasma to regions where it has not previously been observed. This long-range transmission was observed when STOL-infected grapevines were detected in Canada (Rott *et al.*, 2007). A trace-back of the material indicated that the origin of the phytoplasma infection was likely to have come from a nursery in France.

Bois noir disease incidence

Spatial patterns of BN are linked to the host plant of *H. obsoletus*. Nettle, host to both STOL and *H. obsoletus*, is often found on the boundary of vineyards (Credi and Callegari, 1988; Maixner, 2006; Bressan *et al.*, 2007). As a consequence, a gradient of disease is observed, with highest incidence at the edge of the vineyard, where the nettle plants are located, and a decrease in incidence as the distance between the vines and the nettle plants increases. Bindweed often occurs in vineyards and is associated with a random distribution of BN (Maixner, 2006). These differences in the spatial distribution of BN assist in identifying the most likely source of STOL and *H. obsoletus* and therefore in prescribing management plans for disease control.

Hyalesthes obsoletus is attracted by sparse vegetation on open soil where weeds are removed through summer or to newly planted vineyards where a ground cover has not developed (Langer *et al.*, 2003). In this instance, the vectors are attracted into the vineyard and are then forced to feed on vines due to the unavailability of other food sources, which may increase the incidence of disease in vineyards (Maixner, 2006). Consequently, the use of a green cover crop, which is not a host to either the vector or the phytoplasma, could reduce the incidence of the vector in the vineyard.

Bois noir disease progression – recurrence, remission and recovery

BN can reach incidences of over 80% in some years in some locations (Romanazzi *et al.*, 2007; Johannesen *et al.*, 2008). However, the disease is characterized by fluctuations of incidence over time. These are driven by several forces: remission or recovery from disease, expression of disease associated with persistent STOL infection, and new infection events (Osler *et al.*, 1993). Recovery is an almost completely irreversible phenomenon, and the disappearance of symptoms is associated with the inability to detect the phytoplasma in the canopy (Osler *et al.*, 2003; Maixner, 2006; Morone *et al.*, 2007; Romanazzi and Murolo, 2008; Terlizzi and Credi, 2007). However, different cultivars have a different propensity to recover (Romanazzi *et al.*, 2007), and the ability of grapevines to recover is also affected by the rootstock (Romanazzi and Murolo, 2008).

Evidence for persistent expression of BN and persistent STOL infection was shown when infected vines were contained in an insect-proof screen house and some vines displayed symptoms in subsequent years in the absence of new infection events (Osler *et al.*, 1993). This is further supported by PCR tests, which showed that STOL could be detected in the dormant canes, cordons and roots of infected grapevines, although the rate of persistent infection was low (Terlizzi and Credi, 2007). When the rate of infection of the protected vines was compared with the rate in vines that were exposed in the field, the level of infection in the field vines was much greater, indicating that new infections were occurring (Osler *et al.*, 1993).

It is also interesting to note that STOL could not be detected in many asymptomatic shoots from infected vines but was frequently detected in symptomatic shoots (Terlizzi and Credi, 2007). These results, in addition to the lack of detection of STOL in the canopy of recovered vines, suggest that the location and perhaps the titre of the phytoplasma play a significant role in symptom expression.

Partial uprooting, pollarding and pruning are all methods that are thought to assist in the recovery of vines (Romanazzi and Murolo, 2008). It has been suggested that partial uprooting may induce a stress response, which, in turn, reduces phytoplasma concentration.

Case Study 3: Australian Grapevine Yellows – One Disease, Three Phytoplasmas

GY symptoms were first reported in 1976 in Australia and the disease described as Australian grapevine yellows (AGY; Magarey and Wachtel, 1983). A phytoplasma aetiology was inferred due to its similarity to FD disease. Early and recent surveys have shown that AGY and phytoplasmas are found in most viticultural regions of Australia (Magarey and Wachtel, 1986b; Bonfiglioli *et al.*, 1995). Like other GY diseases overseas, Chardonnay and Riesling appear to be most often affected (Magarey and Wachtel, 1986a), but phytoplasmas have also been detected in other white and red varieties (Bonfiglioli *et al.*, 1995).

Three phytoplasmas have been detected in AGY-affected grapevines: '*Ca.* Phytoplasma australiense' (CPA), tomato big bud phytoplasma (TBB) and the Buckland Valley grapevine yellows phytoplasma (BVGY) (Table 11.1; Gibb *et al.*, 1999; Constable *et al.*, 2002). Of the three phytoplasmas, the AGY phytoplasma is most commonly detected in symptomatic grapevines (Gibb *et al.*, 1999). Phytoplasmas closely related to TBB and CPA have been reported in various plant species in other countries, but the three phytoplasmas have been detected in grapevines only in Australia. Both TBB and CPA can occur in the same regions and the same vineyards. Mixed infections of TBB and CPA can occur in the same grapevine. CPA and TBB have been detected in roots, trunks, cordons and shoots, showing that they spread systemically throughout grapevines (Constable *et al.*, 2003a). However, both CPA and TBB may be unevenly distributed or in uneven titre throughout grapevine.

CPA is a member of the 16SrXII-B group and is closely related to, but distinct from, STOL associated with BN in Europe, Lebanon and Israel. Sequence analysis of the *tuf* gene and *rp* operon shows that diversity exists amongst CPA isolates from different hosts, and strains from grapevine fall in the *tuf*-Australia I, *rp*-A subgroup (Streten and Gibb, 2005). Heteroduplex mobility assays of the *tuf* gene indicated some diversity amongst isolates from grapevine (Constable and Symons, 2004).

TBB is a member of the 16SrII-D group phytoplasmas. Some diversity is reported amongst TBBp isolates from different hosts (Streten and Gibb, 2003) but not amongst isolates from grapevine.

BVGY is a unique phytoplasma based on its 16S rRNA gene and 16S–23S spacer region sequences (Constable *et al.*, 2002). The closest sequence similarity observed was 97.1% with clover phyllody phytoplasma from the USA. Constable *et al.* (2002) hypothesized that this phytoplasma represents a new subgroup within the aster yellows group (16SrI) of phytoplasmas. More recently an *in silico* RFLP analysis of phytoplasma 16S rRNA gene sequences suggested that this phytoplasma may represent a distinct 16Sr group (Wei *et al.*, 2007).

Transmission of Australian grapevine phytoplasmas

The insect vector involved in transmission of CPA to grapevine is not known, although CPA has been detected in the common brown leafhopper, *Orosius orientalis* (Matsumura) (=*Orosius argentatus*), using PCR techniques (Beanland *et al.*, 1999). Other phytoplasmas, closely related to CPA, are transmitted by planthopper species from the family Cixiidae, and a strain of CPA is transmitted by *Oliarius atkinsoni* Meyers in New Zealand (Liefting *et al.*, 1997). Additionally, netting experiments to exclude insects from papaya plants have shown that CPA associated with papaya dieback disease is excluded, suggesting the involvement of an insect (Elder *et al.*, 2002).

TBB is transmitted by the common brown leafhopper, *O. orientalis* (Hill, 1943). Recent studies have shown that TBB can be acquired from grapevine by *O. orientalis* and subsequently transmitted to faba bean (Beanland, 2001), but the ability of the leafhopper to transmit TBB back to grapevine has not been confirmed. The transmission of phytoplasmas through grapevine cuttings has not been demonstrated.

Diverse hosts of Australian grapevine phytoplasmas

CPA infects other plant species in Australia and New Zealand. However, many of these are in crops that are not grown in the same areas as grapevine. In a recent survey, several native plant species growing in the vicinity of AGY-affected vineyards tested positive using CPA-specific primers, indicating that these may be alternative hosts for the phytoplasma (Magarey *et al.*, 2005). Plant species in which CPA was detected included yanga bush

(*Maireana brevifolia*), ruby saltbush (*Enchylaena tomentosa*), *Euphorbia terracina* and *Einardia nutans*. However, their role in the epidemiology of AGY is unknown. It is interesting to note that GY disease has not been reported in New Zealand. It is possible that New Zealand strains of CPA do not infect grapevine; it may be that grapevine is not grown where CPA is found in other hosts; or it may be that *O. atkinsoni*, which transmits CPA in New Zealand, does not feed on grapevine.

TBB has a broad plant host range and is found in most parts of Australia where phytoplasma surveys have been conducted (Schneider *et al.*, 1999); consequently, its detection in grapevines was not surprising. It seems likely, given the diversity of host plant species and the extent to which the phytoplasma and its known vector is found throughout Australia, that many potential alternative hosts might be identified near vineyards, which could provide a reservoir of TBB.

No alternative hosts have been identified for BVGY, although the limited distribution of this phytoplasma would suggest that it has origins in some host near the vineyards in which it is found.

Australian grapevine yellows disease progression and possible associations with other syndromes

AGY incidence can fluctuate from year to year (Constable *et al.*, 2004). For example, in one study the incidence of AGY in one vineyard was 20% in 1996, 55% in 1997, 20% in 1998, 44% in 1999, 46% in 2000 and 44% in 2001 (Constable *et al.*, 2004). The pattern of incidence of AGY, i.e. the years in which higher and lower incidence of disease were observed, was different between vineyards observed in the study, indicating that local factors may have had an important influence on disease expression and incidence (Constable *et al.*, 2004). The fluctuation in year-to-year incidence of AGY in this 6-year study was in contrast to a previous 10-year study, in which the incidence of AGY peaked in one year and then declined (Magarey and Wachtel, 1986b). The reason for the differences in disease expression over time between the two studies is unknown but may have been associated with changing viticultural practices, such as pruning techniques, or a change in environmental factors, leading to a change in vector biology.

AGY is characterized by the expression of disease followed by remission in some grapevines, and remission and recurrence of disease in others. The occurrence of AGY in previously unaffected grapevines was also observed (Constable *et al.*, 2004). CPA and TBB can be detected in various grapevine tissues throughout the year, showing that Australian grapevines may be persistently infected from year to year and possibly contributing to recurrent symptom expression in some vines (Constable *et al.*, 2003a). However, the incidence of AGY decreased in grapevines placed in an insect-exclusion house (Magarey *et al.*, 2005). This indicates that new infection events play an important role in recurrent AGY expression. Recurrence of disease might be influenced by both persistent phytoplasma infections and re-inoculation events. Remission of disease might be associated with declining phytoplasma titre in shoots. The decline in phytoplasma titre may be due to plant host defence responses or to environmental factors that reduce the replication of the organism.

In Australia it has been hypothesized that phytoplasmas might also be associated with restricted growth (RG) and late-season leaf curl (LSLC) diseases, as well as AGY. From observations in one vineyard, Bonfiglioli *et al.* (1997) proposed that AGY, RG and LSLC diseases were related and a specific progression of symptoms occurred. It was hypothesized that LSLC was followed by the presence of AGY in the same grapevines in following years and that AGY often led to the presence of RG in subsequent years. PCR testing of RG- and LSLC-affected grapevines from various vineyards was used to determine the possible association of phytoplasmas with both diseases, but the results varied between laboratories, with some studies showing little or no association and others showing a strong association (Bonfiglioli *et al.*, 1995; Padovan *et al.*, 1995; Gibb *et al.*, 1999; Constable *et al.*, 2003a). To further assess the possible association between the three diseases, and therefore a possible association with phytoplasmas, vineyards were visually assessed for the presence and absence of AGY, RG and LSLC (Constable *et al.*, 2004). Analysis of the survey data showed that some grapevines exhibit a combination of AGY and RG or AGY and LSLC but that both RG and LSLC can occur independently of AGY (Constable *et al.*, 2004). Statistical analyses of the survey data using log-linear models also indicated that RG and LSLC were not always associated with AGY. On the basis of these results, it was suggested that phytoplasmas were not the cause of RG or LSLC and their association is coincidental.

Australian grapevine yellows disease incidence in space

A particularly high incidence of AGY occurs in the warmer inland districts of Sunraysia in New South Wales and Victoria, Murrimbidgee Irrigation Area in New South Wales and the Riverland in South Australia. Other Australian grape-growing regions have a lower incidence.

Statistical analyses showed that some clustering of AGY-affected grapevines can occur in some years in some vineyards (Constable *et al.*, 2004; Magarey *et al.*, 2005). Constable *et al.* (2004) observed that the size of the clusters of AGY-affected grapevines was different in the same vineyard from year to year. The change in cluster sizes of AGY from year to year is likely to be a reflection of the observed remission and recurrence in previously affected grapevines and new occurrences of disease in previously unaffected grapevines. In a more recent study, Magarey *et al.* (2005) noted that clustering was more likely to occur in vineyards with high incidence of AGY and that in vineyards with a lower incidence the disease was more likely to be randomly distributed.

Because persistent and asymptomatic phytoplasma infections occur (Constable *et al.*, 2003a), the spatial distribution of AGY does not reflect the spatial distribution of phytoplasmas. Both AGY and TBB can infect grapevines in the same vineyard (Constable *et al.*, 2003a). When surveys for the presence of AGY in each of the vineyards were carried out (Constable *et al.*, 2004), the specific phytoplasma infecting each grapevine was not considered. This could have confounded the interpretation of the statistical analyses for spatial distribution, especially as each phytoplasma might be found in different hosts in different locations and the vectors of each might also be different, with different behaviours and biology.

Recent studies within the Riverland region of South Australia suggested that vineyards with a high incidence of AGY occurred adjacent to wastelands or swamplands (Magarey *et al.*, 2005). Lower levels occurred in vineyards adjacent to native bush lands and lowest levels occurred when vineyards were surrounded by other vineyards. In the regions with high incidences of disease, edge effects were observed, with greatest incidence of disease occurring closest to native bush, wasteland or swampland areas. This suggests that the source of Australian grapevine phytoplasmas and their vectors occurs in these native bush, wasteland or swampland areas and that the vectors occasionally move into the vineyards from these areas. However, very few positive results have been obtained from potential vectors within vineyards (Beanland, 2001).

BVGY has a localized distribution in the Buckland Valley grape-growing region in Australia (Constable *et al.*, 2003b). This phytoplasma has not been detected in any other grape-growing region in Australia or in grape-growing regions in other countries. Prior to 2001, GYd symptoms were observed and BVGY was detected in only one vineyard in the Buckland Valley (Constable *et al.*, 2003b). In 2001, GY and BVGY were detected in two grapevines in a second vineyard, approximately 4 km away from the original vineyard. It is unlikely that the infection in 2001 arose from planting material because the material used to establish the second vineyard in 1998 was obtained from a different source from the other vineyard. These observations suggest that the phytoplasma is endemic to the Buckland Valley area and provide evidence of aerial transmission of the disease and the phytoplasma in the Buckland Valley.

Case Study 4: North American Grapevine Yellows

GY disease occurs in New York and Virginia on the eastern seaboard of the USA (Uyemoto *et al.*, 1977; Pearson *et al.*, 1985; Wolf *et al.*, 1994). The phytoplasma associated with NAGY disease in New York has not been fully characterized, although some serological evidence suggested it may be related to FD phytoplasma (Maixner *et al.*, 1993). In Virginia, two phytoplasmas were associated with NAGY, including an aster yellows (group SrI-A) phytoplasma and an X-disease (group SrIII-I) phytoplasma (Table 11.1). Both phytoplasmas can occur in the same vineyard, and recent studies suggested that they occur with similar prevalence (Tony Wolf, VAES, personal communication). The phytoplasmas infecting grapevines in the USA are distinct and have not been reported in grapevines in other countries. Like all GY diseases, varietal susceptibility differs; Chardonnay and Riesling seem to be most susceptible to NAGY.

Transmission of North American grapevine yellows phytoplasmas

Epidemiological studies in the USA have focused on identification of potential vectors, although none have been confirmed. *S. titanus* is the natural vector of FD in Europe and is a suspected vector of phytoplasmas associated with NAGY. This hypothesis is further supported by results from an Italian study, which showed that *S. titanus* could also harbour and transmit aster yellows phytoplasmas (Alma *et al.*, 1997).

S. titanus is native to the USA and is found in woodland or hedgerow vegetation near vineyards in New York and Virginia (Maixner *et al.*, 1993; Beanland *et al.*, 2006). Phytoplasma bodies were observed by ISEM and detected by ELISA using antibodies raised against FD in *S. titanus* collected in vineyards and also from wild *V. riparia* in hedgerows on the edge of vineyards (Maixner *et al.*, 1993). The phytoplasmas that were detected were not further characterized. Phytoplasma-associated symptoms were observed in *V. faba* plants in transmission studies using *S. titanus* collected from NAGY-affected grapevines (Maixner *et al.*, 1993). However, in similar transmission experiments young, potted Chardonnay vines did not develop NAGY symptoms after 1 year (Maixner *et al.*, 1993). Consequently, further transmission studies are required to determine if this leafhopper is, in fact, a vector of phytoplasmas associated with NAGY.

Several other potential vectors were also identified in epidemiological studies of NAGY, including *Osbornellus auronitens* Provancher and *Jikradia olitorius* (Say) (Beanland *et al.*, 2006). These insects and *S. titanus* are suspected vectors because their movement to and activity within vineyards correlated with the spatial distribution of NAGY, which was clustered and exhibited a gradient effect from the woodlands. Preliminary feeding trials suggested at least three insect species that were found in NAGY-affected vineyards could transmit phytoplasma to grapevines (Beanland *et al.*, 2006). Transmission through vegetative propagation has not been reported.

Alternative hosts of North American grapevine yellows phytoplasmas

Both the 16SrI and 16SrIII phytoplasmas have been detected in wild *V. riparia* (Prince *et al.*, 1993), which can occur in woodland areas and hedgerows near vineyards. A higher incidence of NAGY at the edges of vineyards was associated with the proximity to hedgerow and woodland areas containing *V. riparia*. Potential vectors could also be found in these areas. These observations suggest that *V. riparia* plays an important role in the epidemiology of NAGY and may be an alternative host to the phytoplasma and vector.

The l6SrI-A phytoplasma has also been detected in a range of herbaceous plants and woody hosts and may also play a role in the epidemiology of NAGY.

North American grapevine yellows disease progression

In New York, NAGY-affected vines often showed recurrence of symptoms in following years, and affected shoots were often produced from the basal buds of shoots that had had symptoms in the previous year (Maixner *et al.*, 1993). Recurrent symptom expression was also observed in the subsequent year in three out of ten affected vines that were moved to a glasshouse (Maixner *et al.*, 1993). In Virginia, NAGY-affected vines often die within 2–3 years of the first expression of symptoms (Wolf *et al.*, 1994). These results provide evidence that phytoplasmas can persist and continue to cause disease over time.

The spatial distribution of North American grapevine yellows disease

NAGY-affected grapevines were clustered, and in one vineyard an edge effect was also observed (Beanland *et al.*, 2006). The clustering was attributed to the feeding behaviour of potential vectors. Given that the disease was clustered and occurred more frequently at one end of the vineyard, it was hypothesized that potential vectors of the associated phytoplasmas were living in the woodland area on the vineyard border and occasionally entering the vineyard to feed on several adjacent vines then departing again (Beanland *et al.*, 2006). This hypothesis was also supported by the collection of some potential vectors on sticky traps, which revealed a dispersal behaviour consistent with the insects entering the vineyard from the woodlands (Beanland *et al.*, 2006). Both the 16SrI and 16SrIII phytoplasmas can occur in the one vineyard, but whether the phytoplasmas themselves also occur in a clustered pattern was not reported.

The observed edge effect also coincided with the detection of phytoplasmas in weeds within and near the vineyards and the detection of the 16SrI phytoplasma in asymptomatic native grapevines (*V. riparia*) (Prince *et al.*, 1993; Davis *et al.*, 1998). It is possible that various plant species in the woodland areas are hosts for both the phytoplasma and the vector.

Summary and Conclusions

The comparison of the different GY diseases presented in this chapter illustrates the complexity of phytoplasma-associated disease epidemiology. Each GY disease expresses very similar symptoms, yet each is associated with distinct phytoplasma species. While they seem to fit a similar model, their epidemiologies have different routes, associated with the different phytoplasma species, vectors and alternative hosts. This chapter also highlights the need for understanding every aspect of the disease triangle, so that some part can be modified when designing management practices to reduce the risk of disease. For example, spread of FD is controlled by the application of insecticide when nymphs emerge, but similar control measures are not in place for AGY associated with CPA because the vector, which is unknown, does not live and feed only on grapevines. When describing the epidemiology of phytoplasma-associated diseases the following should be considered:

1. Aetiology – the comparison of different GY diseases clearly shows that different phytoplasmas, both species and strains, can have different biology and life cycles. Identifying the phytoplasma involved may indicate other aspects of the disease triangle, such as alternative hosts and insect vectors.
2. Transmission – the epidemiology of phytoplasma-associated diseases is closely tied to the insect vector of the phytoplasma. Identification of a vector is crucial in determining management strategies, as their control can prevent spread of disease. Transmission of phytoplasmas through propagation material can occur and lead to their long-distance dispersal and introduction into regions where they have not previously been found.
3. Vector biology – the life cycle and behaviour of the vector directly impacts on the rate of spread and incidence of disease in the susceptible crop. Studying life cycle and behaviour may identify some aspect of the vector biology that can be controlled to reduce the risk of phytoplasma transmission.
4. Alternative hosts – alternative hosts can act as reservoirs of phytoplasmas and are often hosts for the vector. Control strategies may best be designed around management of the alternative plant host to reduce the source of phytoplasma inoculum and/or the presence of the insect vector.

Spatial and temporal analyses are used to describe disease distribution and progression. Spatial patterns of disease can indicate how the phytoplasma is being dispersed by indicating where a primary host resides for both the phytoplasma and the vector. The measurement of disease intensity with time defines the disease progress and may assist in identifying its impact on the crop. The disease progression can be linked with other aspects of the disease cycle to better describe the risks associated with plant disease spread and how they might be controlled.

References

Alma, A., Bosco, D., Danielli, A., Bertaccini, A., Vibio, M. and Arzone, A. (1997) Identification of phytoplasmas in eggs, nymphs and adults of *Scaphoideus titanus* Ball reared on healthy plants. *Insect Molecular Biology* 6, 115–121.

Angelini, E., Squizzato, F., Luchetta, G. and Borgo, M. (2004) Detection of a phytoplasma associated with grapevine Flavescence dorée in *Clematis vitalba*. *European Journal of Plant Pathology* 110, 193–201.

Angelini, E., Filippin, L., Michielini, C., Bellotto, D. and Borgo, M. (2006) High occurrence of Flavescence dorée phytoplasma early in the season on grapevines infected with grapevine yellows. *Vitis* 45(3), 151–152.

Arnaud, G., Malembic-Maher, S., Salar, P., Bonnet, P., Maixner, M., Marcone, C., Boudon-Padieu, E. and Foissac, X. (2007) Multilocus sequence typing confirms the close genetic interrelatedness of three distinct Flavescence dorée phytoplasma strain clusters and group 16SrV phytoplasmas infecting grapevine and alder in Europe. *Applied and Environmental Microbiology* 73, 4001–4010.

Arno, C., Alma, A., Bosco, D. and Arzone, A. (1993) Investigations on spatial distribution and symptom fluctuation of Flavescence dorée in 'Chardonnay' vineyards. *Petria* 3, 81–91.

Batlle A., Altabella, N., Sabaté, J. and Laviña A. (2008) Study of the transmission of stolbur phytoplasma to different crop species, by *Macrosteles quadripunctulatus*. *Annals of Applied Biology* 152, 235–242.

Beanland, L. (2001) Developing AGY management strategies. In: *GWRDC 2000–2001 Annual Report*, Grape and Wine Research and Development Corporation, Adelaide, pp. 131–132.

Beanland, L., Kelly, M., Faggian, R., MacFarlane, J. and Glenn, D. (1999) In search of an insect vector of Australian grapevine yellows: species composition and abundance of potential vectors. *The Australian Grapegrower and Winemaker* 430, 41–46.

Beanland, L., Noble, R. and Wolf, T.K. (2006) Spatial and temporal distribution of North American grapevine yellows disease and of potential vectors of the causal phytoplasmas in Virginia. *Environmental Entomology* 35, 332–344.

Belli, G., Fortusini, A., Osler, R. and Amici, A. (1973) The occurrence of a disease of the Flavescence dorée type in vineyards of Oltrepo Pavese. *Rivista di Patologia Vegetale* IV 9, 50–56.

Bellomo, C., Carraro, L., Ermacora, P., Pavan, F., Osler, R., Frausin, C. and Governatori G. (2007) Recovery phenomena in grapevines affected by grapevine yellows in Friuli Venezia Giulia. *Bulletin of Insectology* 60(2), 235–236.

Bertin, S., Guglielmino, C., Karam, N., Gomulski, L., Malacrida, A. and Gasperi, G. (2007) Diffusion of the Nearctic leafhopper *Scaphoideus titanus* Ball in Europe: a consequence of human trading activity. *Genetica* 13, 275–285.

Bonfiglioli, R.G., Magarey, P.A. and Symons, R.H. (1995) PCR confirms an expanded symptomatology for Australian Grapevine Yellows. *Australian Journal of Grape and Wine Research* 1, 71–75.

Bonfiglioli, R.G., Carey, C.T., Schleifert, L.F., Kinnear, A.J. and Symons R.H. (1997) Description and progression of symptoms associated with grapevine yellows disease in young Chardonnay vines in the Sunraysia district. *The Australian Grapegrower and Winemaker* 400, 11–15.

Boudon-Padieu, E. (1999) Grapevine phytoplasmas. In: First internet conference on phytopathogenic mollicutes. http://web.uniud.it/phytoplasma//conf.html

Boudon-Padieu, E. (2003) The situation of Grapevine yellows and current research directions: distribution, diversity, vectors, diffusion and control. *Extended Abstracts 14th Meeting of the ICVG*. Locorotondo (BARI), Italy.

Boudon-Padieu, E. (2005) Phytoplasmes de la vigne et vecteurs potentiels/Grapevine phytoplasmas and potential vectors. *Bulletin O.I.V.* 78, 299–320.

Boudon-Padieu, E. and Maixner, M. (2007) Potential effects of climate change on distribution and activity of insect vectors of grapevine pathogens. *Actes du colloque international et pluridisciplinaire réchauffement climatique, quels impacts probables sur les vignobles?* Dijon, France http://www.u-bourgogne.fr/chaireunesco-vinetculture/Actes%20clima/Actes/Actes.htm

Bressan, A., Spiazzi, S., Girolami, V. and Boudon-Padieu, E. (2005) Acquisition efficiency of Flavescence dorée phytoplasma by *Scaphoideus titanus* Ball from infected tolerant or susceptible grapevine cultivars or experimental host plants. *Vitis* 44, 143–146.

Bressan, A., Larrue, J. and Boudon-Padieu, E. (2006) Patterns of phytoplasma-infected and infective *Scaphoideus titanus* leafhoppers in vineyards with high incidence of Flavescence dorée. *Entomologia Experimentalis et Applicata* 119, 61–69.

Bressan, A., Turata, R., Maixner, M., Spiazzi, S., Boudon-Padieu, E. and Girolami, V. (2007) Vector activity of *Hyalesthes obsoletus* living on nettles and transmitting a stolbur phytoplasma to grapevines: a case study. *Annals of Applied Biology* 150, 331–339.

Caudwell, A. (1957) Deux années d'étude sur la Flavescence dorée, nouvelle maladie grave de la vigne. *Annales d'Amélioration des Plantes* 4, 359–393.

Caudwell, A. (1961) A study of black wood disease of vines: its relationship to Flavescence dorée. *Annales des Epiphyties* 12, 241–262.

Caudwell, A. (1964) Identification d'une nouvelle maladie a virus de la vigne, 'la flavescence dorée'. Etude des phenomènes de localisation des symptomes et de rétablissement. *Annales des Epiphyties* 15 (hors-sér. 1), 193.

Caudwell, A. (1983) Origin of yellows induced by mycoplasma-like organisms (MLO) of plants and the example of grapevine yellows. *Agronomie* 3, 103–111.

Caudwell, A., Larrue, J., Kuzsala, C. and Bachelier, J.C. (1971) Pluralité des jaunisses de la vigne. *Annales de Phytopathologie* 3, 95–105.

Caudwell, A., Larrue, J., Tassart, V., Boidron, R., Grenan, S., Leguay, M. and Bernard, P. (1994) Caractère porteur de la Flavescence dorée chez les vignes porte-greffes en particulier le 3309 C et le Fercal. *Agronomie* 14, 83–94.

Constable, F.E. and Symons, R.H. (2004) Genetic variability amongst isolates of Australian grapevine phytoplasmas. *Australasian Plant Pathology* 33, 115–119.

Constable, F.E., Whiting, J.R., Jones, J., Gibb K.S. and Symons R.H. (2002) A new grapevine yellows phytoplasma from the Buckland Valley of Victoria, Australia. *Vitis* 41, 147–154.

Constable, F.E., Gibb, K.S. and Symons, R.H. (2003a) The seasonal distribution of phytoplasmas in Australian grapevines. *Plant Pathology* 52, 267–276.

Constable, F.E., Whiting, J.R., Jones, J., Gibb, K.S. and Symons, R.H. (2003b) The distribution of grapevine yellows disease associated with the Buckland Valley grapevine yellows phytoplasma. *Journal of Phytopathology* 151, 65–73.

Constable, F.E., Jones, J., Gibb, K.S., Chalmers, Y.M. and Symons, R.H. (2004) The incidence distribution and expression of Australian grapevine yellows, restricted growth and late season curl diseases in selected Australian vineyards. *Annals of Applied Biology* 144, 205–218.

Credi, R. and Callegari, D. (1988) Epidemiological profile of grapevine Flavescence dorée in Emilia Romagna: temporal spread, spatial distribution of diseased plants and incidence gradients. *Phytopathologia Mediterranea* 27, 90–98.

Davis, R.E., Jomantiene, R., Dally, E.L. and Wolf, T.K. (1998) Phytoplasmas associated with grapevine yellows in Virginia belong to group 16SrI, subgroup A (tomato big bud phytoplasma subgroup) and group 16SrIII, new subgroup I. *Vitis* 37, 131–137.

Delić, D., Selak, G., Martini, M., Ermacora, P., Carraro, L., Myrta, A. and Durić, G. (2007) Surveys for grapevine yellows phytoplasmas in Bosnia and Herzegovina. *Bulletin of Insectology* 60, 197–198.

Dér, Z., Koczor, S., Zsolnai, B., Ember, I., Kölber, M., Bertaccini, A. and Alma, A. (2007) *Scaphoideus titanus* identified in Hungary. *Bulletin of Insectology* 60, 199–200.

Elder, R.J., Milne, J.R., Reid, D.J., Guthrie, J.N. and Persley, D.M. (2002) Temporal incidence of three phytoplasma-associated diseases of *Carica papaya* and their potential hemipteran vectors in central and south-east Queensland. *Australasian Plant Pathology* 31, 165–176.

Esau, K. (1948) Phloem structure in the grapevine and its seasonal changes. *Hilgardia* 18, 217–296.

Filippin, L., Jović, J., Forte, V., Cvrković, T., Toševski, I., Borgo M. and Angelini, E. (2007) Occurrence and diversity of phytoplasmas detected in clematis and their relationships with grapevine 'flavescence dorée' phytoplasmas. *Bulletin of Insectology* 60, 327–328.

Firrao, G., Gibb K. and Streten C. (2005) Short taxonomic guide to the genus '*Candidatus* Phytoplasma'. *Journal of Plant Pathology* 87, 249–263.

Gibb, K.S., Constable, F.E., Moran, J.R. and Padovan A.C. (1999) Phytoplasmas in Australian grapevines – detection, differentiation and associated diseases. *Vitis* 38, 107–114.

Hill, A.V. (1943) Insect transmission and host plants of virescence (big bud of tomato). *Journal of the Council for Scientific and Industrial Research, Australia* 16, 85–90.

Johannesen, J., Lux, B., Michel, K., Seitz, A. and Maixner, M. (2008) Invasion biology and host specificity of the grapevine yellows disease vector *Hyalesthes obsoletus* in Europe. *Entomologia Experimentalis et Applicata* 126, 217–227.

Kranz, J. (1990) Epidemics, their mathematical analysis and modelling. In: Kranz, J. (ed.) *Epidemics of Plant Diseases: Mathematical Analysis and Modeling*. Springer-Verlag, Berlin, Germany, pp. 1–11.

Langer, M. and Maixner, M. (2004) Molecular characterisation of grapevine yellows associated phytoplasmas of the stolbur-group based on RFLP-analysis of non-ribosomal DNA. *Vitis* 43, 191–200.

Langer, M., Darimont, H. and Maixner, M. (2003) Control of phytoplasma vectors in organic viticulture. *Bulletin OILB/SROP* 26, 197–202.

Lee, I.M., Gundersen-Rindal, D.E. and Bertaccini, A. (1998) Phytoplasma: ecology and genomic diversity. *Phytopathology* 88, 1359–1366.

Lessio, F. and Alma, A. (2004) Seasonal and daily movement of *Scaphoideus titanus* Ball (Homoptera:Cicadellidae). *Environmental Entomology* 33, 1689–1694.

Lessio, F., Tedeschi, R. and Alma, A. (2007) Presence of *Scaphoideus titanus* on American grapevine in woodlands, and infection with 'flavescence dorée' phytoplasmas. *Bulletin of Insectology* 60, 373–374.

Liefting, L.W., Beever, R.E., Winks, C.J., Pearson M.N. and Forster R.L.S. (1997) Planthopper transmission of Phormium yellow leaf phytoplasma. *Australasian Plant Pathology* 26, 148–154.

Madden, L.V., Hughes, G. and Van Den Bosch, F. (2007) *The Study of Plant Disease Epidemics*. The American Phytopathological Society, St Paul, Minnesota.

Magarey, P.A. and Wachtel, M.F. (1983) Australian vine yellows, I. Inference of a mycoplasma-like aetiology for a new disease of Australian grapevine. *4th International Congress of Plant Pathology, Melbourne, Australia,* p. 118.

Magarey, P.A. and Wachtel, M.F. (1986a) Grapevine yellows – a widespread, apparently new disease in Australia. *Plant Disease* 70, 694.

Magarey, P.A. and Wachtel, M.F. (1986b) Australian grapevine yellows. *International Journal of Tropical Plant Disease* 4, 1–14.

Magarey, P.A., Gillett, S.R., Gurr, G.M., Cass, J., Habili, N., Magarey, C.C. and Thiele, S.A. (2005) Australian grapevine yellows: searching for the source and vector of disease. *The 15th Biennial Australasian Plant Pathology Society Conference Geelong, Australia*, p. 312.

Maixner, M. (2006) Grapevine yellows – current developments and unsolved questions. *Extended abstracts 15th Meeting of the ICVG*. Stellenbosch, South Africa.

Maixner, M. (2007) Biology of *Hyalesthes obsoletus* and approaches to control this soilborne vector of Bois noir disease. *Bulletin IOBC/WPRS* 30, 3–9.

Maixner, M., Pearson, R.C., Boudon-Padieu, E. and Caudwell, A. (1993) *Scaphoideus titanus*, a possible vector of grapevine yellows in New York. *Plant Disease* 77, 408–413.

Maixner, M., Reinert, W. and Darimont, H. (2000) Transmission of grapevine yellows by *Oncopsis alni* (Schrank) (*Auchenorrhyncha*: *Macropsinae*). *Vitis* 39, 83–84.

Maixner, M., Darimont, H. and Mohr, H.D. (2001) Studies on the transmission of bois noir to weeds and potential ground-cover plants by *Hyalesthes obsoletus* Signoret (Auchenorrhyncha: Cixiidae). *Bulletin OILB/SROP* 24, 249–251.

Morone, C., Boveri, M., Giosue, S., Gotta, P., Rossi, V., Scapin, I. and Marzachì, C. (2007) Epidemiology of flavescence dorée in vineyards in northwestern Italy. *Phytopathology* 97, 1422–1427.

Musetti, R., Marabottini, R., Badiani, M., Martini, M., Sanità di Toppi, L.S., di Borselli, S., Borgo, M. and Osler, R. (2007) On the role of H_2O_2 in the recovery of grapevine (*Vitis vinifera* cv. Prosecco) from Flavescence dorée disease. *Functional Plant Biology* 34, 750–758.

Osler, R., Carraro, L., Loi, N. and Refatti, E. (1993) Symptom expression and disease occurrence of a yellows disease of grapevine in northeastern Italy. *Plant Disease* 77, 496–498.

Osler, R., Vindimian, M.E., Filippi, M., Carraro, L. and Refatti, E. (1997) Possibility of propagation of grapevine yellows (black wood) by grafting. *Informatore Fitopatologico* 47, 61–63.

Osler, R., Carraro, L., Ermacora, P., Ferrini, F., Loi, N., Loschi, A., Martini, M., Mutton, P.B. and Refatti E. (2003) Roguing: a controversial practice to eradicate grape yellows caused by phytoplasmas. *Extended Abstracts 14th Meeting of the International Council for the Study of Virus and Virus-like Diseases of the Grapevine (ICVG)*, Locorotondo (BA), Italy, p. 68.

Pacifico, D., Foissac, X., Veratti, F. and Marzachì, C. (2007) Genetic diversity of Italian and French 'bois noir' phytoplasma isolates. *Bulletin of Insectology* 60, 345–346.

Padovan, A.C., Gibb, K.S., Bertaccini, A., Vibio, M., Bonfiglioli, R.E., Magarey, P.A. and Sears, B.B. (1995) Molecular detection of the Australian grapevine yellows phytoplasma and comparison with grapevine yellows phytoplasmas from Italy. *Australian Journal of Grape and Wine Research* 1, 25–31.

Pearson, R.C., Pool, R.M., Gonsalves, D. and Goffinet, D.C. (1985) Occurrence of Flavescence dorée-like symptoms on White Riesling grapevines in New York, U.S.A. *Phytopathologia Mediterranea* 24, 82–87.

Prince, J.P., Davis, R.E., Wolf, T.K., Lee, I.-M., Mogen, B.D., Dally, E.L., Bertaccini, A., Credi, R. and Barba, M. (1993). Molecular detection of diverse mycoplasma-like organisms (MLOs) associated with grapevine yellows and their classification with aster yellows, X-disease, and elm yellows MLOs. *Phytopathology* 83, 1130–1137.

Reidle-Bauer, M., Sára, A. and Regner, F. (2008) Transmission of a stolbur phytoplasma by the Agalliinae leafhopper *Anaceratagallia ribauti* (Hemiptera, Auchenorrhyncha, Cicadellidae). *Journal of Phytopathology* 156, 687–690.

Romanazzi, G. and Murolo, S. (2008) Partial uprooting and pulling to induce recovery in Bois noir-infected grapevines. *Journal of Phytopathology* 156, 747–750.

Romanazzi, G., Prota, V.A., Casati, P., Murolo, S., Silletti, M.R., Di Giovanni, R., Landi, L., Zorloni, A., D'Ascenzo, D., Virgili, S., Garau, R., Savini, V. and Bianco P.A. (2007) Incidence of recovery in grapevines infected by phytoplasma in different Italian climatic and varietal conditions and attempts to understand and promote the phenomenon. *Proceedings Workshop 'Innovative strategies to control grapevine and stone fruit phytoplasma based on recovery, induced resistance and antagonists'*, Ancona, Italy, pp. 9–11.

Rott, M., Johnson, R., Masters, C. and Green, M. (2007) First report of bois noir phytoplasma in grapevine in Canada. *Plant Disease* 91, 1682.

Schneider, B., Padovan, A., De La Rue, S., Eichner, R., Davis, R., Bernuetz, A. and Gibb, K. (1999) Detection and differentiation of phytoplasmas in Australia: an update. *Australian Journal of Agricultural Research* 50, 333–342.

Schvester, D., Carle, P. and Moutous, G. (1969) Nouvelles données sur la transmission de la flavescence dorée de la vigne par *Scaphoideus littoralis* Ball. *Annales de Zoologie et Ecologie Animale* 1, 445–465.

Sforza, R., Clair, D., Daire, X., Larrue, J. and Boudin-Padieu, E. (1998) The role of *Hyalesthes obsoletus* (Hemiptera: Cixiidae) in the occurrence of Bois noir of grapevines in France. *Journal of Phytopathology* 146, 549–556.

Sharon, R., Soroker, V., Wesley, S.D., Zahavi, T., Harari, A. and Weintraub P.G. (2005) *Vitex agnus-castus* is a preferred host plant for *Hyalesthes obsoletus*. *Journal of Chemical Ecology* 31, 1051–1063.

Streten, C. and Gibb, K.S. (2003) Identification of genes in the tomato big bud phytoplasma and comparison to those in sweet potato little leaf-V4 phytoplasma. *Microbiology* 149, 1797–1805.

Streten, C. and Gibb, K.S. (2005) Genetic variation in *Candidatus* Phytoplasma australiense. *Plant Pathology* 54, 8–14.

Terlizzi, F. and Credi, R. (2007) Uneven distribution of stolbur phytoplasma in Italian grapevines as revealed by nested-PCR. *Bulletin of Insectology* 60, 365–366.

Uyemoto, J.K., Cummins, J.R. and Abawi, G.S. (1977) Virus and virus-like diseases affecting grapevines in New York vineyards. *American Journal of Enology and Viticulture* 28, 131–136.

Wei, W., Davis R.E., Lee, I.-M. and Zhao, Y. (2007) Computer-simulated RFLP analysis of 16S rRNA genes: identification of ten new phytoplasma groups. *International Journal of Systematic and Evolutionary Microbiology* 57, 1855–1867.

Wolf, T.K., Prince, J.P. and Davis R.E. (1994) Occurrence of grapevine yellows in Virginia vineyards. *Plant Disease* 78, 208.

Zahavi, T., Peles, S., Harari, A.R., Soroker, V. and Sharon, R. (2007) Push and pull strategy to reduce *Hyalesthes obsoletus* population in vineyards by *Vitex agnus castus* as trap plant. *Bulletin of Insectology* 60(2), 297–298.

12 Phytoplasma Epidemiological Systems with Multiple Plant Hosts

Michael Maixner

Julius Kühn Institute, Germany

Introduction

Phytoplasmas are associated with several hundred plant species. As obligate parasites, they are adapted to live in two ecological niches, plant phloem and insects. They depend on transmission by phloem-feeding hemipteran vectors, because they are, in general, not able to pass vertically to plant seed or insect eggs. Phytoplasmas differ with respect to host specificity, which is usually higher for vectors than for the plant hosts (Lee *et al.*, 1998). For example, whereas phytoplasmas of the aster yellows (16SrI) phylogenetic group infect various plant species and are transmitted by a wide range of leafhopper species, phytoplasmas of the apple proliferation group (16SrX) are restricted to only one or a few closely related plant hosts and vectors (Lee *et al.*, 1998). The reasons for plant host specificity are still not completely understood, but the existence of resistant plant taxa and the variation of phytoplasma titres in different host plants indicate that plant susceptibility is not only a question of successful inoculation by vectors but also a result of complex phytoplasma–plant host interactions. Therefore, phytoplasmas of different phylogenetic groups show considerable differences with respect to their plant host range and their vector specificity (Weintraub and Beanland, 2006). This chapter deals with phytoplasmas using multiple plant hosts. The topic will be restricted to such phytoplasmas that regularly use more than one plant–vector system to exploit different ecological niches, and to examples where the specific host association of genetically diverse phytoplasma strains leads to separated epidemiological systems in the field. The relationship between these phytoplasmas, their hosts and the vectoring insects, as well as the consequences of the specific plant–phytoplasma–vector interactions for the management of the respective plant diseases, will be discussed.

The full understanding of the epidemiological systems is often impeded by the lack of knowledge of phytoplasma epidemiology. Not much is known about phytoplasma infection of wild plants that frequently serve as sources of inoculum for cultivated crops. Natural host plants may stay free from obvious disease symptoms due to a long co-evolution with their phytoplasmas (Caudwell, 1983). Where more than one plant species or vector are involved, the systems are often too complex for the identification of proper targets for control strategies, or the biology of the species involved prevents appropriate measures. Furthermore, care has to be taken to avoid rash conclusions about the phytoplasma aetiology of particular diseases, since other pathogens may be involved (Danet *et al.*, 2003; Streten *et al.*, 2005).

Studies of the aetiology and epidemiology of phytoplasma diseases depend on sensitive and reliable techniques for the detection of the pathogens in plant and insect hosts (Firrao *et al.*, 2007). Routine diagnosis is based on PCR, using primers targeting phylogenetic-group-specific fragments of the ribosomal RNA operon. Other, more variable, genetic loci are analysed in order to differentiate and characterize closely related phytoplasma strains and isolates (e.g. Arnaud *et al.*, 2007). Woody plants especially often require particular efforts to increase sensitivity, such as nested PCR or real-time PCR. Phytoplasma detection in insects is easier; however, it is not sufficient to identify vectoring species, since non-vector, phloem-feeding species may also acquire the pathogens from infected plants. Therefore, the vectoring ability has to be confirmed by the experimental transmission of the phytoplasma to healthy plants.

The basic epidemiological system of phytoplasma diseases consists of at least three components: the phytoplasma itself, a susceptible host plant and a competent vector feeding on the host plant. Increasing complexity evolves from either a wide host range of the phytoplasma or differences in the host specificity due to genetically diverse strains or isolates. On the side of the vector, multiple vector species, vector populations with different host plant affiliation or the polyphagous feeding habit of vectors need to be considered. Consequently, the analysis of the epidemiological systems of phytoplasmas with multiple host plants needs to take account of the existence of additional natural plant hosts and their significance as alternative sources of inoculum as well as the host range and feeding preferences of vector species or particular vector populations. The genetic variability of phytoplasmas as well as their vectors with respect to plant host specificity plays a role, too. A principal element that affects the efficiency of an epidemiological system is the life history of the particular vector(s), mainly the number of generations, the hibernation strategy, feeding preferences and mobility or seasonal migration activity. Other factors that determine infection pressure are the time and efficiency of phytoplasma acquisition in relation to host phenology and the temporal fluctuations of phytoplasma titres, the host-plant-related infestation of vector populations, and vector propensity to feed on host plants and their transmission efficiency, as well as the temporal and spatial synchronization of vectors and particular host plants.

A hypothesis for the emergence of phytoplasma diseases of cultivated plants was developed by Caudwell (1983), who proposed the existence of natural cycles of phytoplasmas, consisting of more or less symptomless wild hosts and vectors that are not affected by the phytoplasma infection. The low virulence is explained as the result of a long co-evolution of a pathogen and its hosts (Elliott *et al.*, 2003). Outbreaks of new disease are induced either by the extension of a natural cycle to a newly occurring plant as a suitable feeding host for the natural vector (direct mode, according to Caudwell, 1983) or by the introduction of a new competent vector into a habitat where both the natural and the cultivated host already exist (indirect mode). Disease outbreaks in cultivated crops should therefore be the consequence of the branching of a natural epidemiological cycle to a new plant species, enabling the phytoplasmas to exploit a new ecological niche (Lee *et al.*, 1998). This can lead to genetically distinct phytoplasma strains if the secondary epidemiological cycles are isolated from the original system (Lee *et al.*, 1998). Three levels of epidemiological systems of phytoplasmas that include multiple host plants will be discussed further: natural epidemiological cycles branching to cultivated plants as dead-end hosts, phytoplasmas with parallel but interconnected epidemiological cycles, and crop-specific epidemic systems that are isolated from the original plant–vector systems.

Natural Epidemiological Cycles Branching to Cultivated Plants as Dead-end Hosts

Many phytoplasmas are maintained by natural disease cycles that include only wild plants. If cultivated plants are grown in the same environment, vector feeding on these crops can result in the emergence of new diseases of economic importance. The feeding preference of the vector species decides whether a newly emerged disease is transmitted independently from the original epidemiological system or depends constantly on the natural sources of inoculum. The latter is true if vectors are unable to acquire the phytoplasma from the cultivated plants, which then are considered dead-end hosts for the phytoplasma, as long as no other vectors exist. Some epidemiological systems of phytoplasmas of the 16SrXII phylogenetic group (stolbur group or STOL) are examples of such an erratic branching of natural cycles to cultivated plants. In spite of this rather inefficient mode of transmission, they are causing diseases of high economic impact, e.g. in viticulture and solanaceous crops. Phytoplasmas of this group occur in Europe, Asia and Australasia. Two major subgroups, 16SrXII-A (STOL, proposed as '*Ca.* Phytoplasma solani') and 16SrXII-B ('*Ca.* Phytoplasma australiense'), can be distinguished. Recently, additional 16SrXII-subgroups were identified in strawberry, *Hydrangea* spp. and grapevine (Quaglino *et al.*, 2009). Additional markers besides the 16S rRNA gene have been used to differentiate further the subgroup XII-A phytoplasmas, namely the *tuf* gene (Langer and Maixner, 2004) and *stol-1H10*, a gene encoding a putative membrane protein of STOL (Pacifico *et al.*, 2007). The *tuf* marker proved a useful tool for epidemiological studies, since its genetic variability was found to be linked to host plant specificity.

Bois noir of grapevine

Stolbur phytoplasmas of subgroup 16SrXII-A are endemic to Europe and the Mediterranean area. They are associated with diseases of woody plants such as grapevine (bois noir) or *Lavandula* spp. (lavender decline), and of various solanaceous crops, strawberry, celery, maize and sugarbeet (Jovic *et al.*, 2007; Semetey *et al.*, 2007; Marzachi, 2008). A wide variety of herbaceous as well as woody plants have been reported as regular or occasional hosts (Credi *et al.*, 2006). Bois noir (BN) of grapevine is the most widespread grapevine yellows disease in Europe and the Mediterranean area and occurs in almost all viticultural regions (Boudon-Padieu, 2003). It developed to a major disease in viticulture because of new outbreaks and increasing damage in many regions.

Typing of STOL isolates from grapevine, wild hosts and vectors using the tufAY marker revealed the existence of three different strains of the pathogen associated with bois noir (Langer and Maixner, 2004). They are associated with different natural hosts: tuf type 'a' is confined to *Urtica dioica* (stinging nettle), while type 'b' was found in *Convolvulus arvensis* (field bindweed) and *Calystegia sepium* (hedge bindweed) only. Type 'c' is restricted to *C. sepium* and limited to a small geographic area. The specific association of tuf types 'a' and 'b' (or BN types I and II) with their respective host plants was confirmed in various areas, and regional differences of their prevalence were found (Bressan *et al.*, 2007; Pacifico *et al.*, 2007; Pasquini *et al.*, 2007; Mori *et al.*, 2008; Riedle-Bauer *et al.*, 2008).

Different leafhoppers and planthoppers have been reported to transmit STOL to either herbaceous or woody plants or both (Weintraub and Beanland, 2006). Planthoppers of the family Cixiidae are the most important vectors. Three cixiid species transmit 16SrXII-A phytoplasmas: *Hyalesthes obsoletus* Signoret (Fos *et al.*, 1992; Maixner *et al.*, 1995), *Pentastiridius leporinus* (L.) (Bressan *et al.*, 2008) and *Reptalus panzeri* (Löw) (Jovic *et al.*, 2007). Another species, *Reptalus quinquecostatus* (Dufour), was found carrying this phytoplasma and being able to transmit it to an experimental feeding medium (Pinzauti *et al.*, 2008). Its vectoring ability to plants needs to be confirmed. The non-cixiid species *Macrosteles quadripunctulatus* (Kirschbaum) and *Anaceratagallia ribauti* Ossiannilsson have been reported as vectors of STOL to herbaceous plants (Batlle *et al.*, 2008; Riedle-Bauer *et al.*, 2008).

H. obsoletus is the only known vector of BN and the major vector to most of the other crops. As in many species of the family Cixiidae, all developmental stages of this planthopper live in the soil, feeding on the roots of their host plants. Eggs are deposited close to host plants and the first larval instars move to the roots for feeding. Major herbaceous host plants of this planthopper are *C. arvensis* and *C. sepium*, perennial species of *Ranunculus*, and *U. dioica*. Woody hosts are lavender (*Lavandula agustifolia*) in southern France (Sforza *et al.*, 1999) and monks pepper (*Vitex agnus-castus*) in Israel (Sharon *et al.*, 2005). Larval instars do not feed on roots of grapevine, which is only an erroneous host of adult *H. obsoletus*. In central Europe at the northern range of this species, second and third instars move within the soil to a depth of 20–25 cm to avoid frost damage during winter and return to the surface during April and May. Since STOL is acquired during larval development, the adult planthoppers are already infective when they emerge from the soil in June/July for a flight period of 6–8 weeks. Infestation levels do not change considerably during the flight period, which underlines the significance of larval acquisition feeding. *H. obsoletus* is able to inoculate grapevine quite efficiently, in spite of its irregular feeding activity on this plant. This can be explained by the short minimum inoculation access period and the usually high proportions of infected vectors (Bressan *et al.*, 2007). There is no evidence so far that *H. obsoletus* could be able to acquire STOL from grapevine, because of the only erratic feeding and the short lifespan of the adult vectors, the only stage that feeds occasionally on vines. Therefore, as long as *H. obsoletus* is the sole vector of BN, grapevine is indeed a classical dead-end host for STOL, being affected only by erroneous feeding of *H. obsoletus*, which causes the one-directional extension of the natural epidemiological cycles to grapevine.

There is evidence for a host plant specialization not only of STOL but also of its vector *H. obsoletus*, which uses either bindweed or nettle as host plants. For instance, specimens of *H. obsoletus* that were force-fed on their homologous host plants (e.g. vectors from nettle fed on nettle) lived significantly longer than those fed on the heterologous species (Maixner, 2007). Furthermore, the populations from the two hosts differ in flight phenology. The period of adult flight activity of nettle populations is delayed by 2–4 weeks compared with bindweed populations.

The host plant utilization changed in northern viticultural regions such as Germany during the last decade, as *H. obsoletus* started to exploit nettle as a new host plant besides the traditional bindweed host. A genetic population analysis of *H. obsoletus* populations based on mitochondrial DNA revealed a high correlation of their genealogy with geography (Johannesen *et al.*, 2008). The study provided evidence for a recent geographic range expansion of this vector and a circum-Alpine immigration to central Europe, which could explain the emergence of BN in central Europe in the 20th century. A lack of haplotype–host plant affiliation suggests that both host plants can be used by *H. obsoletus*; however, a slight genetic differentiation between host plant populations was found using random amplified polymorphic DNA analysis (Johannesen *et al.*, 2008). Environmental factors are probably playing an additional role in the host shift of the vector, since higher temperatures are required by *H. obsoletus* to complete its life cycle on nettle than on bindweed (Maixner, unpublished).

The existence of different strains of STOL associated with different natural plant hosts, together with the host affiliation of vector populations, led to the assumption that distinct epidemiological cycles of STOL with different regional prevalence exist in the field (Langer and Maixner, 2004). The 'nettle cycle', i.e. the system including BN type I of the phytoplasma, *U. dioica* as plant host and nettle-adapted *H. obsoletus* populations, seems to be the predominant one in northern Italy (Alma *et al.*, 2002) and extended to the northern viticultural areas recently, causing new outbreaks of BN in various regions (Maixner *et al.*, 2006). The second epidemiological cycle ('bindweed cycle'), based on BN type II of STOL, *C. arvensis* and vector populations affiliated with this host plant, is the traditional system in central Europe and the predominant one in central and southern Italy (Pasquini *et al.*, 2007). Transmission experiments revealed differences in the inoculation efficiency of *H. obsoletus* for the two types of STOL (Maixner, unpublished). BN type II was transmitted with higher efficiency to *C. arvensis, C. sepium* and grapevine, whereas *U. dioica* was more efficiently inoculated with BN type I (Maixner *et al.*, 2006). No double infections with both strains in herbaceous hosts or grapevine have yet been observed. While the available data support the hypothesis of separate epidemiological systems of STOL, the mechanisms of this host specificity are still unknown. Studies of the genetic variability of genes of putative membrane proteins that might be involved in host recognition (Pacifico *et al.*, 2007) could be appropriate tools to investigate this question. Which of the major host plants was the original one is an interesting question. BN type II infecting *C. arvensis* appears to be the more widespread strain; however, bindweed is more severely affected by STOL infection than *U. dioica,* which shows no or only weak disease symptoms. Moreover, *H. obsoletus* seems to be well adapted to this strain, as no influence of BN type I infection on the size of *H. obsoletus* nymphs was observed (Kaul *et al.*, 2009). Both observations point to a low virulence of this BN type due to a long phytoplasma–host–vector relationship (Elliott *et al.*, 2003), while branching of this cycle to grapevine initiates severe symptoms in this new, non-adapted host. Nothing is known yet about possible interactions of the two strains because of the lack of plants that are simultaneously infected.

The dead-end characteristic of the grapevine host for STOL is a consequence of the life history and feeding preferences of *H. obsoletus*. Key factors for disease pressure are therefore uncoupled from the incidence of BN in grapevine but related to the predominant host species of both the vector and the pathogen, as well as their density, distribution patterns and infestation. Additional aspects are the propensity of *H. obsoletus* to move to grapevine and the vector's transmission efficiency (Bressan *et al.*, 2007). Some plants are highly suitable hosts for the vector but not for the phytoplasma. Whereas *H. obsoletus* achieves high population densities on perennial *Ranunculus* species in Germany, infected vectors were scarcely found. However, STOL was transmitted to *Ranunculus bulbosus* by *H. obsoletus* with high efficiency (Ge and Maixner, 2003), but infected plants died rapidly. Consequently, only healthy *Ranunculus* are available as hosts for the larval development of *H. obsoletus*. Another common weed, *Hieracium pilosella*, proved to be a good feeding host for adult *H. obsoletus*, but repeated attempts to inoculate it with STOL failed. Furthermore, STOL was not detected in *V. agnus-castus* in Israel, although this shrub supported high populations of *H. obsoletus* (Sharon *et al.*, 2005). On the other hand, more than 50% of *C. arvensis* and *U. dioica* plants were found to be infected at sites with high infection pressure in Germany. The proportion of infected vectors on bindweed and nettle at different locations of the same viticultural site varied from 26 to 62% and from 6 to 25%, respectively. A wide variety of additional plant species were found to be occasionally infected by STOL (e.g. Credi *et al.*, 2006), but their role in bois noir epidemiology is unclear. Only perennial plants would be able to serve as sources of inoculum for *H. obsoletus* nymphs. Due to the polyphagous behaviour of *H. obsoletus* adults, but the restricted host range of nymphs that acquire the phytoplasma, most of those plants might be dead-end hosts as well. However, they could probably play a role in secondary natural epidemiological cycles with minor vector species (Riedle-Bauer *et al.*, 2008).

The level of bois noir shows extensive spatio-temporal fluctuations, with epidemic outbreaks followed by periods of low infection pressure and decreasing disease incidence (Maixner *et al.*, 2006). The relative significance of antagonistic phenomena like new infection and recovery of infected vines, as well as remission and reoccurrence of symptoms, is influenced by the cultivar and modified by abiotic factors or cultural practice. However, only the rate of new infection was found to vary significantly between epidemic and endemic periods. The spatial patterns of BN-diseased vines are influenced by the distribution of the alternative host plants. Random distribution and clusters of diseased vines are often associated with patches of bindweed, which frequently grows within the vineyards. On the other hand, disease gradients due to aggregations of infected vines along the vineyard borders are more common with BN type I, since *U. dioica* is frequently growing on uncultivated land in the vicinity of vineyards (Bressan *et al.*, 2007; Mori *et al.*, 2008).

The management of diseases like BN, where the affected crop is not involved in the epidemiological cycle, is severely hampered, because both the phytoplasma and its vector are endemic and widely dispersed in the natural vegetation. Eradication of infected vines does not hold up disease progress and the application of insecticide in vineyards hardly affects the non-ampelophagous vectors. Infected vines may be tolerated in the vineyards, but appropriate pruning of partially infected vines or cutting of the trunks of systemically infected plants promotes the recovery phenomenon and helps to decrease disease incidence (Stark-Urnau and Kast, 2008). The life history of *H. obsoletus*, its affiliation to wild host plants and its erratic feeding behaviour on grapevine impede the effective control of this species. Insecticide treatments, which are a routine measure to control the vector of flavescence dorée, another grapevine yellows disease, are not effective against *H. obsoletus* (Mori *et al.*, 2008) because the planthopper is not restricted to grapes. Cultural methods such as ploughing densely populated plots during winter can help to reduce the abundance of hibernating vectors (Maixner, 2007). Zahavi *et al.* (2007) showed that caged *V. agnus-castus* plants situated around the vineyards attracted *H. obsoletus*, which thereafter were caught on sticky traps. The vector's density in adjacent vineyard rows was significantly reduced.

The alternative host plants play a key role in BN epidemiology. Weed management is the most effective method to reduce disease pressure, by lowering density and infestation levels of vector populations. Herbicide treatments of nettle stands in early spring reduced the numbers of emerging adult *H. obsoletus* as efficiently as insecticides (Maixner, unpublished). However, the timing of weed control measures is crucial. If the natural hosts are extinguished during the flight period, the search of adult *H. obsoletus* for alternative food sources would increase infection pressure to grapevine. Most appropriate is the control of host plants after the flight period of adult vectors. Open soil with sparse vegetation is highly attractive for *H. obsoletus*. This may explain the often rapid disease progress of BN in young vineyards (Caudwell, 1983). Covering the soil with competitive plants can help to reduce the density of bindweed and nettle at sensitive sites, e.g. the embankments of terraced vineyards, vineyard borders or adjacent uncultivated areas (Maixner, 2007).

Stolbur diseases of other crops

Stolbur phytoplasma is also associated with diseases of potato and tomato and other solanaceous crops with great economic importance (Marzachi, 2008). As long as *H. obsoletus* is involved as a vector, the epidemiological cycle is similar to BN, with the same natural disease cycles branching in direct mode (Caudwell, 1983) to these crops. However, additional vectors are probably involved in STOL transmission to those plants (Batlle *et al.*, 2008; Riedle-Bauer *et al.*, 2008). The syndrome 'basses richesses' of sugarbeet is induced by either infection with a γ-3 proteobacterium or infection with STOL (Bressan *et al.*, 2008). The former is transmitted by *P. leporinus* and the latter by *H. obsoletus*, which spreads it from wild hosts to sugarbeet. Interestingly, only tuf type 'b' associated with bindweed is transmissible to sugarbeet (Bressan *et al.*, 2008). In two crop species, the phytoplasma–vector–plant system obviously evolved to complete epidemiological cycles that do not necessarily depend on alternative sources of inoculum. Decline of lavender (*Lavandula* spp.) is an economically important disease in southern France, which is probably caused by an STOL strain distinct from the BN isolates (Langer and Maixner, 2004). Lavender is a preferred host of *H. obsoletus*, and the vector fulfils its complete life cycle on this plant (Sforza *et al.*, 1999). Although bindweed is considered a source of infection in newly planted fields, the phytoplasma can subsequently be maintained by a cyclic change between lavender and *H. obsoletus*. This simple, hence effective, system causes severe damage to lavender cultivation. STOL was also identified as the causal agent of maize redness (MR), a disease of *Zea mays* in the Balkans (Duduk and Bertaccini, 2006). *H. obsoletus* occurs in low numbers in affected maize fields, but another cixiid planthopper, *R. panzeri*, is much more abundant, with high proportions of infected specimens and it is able to transmit STOL to maize (Jovic *et al.*, 2007). Whether the natural disease cycles play a role in the MR epidemiology, probably by occasional transmission of STOL to maize by *H. obsoletus* as a start-off for the *Reptalus*–maize cycle, is unclear. STOL-infected *R. panzeri* were also found in vineyards (Palermo *et al.*, 2004), but their capability to transmit BN to grapevine has not yet been shown.

Australian grapevine yellows

Australian grapevine yellows (AGY) is associated with '*Ca.* Phytoplasma australiense' (16SrXII-B phylogenetic group). This disease occurred with the increased cultivation of cv. Riesling in Australia ('Rhine Riesling problem'; Magarey and Wachtel (1978)) and gained high economic importance in Australian viticulture. The epidemiology of AGY is unclear, since a vector is still unknown. The spatio-temporal behaviour of AGY hints at an epidemiological system similar to BN: a natural disease cycle including indigenous alternative hosts and an occasional branching of the cycle to grapevine (Magarey *et al.*, 2006). Endemic and introduced wild host plants of '*Ca.* Phytoplasma australiense' have been identified (Streten *et al.*, 2005), and some of these plants are growing in the same habitat as grapevine (Magarey *et al.*, 2006). Environmental conditions might force the (still unknown) vectors to migrate to vineyards, where they inoculate grapevine with AGY (Magarey *et al.*, 2006). Such an irregular short-term occurrence of migrating vectors could explain the hitherto lack of success in identifying the vector of AGY.

Phytoplasmas with Parallel but Interconnected Epidemiological Cycles

Some phytoplasmas infect different plant hosts but do not depend on changing between them because their vector(s) are able to acquire and transmit them from and to each of those plants. If the same vectors feed on both plants, the two systems are interlinked and the phytoplasmas can be exchanged between the two hosts. Vector feeding habits and mobility, as well as the synchrony of host plant and vector phenology, determine whether the relationship is well balanced or one plant is the predominant source for phytoplasma acquisition. Phytoplasma diseases of fruit trees such as apple proliferation (AP) and European stone fruit yellows (ESFY) or Phormium yellow leaf (PYL) disease in New Zealand are examples of such systems.

Phytoplasmas of the 16Sr-X phylogenetic group (apple proliferation group) are economically important pathogens of fruit trees in Europe and North America. The causal agents of the European fruit tree diseases have been described recently as '*Ca.* Phytoplasma mali' (AP), '*Ca.* Phytoplasma pyri' (pear decline, PD) and '*Ca.* Phytoplasma prunorum' (ESFY) (Seemüller and Schneider, 2004). Phytoplasmas of this group not only are closely related but also share epidemiological traits. In contrast to all other phytoplasmas, they are vectored by psyllids (see Jarausch and Jarausch, Chapter 14, this volume). These vectors show a high host plant specificity, being either monophagous or oligophagous on closely related plant species. Dispersal activity of summer generations, as well as the migration to and from specific hibernation sites at elevated areas, favours disease dissemination (Sauvion *et al.*, 2007). Although hibernating adult psyllids were found to be infected, there is no evidence that the conifers used as hibernation hosts have any function for the epidemiology of the phytoplasmas. Besides cultivated fruit trees of different species and cultivars, wild host plants of both vectors and pathogens exist for '*Ca.* Phytoplasma mali' and '*Ca.* Phytoplasma prunorum' (Seemüller and Schneider, 2004). They enable additional epidemiological cycles and may serve as sources of inoculum.

Apple proliferation

Apple proliferation is one of the most important phytoplasma diseases in Europe (Seemüller and Schneider, 2004). Symptoms include proliferation of side shoots, enlarged stipules, discoloration of leaves and small fruit. Diseased trees usually do not die and may recover, but their roots at least stay constantly infected (Seemüller *et al.*, 2008). Due to the blockage of the phloem of aerial parts of apple trees, the pathogen is restricted to the roots during winter, from which they recolonize the trees in spring (Seemüller *et al.*, 2008).

Two psyllid species have been identified as vectors of AP, *Cacopsylla picta* (Forster) (Frisinghelli *et al.*, 2000) and *C. melanoneura* (Forster) (Tedeschi *et al.*, 2002). Their significance as vectors of AP varies due to their regional distribution and prevalence, their host preferences and their natural transmission rates (Jarausch *et al.*, 2007b). The vectors recolonize their feeding hosts in early spring, moving back from distant hibernation sites (Tedeschi *et al.*, 2002). This migratory behaviour leads to dissemination of infective individuals within the range of the vector populations. Hibernated psyllids are able to inoculate apple trees efficiently (Tedeschi *et al.*, 2002), while acquisition of AP by overwintered vectors appears to be negligible because of the still low phytoplasma titres in the aerial parts of apple trees in early spring (Seemüller *et al.*, 2008). However, the overwintered populations are considered to be more significant for AP transmission to apple, because they stay longer on the trees than the summer populations (Tedeschi *et al.*, 2003). The developmental stages of the offspring have a better chance to acquire the phytoplasmas during late spring and summer, but they leave the apple orchards early in the adult stage for hibernation (Tedeschi *et al.*, 2003).

The principal host plant of *C. melanoneura* is not apple but *Crateagus monogyna* (hawthorn), a common shrub. It was also found to be infected by AP phytoplasma (Tedeschi *et al.*, 2009). The host range extension of *C. melanoneura* to apple could have opened the *Crateagus*–*C. melanoneura* system for '*Ca.* Phytoplasma mali' as a new ecological niche. The two epidemiological cycles on apple and hawthorn can exist independently, but they are connected by *C. melanoneura* as the common vector. This situation, however, appears to be restricted to north-western Italy (Tedeschi *et al.*, 2003), while *C. picta* is the predominant vector species in other regions (Jarausch *et al.*, 2007b). Previous reports of the ability of the leafhopper *Fieberiella florii* (Stål) to transmit '*Ca.* Phytoplasma mali' were recently confirmed (Tedeschi and Alma, 2007). Although the significance of this species as a vector of AP is considered to be inferior, the leafhopper could disseminate '*Ca.* Phytoplasma mali' to other host plants due to its polyphagous feeding habit (Tedeschi and Alma, 2007), and thereby further extend its host range.

Management of apple proliferation is difficult, as with all phytoplasma diseases with a complex epidemiology. Alternative host plants, different vectors and phytoplasma strains are involved, and vectors are highly mobile, with seasonal changes of the host plants. Insecticide control of the vectors and uprooting of diseased trees are the most common management tools. Infected trees of tolerant cultivars or recovered trees might be tolerated in orchards, because vectors cannot acquire the phytoplasma if the aerial parts of the trees are free from the phytoplasma (Seemüller *et al.*, 2008). Transmission through root bridges, however, is still possible (Ciccotti *et al.*, 2007). Disease progress of AP might be reduced by control measures, but it cannot be eliminated completely from orchards wherever mobile aerial vectors are involved in the transmission. An alternative approach to reduce the impact of AP is the use of resistant rootstocks (see Seemüller and Harries, Chapter 9, this volume). Their use could prevent the overwintering of '*Ca.* Phytoplasma mali' in the roots and lead to complete recovery of inoculated trees during winter.

European stone fruit yellows

European stone fruit yellows (ESFY) is a common name for diseases of European stone fruit caused by '*Ca.* Phytoplasma prunorum'. The disease is of high economic importance, causing decline of apricot (apricot chlorotic leaf roll), Japanese plum (plum leptonecrosis) and peach (peach decline). '*Ca.* Phytoplasma prunorum' was also detected in wild *Prunus* species, especially the widely distributed *Pr. spinosa* and *Pr. serratina* (Carraro *et al.*, 2002). Some species, such as European plum (*Pr. domestica*) or Myrabolan plum (*Pr. cerasifera*), appear to be tolerant and symptomless carriers of the phytoplasma, while resistant species (*Pr. avium*, *Pr. cerasus*) have also been identified (Carraro *et al.*, 2002).

The only known vector of ESFY is *C. pruni* Scopoli, a strictly oligophagous species on *Prunus* (Carraro *et al.*, 1998). Like the vectors of AP, *C. pruni* hibernates on conifers and returns to its feeding hosts in early spring. The hibernated psyllids are already infective, but the frequency of infection increases with continuing acquisition feeding. The summer generation also acquires the phytoplasma from infected trees in orchards, but many vectors move to secondary hosts before the latency period is completed (Carraro *et al.*, 2004).

The significance of wild *Pr. spinosa* for the epidemiology of ESFY varies between regions. It is a principal host of *C. pruni*, which reaches higher populations on this species than on cultivated *Prunus* species (Carraro *et al.*, 2002; Jarausch *et al.*, 2007a). *Pr. cerasifera*, *Pr. domestica* and *Pr. spinosa* were identified as regular host plants of both *C. pruni* and '*Ca.* Phytoplasma prunorum' in northern Italy (Carraro *et al.*, 2002). This allows the completion of the epidemiological cycle of ESFY phytoplasma apart from cultivated trees in orchards. Nevertheless, the different cycles are overlapping by the vector's moving and feeding behaviour. Some cultivated as well as uncultivated *Pr. domestica* are tolerant carriers of ESFY phytoplasma and, as additional sources of inoculum, could be an unrecognized disease reservoir and a threat to orchards with susceptible fruit trees (Carraro *et al.*, 2002).

Like AP, the ESFY phytoplasma overwinters in the roots of infected trees. Hibernation of the pathogen in the root system could be prevented by the use of resistant rootstocks. The efficient control of *C. pruni* is hampered by its wide distribution on cultivated and wild *Prunus* host species and its movement between different hosts. To remove inoculum inside the orchards, infected trees need to be destroyed. However, the existence of parallel epidemic cycles, including wild and tolerant *Prunus* hosts, impedes the efficient management of disease pressure.

It is interesting from an epidemiological point of view not only that the European diseases caused by phytoplasmas of the 16SrX phylogenetic group are closely related but that their host plants are interconnected by the host affiliation and the feeding preferences of their psyllid vectors (see Jarausch and Jarausch, Chapter 14, this volume). While *C. pruni* feeds only on *Prunus* spp., the 'apple species', *C. picta*, is also found on *Pr. armeniaca*, and *C. melanoneura*, on the other hand, feeds on *Pyrus communis* in addition to *Malus* and *C. monogyna*.

Flax yellows

Phormium yellow leaf (PLY) or flax yellows is a lethal disease of New Zealand flax (*Phormium tenax*) and mountain flax (*Ph. cookianum*), two monocotyledon fibre plants endemic to New Zealand (Liefting *et al.*, 1997). *Ph. tenax* was of high importance for the New Zealand fibre industry. '*Ca.* Phytoplasma australiense' was found to be associated with PLY (Liefting *et al.*, 1998). It is transmitted by a cixiid, the flax planthopper *Zeoliarus atkinsoni* (syn. *Oliarus atkinsoni*), which is oligophagous on the genus *Phormium* (Beever *et al.*, 2004) and completes its whole life cycle on or close to this plant (Liefting *et al.*, 1997). Therefore, the epidemiological cycle of PLY includes only *Ph. tenax* and the vectoring planthopper. '*Ca.* Phytoplasma australiense' is also associated with other diseases of endemic New Zealand plants, while in Australia it is known from various introduced species. This led to the assumption that the pathogen is endemic to New Zealand (Liefting *et al.*, 1998). While strawberry (strawberry lethal yellows) and cabbage tree (*Cordyline australis*; cabbage tree sudden decline) are considered to be secondary hosts of the pathogen, an indigenous and widespread shrub (*Coprosoma robusta*) was found to be frequently infected by '*Ca.* Phytoplasma australiense'. It is suspected to be a natural source plant of the phytoplasma (Beever *et al.*, 2004). Several endemic leafhopper and planthopper species on *C. robusta* are considered as possible vectors of '*Ca.* Phytoplasma australiense' (Beever *et al.*, 2004). The confirmation of their vector status would imply the existence of a second specific natural epidemiological cycle of '*Ca.* Phytoplasma australiense' in New Zealand besides the *Phormium–Z. atkinsoni* system. Interestingly, *Zeoliarus oppositus*, a species closely related to the latter, is more polyphagous and was found on both *C. robusta* and *Ph. tenax* (Beever *et al.*, 2004). This species could provide the link between the two epidemiological systems; however, its ability to vector '*Ca.* Phytoplasma australiense' needs to be confirmed.

Crop-specific Epidemic Systems that are Isolated from the Original Plant–Vector Systems

New epidemiological cycles can evolve from sporadic events of phytoplasma inoculation into erratic hosts if potentially competent vectors exist and are regularly affiliated to the new host plants. This can enable new disease systems to be completely separated from the original pathosystem. In the case of cultivated plants, crop-specific diseases may emerge, whose relationship with naturally occurring diseases of wild plants can only be recognized by DNA-based analyses.

Examples of divergent plant host and vector specificity can be found in the 16SrV phylogenetic group (elm yellows group). Phytoplasmas of this group are closely related, but they constitute several strain clusters that use different ecological niches (Lee *et al.*, 2004). Five subgroups with different host plants, vectors and geographic distribution can be distinguished within the EY group, based on 16S rDNA sequences. The known vectors are representatives of two distinct leafhopper taxa: Nearctic vectors are Deltocephalinae of the genus *Scaphoideus* while Macropsidae of the genera *Macropsis* and *Oncopsis* are of European origin (see Weintraub and Beanland, 2006, for references). Vectors of Asian isolates (subgroup 16SrV-B) have not yet been identified. Pathogens of subgroup 16SrV-A are associated with elm yellows. They are transmitted by *Scaphoideus luteolus* van Duzee in North America and *Macropsis mendax* (Fieber) in Europe, while *M. fuscula* (Zetterstedt) is the vector of phytoplasmas of subgroup 16SrV-E, which are associated with Rubus stunt disease. Phytoplasmas of the subgroup 16SrV-C have a wider range of host plants. They are known from wild woody plants in North America and Europe (Arnaud *et al.*, 2007), and together with subgroup D they are associated with grapevine flavescence dorée in Europe, which is transmitted by *S. titanus* Ball.

Flavescence dorée and alder yellows

Grapevine flavescence dorée (FD) is the most important type of grapevine yellows and classified as a quarantine disease. It is spread over a large geographic region in southern Europe (Boudon-Padieu, 2003). Symptoms resemble those of other grapevine yellows. Infected vines show leaf discoloration, incomplete lignification of shoots and abortion of inflorescences and clusters. FD is graft transmissible and can be disseminated with vegetative propagation material (Boudon-Padieu, 2003).

The Nearctic leafhopper *S. titanus* is the only vector of FD. It was introduced to Europe, presumably at the beginning of the 20th century, probably in the egg stage, hidden in the bark of imported vines (Bertin *et al.*, 2007). The occurrence of FD is connected to the distribution range of this insect. It is strictly ampelophagous and closely associated to grapevine throughout its life cycle, although a sporadic occurrence on *Clematis vitalba* has been reported (Angelini *et al.*, 2004). Larval instars of this univoltine species acquire the FD phytoplasma from infected vines in spring. The risk of infection is highest in late summer and is a function of vector density and degree-days (Bressan *et al.*, 2006), but phytoplasma acquisition and the infestation level of vectors are also influenced by the grapevine cultivar (Bressan *et al.*, 2005b). The epidemiological system is simple and hence very efficient. Epidemic outbreaks of FD may originate from small disease foci or even from single infected vines. Both elements of the disease cycle, infected grapevine and the vector, can be targets for control measures. Therefore, management of FD is more efficient than that of bois noir or other grapevine yellows with alternative, non-grapevine sources of infection. Infected vines are destroyed and the vector, living exclusively on vines, is regularly controlled by insecticides (Boudon-Padieu, 2003). Both measures are compulsory in countries or regions where FD occurs.

Phytoplasmas of two subgroups (16SrV-C and -D) are associated with FD and transmitted by *S. titanus*. FD of type C is only known from north-eastern Italy, but phytoplasmas of subgroup D are widespread in Europe, presumably because of the dissemination of infected grapevine wood (Arnaud *et al.*, 2007). Genetic analyses suggest that *S. titanus*, whose eggs are deposited in the bark of shoots and canes, was also spread with grapevine wood (Bertin *et al.*, 2007).

The origin of FD is still unclear. It was suspected to come from North America, like many other pathogens in viticulture, because of its very close association with the Nearctic vector (Caudwell, 1983). However, FD has not been detected in North America so far, although subgroup 16SrV-C phytoplasmas were found in *Apocynum cannabinum* and *Parthenocissus quinquefolia* (Lee *et al.*, 2004), and *S. titanus* collected from *Vitis riparia* in New York reacted positively with antibodies against FD (Maixner *et al.*, 1993). The tolerance of American *Vitis* species and rootstocks to FD (Caudwell, 1983), on the other hand, could be taken as a result of a long co-evolution with the pathogen (Elliott *et al.*, 2003), while the observation of adverse effects of FD infection on *S. titanus* (Bressan *et al.*, 2005a) supports the hypothesis of a rather new association of the phytoplasma with this vector. According to the idea of an American origin of FD, a complete epidemiological cycle was transferred to Europe and was able to branch to *Vitis vinifera* as an alternative, although extremely susceptible, *Vitis* host, due to the oligophagous feeding habit of *S. titanus*. The affiliation of FD phytoplasmas with two distinct 16SrV groups implies at least two introduction events of this pathogen into Europe.

Recent data support an alternative hypothesis of a European origin of FD, because they show a close phylogenetic relationship of FD and phytoplasmas associated to alder yellows (Angelini *et al.*, 2001; Arnaud *et al.*, 2007). Alder yellows (AldY) phytoplasmas belong to group 16SrV-C and are frequently associated with *Alnus glutinosa* and *A. incana* in Europe (Lederer and Seemüller, 1991). Different isolates of AldY have been described (Maixner and Reinert, 1999; Marcone *et al.*, 1997), often together in the same trees. *Oncopsis alni* Shrank, a leafhopper of the family Macropsidae, is the vector of AldY (Maixner and Reinert, 1999). This univoltine species is strictly oligophagous on *Alnus*.

Palatinate grapevine yellows (PGY) emerged as a new grapevine yellows disease in the 1990s in Germany. Unlike BN it did not occur in xerothermic habitats but was mainly found in vineyards close to waterbodies lined by alder trees. Three different phytoplasma isolates were found to be associated with PGY, and they proved to be indistinguishable from the AldY strains that infected alder in the same area. Their experimental transmission from alder to grapevine by *O. alni* was successful, but the transmission efficiency was low (Maixner *et al.*, 2000). In spite of the strict oligophagy of *O. alni*, a few specimens of this vector are usually found in vineyards close to alders, probably due to passive wind drift. Once in a vineyard, the vector may inoculate grapevine by erratic feeding on this atypical host. PGY is therefore another example, besides BN, for a natural epidemiological cycle branching to a cultivated plant by irregular feeding of the vector. However, the disease pressure is very low compared with BN, because of the extreme host restriction of *O. alni* and its low transmission efficiency of AldY/PGY to grapevine. Corresponding to the low risk of infection, PGY is almost exclusively found in rather old vineyards and with a low incidence (<1%), but it seems to occur wherever grapevine is grown in close association with alder trees.

A phylogenetic analysis of non-ribosomal gene sequences proved the presence of three distinct FD gene clusters and revealed a monophyletic origin of FD, AldY and PGY phytoplasmas (Arnaud *et al.*, 2007). The very close relationship of an AldY phytoplasma isolate (ALY) with FD (Angelini *et al.*, 2001) was confirmed. Some isolates of AldY, for example, proved to be more closely related to particular FD isolates than to each other. This genetic evidence for a common origin of the alder phytoplasmas and FD, as well as the occurrence of FD-C in *C. vitalba* in Italy (Angelini *et al.*, 2004), supports the hypothesis of a European origin of FD.

The occasional transmission of AldY phytoplasmas from the original alder hosts to grapevine by *O. alni* is most probably a common phenomenon that, under normal circumstances, has no further consequences except of single infected vines. However, in the case of a coincidental presence of a competent and ampelophagous vector like *S. titanus* on such an infected vine, a new disease cycle could be initiated. Given the existence of three distinct FD strain clusters in Europe (Arnaud *et al.*, 2007), this must have happened at least three times. There is probably a constant risk of the emergence of new FD isolates wherever *S. titanus* occurs and AldY is occasionally transmitted to grapevine.

Flavescence dorée is, irrespective of the origin, a good example of a new epidemiological system of a phytoplasma that could become independent and isolated from the original cycle as the result of the introduction of a new, competent vector species. This underlines the importance of phytosanitary measures to prevent the dissemination of both phytoplasmas and their vectors. Both the introduction of a competent vector into a new area and the movement of infected plants to disease-free areas pose the risk of new phytoplasma–plant–vector combinations. Plants such as grapevine, which are propagated vegetatively and are subject to supraregional trade, bear a high risk of dissemination of the graft-transmissible phytoplasmas. The widespread occurrence and clonal structure of one of the FD strain clusters identified by Arnaud *et al.* (2007) and the lack of genetic structuring within the European populations of *S. titanus* (Bertin *et al.*, 2007) underline the role of grapevine propagation material for both the dissemination of FD and the spread of its vector.

Concluding Remarks

The examples of phytoplasma disease cycles presented in this chapter underline the importance of the specific interrelationship of phytoplasmas with their particular host plants and vectors but also of the vector's biology and life history for disease epidemiology. The incursion of cultivated plants in previously unknown natural disease cycles may trigger sudden outbreaks of economically significant diseases. The identification of vectors and natural sources of inoculum is a prerequisite for the development of control strategies but does not assure that specific and effective management solutions are found, for example if distribution or behaviour of vectors impede efficient control measures.

Current research activities on the molecular background of host specificity and the genetic variability of phytoplasmas as well as vectors could help to improve our understanding of the mechanisms of host specificity and virulence and thereby provide new means of prevention or control of phytoplasma diseases. However, they are a complement to, but not a substitute for, comprehensive field studies of the epidemiological parameters and their dependence on environmental factors and cultural practice. Investigations of the influence of phytoplasma–plant interactions on the attraction of infected plants for vectors might open additional perspectives for specific control strategies.

References

Alma, A., Soldi, G., Tedeschi, R. and Marzachì, C. (2002) Role of *Hyalesthes obsoletus* Signoret (Homoptera Cixidae) in the transmission of grapevine bois noir in Italy. *Petria* 12, 411–412.

Angelini, E., Clair, D., Borgo, M., Bertaccini, A. and Boudon-Padieu, E. (2001) Flavescence doree in France and Italy – occurrence of closely related phytoplasma isolates and their near relationships to Palatinate grapevine yellows and an alder yellows phytoplasma. *Vitis* 40, 79–86.

Angelini, E., Squizzato, F., Lucchetta, G. and Borgo, M. (2004) Detection of a phytoplasma associated with grapevine flavescence doree in *Clematis vitalba*. *European Journal of Plant Pathology* 110, 193–201.

Arnaud, G., Malembic-Maher, S., Salar, P., Bonnet, P., Maixner, M., Marcone, C., Boudon-Padieu, E. and Foissac, X. (2007) Multilocus sequence typing confirms the close genetic interrelatedness of three distinct flavescence dorée phytoplasma strain clusters and group 16SrV phytoplasmas infecting grapevine and alder in Europe. *Applied and Environmental Microbiology* 73, 4001–4010.

Batlle, A., Altabella, N., Sabate, J. and Lavina, A. (2008) Study of the transmission of stolbur phytoplasma to different crop species, by *Macrosteles quadripunctulatus*. *Annals of Applied Biology* 152, 235–242.

Beever, R.E., Wood, G.A., Andersen, M.T., Pennycook, S.R., Sutherland, P.W. and Forster, R.L.S. (2004) '*Candidatus* phytoplasma australiense' in *Coprosoma robusta* in New Zealand. *New Zealand Journal of Botany* 42, 663–675.

Bertin, S., Guglielmino, C.R., Karam, N., Gomulski, L.M., Malacrida, A. and Gasperi, G. (2007) Diffusion of the Nearctic leafhopper *Scaphoideus titanus* Ball in Europe: a consequence of human trading activity. *Genetica* 131, 275–285.

Boudon-Padieu, E. (2003) The situation of grapevine yellows and current research directions: distribution, diversity, vectors, diffusion and control. *Extended Abstracts 14th Meeting of the ICVG*, 12–17 September 2003, Locorotondo (Bari), Italy, pp. 47–53.

Bressan, A., Girolami, V. and Boudon-Padieu, E. (2005a) Reduced fitness of the leafhopper vector *Scaphoideus titanus* exposed to flavescence dorée phytoplasma. *Entomologia Experimentalis et Applicata* 115, 283–290.

Bressan, A., Spiazzi, S., Girolami, V. and Boudon-Padieu, E. (2005b) Acquisition efficiency of flavescence dorée phytoplasma by *Scaphoideus titanus* Ball from infected tolerant or susceptible grapevine cultivars or experimental host plants. *Vitis* 44, 143–146.

Bressan, A., Larrue, J. and Boudon-Padieu, E. (2006) Patterns of phytoplasma infected and infective *Scaphoideus titanus* leafhoppers in vineyards with high incidence of flavescence dorée. *Entomologia Experimentalis et Applicata* 119, 61–69.

Bressan, A., Turata, R., Maixner, M., Spiazzi, S., Boudon-Padieu, E. and Girolami, V. (2007) Vector activity of *Hyalesthes obsoletus* living on nettles and transmitting a stolbur phytoplasma to grapevines: a case study. *Annals of Applied Biology* 150, 331–339.

Bressan, A., Semetey, O., Nusillard, B., Clair, D. and Boudon-Padieu, E. (2008) Insect vectors (Hemiptera : Cixiidae) and pathogens associated with the disease syndrome 'Basses Richesses' of sugar beet in France. *Plant Disease* 92, 113–119.

Carraro, L., Osler, R., Loi, N., Ermacora, P. and Refatti, E. (1998) Transmission of European stone fruit yellows phytoplasma by *Cacopsylla pruni*. *Journal of Plant Pathology* 80, 233–239.

Carraro, L., Ferrini, F., Ermacora, P. and Loi, N. (2002) Role of wild *Prunus* species in the epidemiology of European stone fruit yellows. *Plant Pathology* 51, 513–517.

Carraro, L., Ferrini, F., Labonne, G., Ermacora, P. and Loi, N. (2004) Seasonal infectivity of *Cacopsylla pruni*, vector of European stone fruit yellows phytoplasma. *Annals of Applied Biology* 144, 191–195.

Caudwell, A. (1983) L'origine des jaunisses a mycoplasmes (MLO) des plantes et l'exemple des jaunisses de la vigne. *Agronomie* 3, 103–111.

Ciccotti, A.M., Bianchedi, P.L., Biragagna, P., Deromedi, M., Filippi, M., Forno, F. and Mattedi, L. (2007) Transmission of 'Candidatus Phytoplasma mali' by root bridges under natural and experimental conditions. *Bulletin of Insectology* 60(2), 387–388.

Credi, R., Terlizzi, F., Milanesi, L., Bondavalli, R., Cavallini, G., Montermini, A. and Dradi, D. (2006) Wild host plants of stolbur phytoplasma and its vector, *Hyalesthes obsoletus*, at sites of grapevine bois noir occurrence in Emilia-Romagna, Italy. *Extended Abstracts 15th Meeting ICVG*, 3–7 April 2006, Stellenbosch, South Africa, pp. 182–183.

Danet, J.L., Foissac, X., Zreik, L., Salar, P., Verdin, E., Nourrisseau, J.G. and Garnier, M. (2003) '*Candidatus* Phlomobacter fragariae' is the prevalent agent of marginal chlorosis of strawberry in French production fields and is transmitted by the planthopper *Cixius wagneri* (China). *Phytopathology* 93, 644–649.

Duduk, B. and Bertaccini, A. (2006) Corn with symptoms of reddening: new host of stolbur phytoplasma. *Plant Disease* 90, 1313–1319.

Elliott, S.L., Adler, F.R. and Sabelis, M.W. (2003) How virulent should a parasite be to its vector? *Ecology* 84, 2568–2574.

Firrao, G., Garcia-Chapa, M. and Marzachì, C. (2007) Phytoplasmas: genetics, diagnosis and relationships with the plant and insect hosts. *Frontiers in Bioscience* 12, 1353–1375.

Fos, A., Danet, J.L., Zreik, L., Garnier, M. and Bove, J.M. (1992) Use of a monoclonal antibody to detect the stolbur mycoplasmalike organism in plants and insects and to identify a vector in France. *Plant Disease* 76, 1092–1096.

Frisinghelli, C., Delaiti, L., Grando, M.S., Forti, D. and Vindimian, M.E. (2000) *Cacopsylla costalis* (Flor 1861), as a vector of apple proliferation in Trentino. *Journal of Phytopathology* 148, 425–431.

Ge, Q. and Maixner, M. (2003) Comparative experimental transmission of grapevine yellows phytoplasmas to plants and artificial feeding medium. *Extended Abstracts 14th Meeting of the ICVG*, 12–17 September 2003, Locorotondo (Bari), Italy, pp. 109–110.

Jarausch, B., Fuchs, A., Muhlenz, I., Lampe, I., Harzer, U. and Jarausch, W. (2007a) Research on European stone fruit yellows (ESFY) in Germany. *Bulletin of Insectology* 60(2), 389–390.

Jarausch, B., Fuchs, A., Schwind, N., Krcal, G. and Jarausch, W. (2007b) *Cacopsylla picta* as most important vector for '*Candidatus* Phytoplasma mali' in Germany and neighbouring regions. *Bulletin of Insectology* 60, 189–190.

Johannesen, J., Lux, B., Michel, K., Seitz, A. and Maixner, M. (2008) Invasion biology and host specificity of the grapevine yellows disease vector *Hyalesthes obsoletus* in Europe. *Entomologia Experimentalis et Applicata* 126, 217–227.

Jovic, J., Cvrkovic, T., Mitrovic, M., Krnjajic, S., Redinbaugh, M.G., Pratt, R.C., Gingery, R.E., Hogenhout, S.A. and Tosevski, I. (2007) Roles of stolbur phytoplasma and *Reptalus panzeri* (Cixiinae, Auchenorrhyncha) in the epidemiology of maize redness in Serbia. *European Journal of Plant Pathology* 118, 85–89.

Kaul, C., Seitz, A., Maixner, M. and Johannesen, J. (2009) Time of infection by Bois Noir tuf-type-I stolbur phytoplasma in *Hyalesthes obsoletus* (Homoptera: Cixiidae) and influence on larval size. *Journal of Applied Entomology* 133, 596–601.

Langer, M. and Maixner, M. (2004) Molecular characterisation of grapevine yellows associated phytoplasmas of the stolbur-group based on RELP-analysis of non-ribosomal DNA. *Vitis* 43, 191–199.

Lederer, W. and Seemüller, E. (1991) Occurrence of mycoplasma-like organisms in diseased and non-symptomatic alder trees (*Alnus* spp.). *European Journal of Forest Pathology* 21, 90–96.

Lee, I.M., Gundersen-Rindal, D.E. and Bertaccini, A. (1998) Phytoplasma: ecology and genomic diversity. *Phytopathology* 88, 1359–1366.

Lee, I.M., Martini, M., Marcone, C. and Hu, S.F. (2004) Classification of phytoplasma strains in the elm yellows group (16SrV) and proposal of '*Candidatus* Phytoplasma ulmi' for the phytoplasma associated with elm yellows. *International Journal of Systematic and Evolutionary Microbiology* 54, 337–347.

Liefting, L.W., Beever, R.E., Winks, C.J., Pearson, M.N. and Forster, R.L.S. (1997) Planthopper transmission of Phormium yellow leaf phytoplasma. *Australasian Plant Pathology* 26, 148–154.

Liefting, L.W., Padovan, A.C., Gibb, K.S., Beever, R.E., Andersen, M.T., Newcomb, R.D., Beck, D.L. and Forster, R.L.S. (1998) '*Candidatus* Phytoplasma australienses' is the phytoplasma associated with Australian grapevine yellows, papaya dieback and Phormium yellow leaf diseases. *European Journal of Plant Pathology* 104, 619–623.

Magarey, P.A. and Wachtel, M.F. (1978) The Rhine Riesling problem – most severe in the Riverland. *Australian Grapegrower and Winemaker* 172, 24–26.

Magarey, P.A., Gillett, S.R., Magarey, C.C., Thiele, S.A., Bogacz, H. and Habili, N. (2006) Australian grapevine yellows: first detection of the phytoplasma in non-vitis hosts in South Australia. *Extended Abstracts 15th Meeting ICVG*, 3–7 April 2006, Stellenbosch, South Africa, pp. 205–206.

Maixner, M. (2007) Biology of *Hyalesthes obsoletus* and approaches to control this soil-borne vector of bois noir disease. *IOBC/WPRS Bulletin* 30(7), 3–9.

Maixner, M. and Reinert, W. (1999) *Oncopsis alni* (Schrank) (Auchenorrhyncha: Cicadellidae) as a vector of the alder yellows phytoplasma of *Alnus glutinosa* (L.). Gaertn. *European Journal of Plant Pathology* 105, 87–94.

Maixner, M., Pearson, R.C., Boudon-Padieu, E. and Caudwell, A. (1993) *Scaphoideus titanus* Ball, a possible vector of grapevine yellows in New York. *Plant Disease* 77, 408–413.

Maixner, M., Ahrens, U. and Seemüller, E. (1995) Detection of the German grapevine yellows (Vergilbungskrankheit) MLO in grapevine, alternative hosts and a vector by a specific PCR procedure. *European Journal of Plant Pathology* 101, 241–250.

Maixner, M., Reinert, W. and Darimont, H. (2000) Transmission of grapevine yellows by *Oncopsis alni* (Schrank) (Auchenorrhyncha: Macropsinae). *Vitis* 39, 83–84.

Maixner, M., Langer, M. and Gerhard, Y. (2006) Epidemiological characteristics of bois noir type I. *Extended Abstracts 15th Meeting ICVG*, 3–7 April 2006, Stellenbosch, South Africa, pp. 86–87.

Marcone, C., Ragozzino, A. and Seemüller, E. (1997) Identification and characterization of the phytoplasma associated with elm yellows in southern Italy and its relatedness to other phytoplasmas of the elm yellows group. *European Journal of Forest Pathology* 27, 45–54.

Marzachì, C. (2008) Stolbur phytoplasma. In: Harrison, N.A., Rao, G.P. and Marcone, C. (eds) *Characterization, Diagnosis and Management of Phytoplasmas*. Studium Press LLC, Houston, Texas, pp. 365–386.

Mori, N., Pavan, F., Bondavalli, R., Reggiani, N., Paltrinieri, S. and Bertaccini, A. (2008) Factors affecting the spread of 'Bois Noir' disease in north Italy vineyards. *Vitis* 47, 65–72.

Pacifico, D., Foissac, X., Veratti, F. and Marzachì, C. (2007) Genetic diversity of Italian phytoplasma and French 'bois noir' isolates. *Bulletin of Insectology* 60(2), 345–346.

Palermo, S., Elekes, M., Botti, S., Ember, I., Alma, A., Orosz, A., Bertaccini, A. and Kolber, M. (2004) Presence of stolbur phytoplasma in Cixiidae in Hungarian vineyards. *Vitis* 43, 201–203.

Pasquini, G., Ferretti, L., Gentili, A., Bagnoli, B., Cavalieri, V. and Barba, M. (2007) Molecular characterization of stolbur isolates collected in grapevines, weeds and insects in central and southern Italy. *Bulletin of Insectology* 60(2), 355–356.

Pinzauti, F., Trivellone, V. and Bagnoli, B. (2008) Ability of *Reptalus quinquecostatus* (Hemiptera: Cixiidae) to inoculate stolbur phytoplasma to artificial feeding medium. *Annals of Applied Biology* 153, 299–305.

Quaglino, F., Zhao, Y., Bianco, P.A., Wei, W., Casati, P., Durante, G. and Davis, R.E. (2009) New 16Sr subgroups and distinct SNP lineages among grapevine bois noir phytoplasma populations. *Annals of Applied Biology* 154, 279–289.

Riedle-Bauer, M., Sara, A. and Regner, F. (2008) Transmission of a stolbur phytoplasma by the Agalliinae leafhopper *Anaceratagallia ribauti* (Hemiptera, Auchenorrhyncha, Cicadellidae). *Journal of Phytopathology* 156, 687–690.

Sauvion, N., Lachenaud, O., Genson, G., Rasplus, J.Y. and Labonne, G. (2007) Are there several biotypes of *Cacopsylla pruni*? *Bulletin of Insectology* 60(2), 185–186.

Seemüller, E. and Schneider, B. (2004) '*Candidatus* Phytoplasma mali', '*Candidatus* Phytoplasma pyri' and '*Candidatus* Phytoplasma prunorum', the causal agents of apple proliferation, pear decline and European stone fruit yellows, respectively. *International Journal of Systematic and Evolutionary Microbiology* 54, 1217–1226.

Seemüller, E., Moll, E. and Schneider, B. (2008) Apple proliferation resistance of *Malus sieboldii*-based rootstocks in comparison to rootstocks derived from other *Malus* species. *European Journal of Plant Pathology* 121, 109–119.

Semetey, O., Gatineau, F., Bressan, A. and Boudon-Padieu, E. (2007) Characterization of a gamma-3 proteobacteria responsible for the syndrome 'basses richesses' of sugar beet transmitted by *Pentastiridius* sp. (Hemiptera, Cixiidae). *Phytopathology* 97, 72–78.

Sforza, R., Bourgoin, T., Wilson, S.W. and Boudon-Padieu, E. (1999) Field observations, laboratory rearing and descriptions of immatures of the planthopper *Hyalesthes obsoletus* (Hemiptera : Cixiidae). *European Journal of Entomology* 96, 409–418.

Sharon, R., Soroker, V., Wesley, S.D., Zahavi, T., Harari, A. and Weintraub, P.G. (2005) *Vitex agnus-castus* is a preferred host plant for *Hyalesthes obsoletus*. *Journal of Chemical Ecology* 31, 1051–1063.

Stark-Urnau, M. and Kast, W.K. (2008) Control methods to reduce bois noir disease (Stolbur Type I) in grape vine (*Vitis vinifera*). *Gesunde Pflanzen* 60, 85–89.

Streten, C., Herrington, M.E., Hutton, D.G., Persley, D.M., Waite, G.K. and Gibb, K.S. (2005) Plant hosts of the phytoplasmas and rickettsia-like-organisms associated with strawberry lethal yellows and green petal diseases. *Australasian Plant Pathology* 34, 165–173.

Tedeschi, R. and Alma, A. (2007) '*Candidatus* Phytoplasma mali': the current situation of insect vectors in northwestern Italy. *Bulletin of Insectology* 60(2), 187–188.

Tedeschi, R., Bosco, D. and Alma, A. (2002) Population dynamics of *Cacopsylla melanoneura* (Homoptera : Psyllidae), a vector of apple proliferation phytoplasma in northwestern Italy. *Journal of Economic Entomology* 95, 544–551.

Tedeschi, R., Visentin, C., Alma, A. and Bosco, D. (2003) Epidemiology of apple proliferation (AP) in northwestern Italy: evaluation of the frequency of AP-positive psyllids in naturally infected populations of *Cacopsylla melanoneura* (Homoptera: Psyllidae). *Annals of Applied Biology* 142, 285–290.

Tedeschi, R., Lauterer, P., Brusetti, L., Tota, F. and Alma, A. (2009) Composition, abundance and phytoplasma infection in the hawthorn psyllid fauna of northwestern Italy. *European Journal of Plant Pathology* 123, 301–310.

Weintraub, P.G. and Beanland, L. (2006) Insect vectors of phytoplasma. *Annual Review of Entomology* 51, 91–111.

Zahavi, T., Peles, S., Harari, A.R., Soroker, V. and Sharon, R. (2007) Push and pull strategy to reduce *Hyalesthes obsoletus* population in vineyards by *Vitex agnus-castus* as trap plant. *Bulletin of Insectology* 60(2), 297–298.

13 Control of Phytoplasma Diseases and Vectors

PHYLLIS G. WEINTRAUB[1] AND MICHAEL R. WILSON[2]

[1]*Agricultural Research Organization, Israel;* [2]*National Museum of Wales, UK*

Introduction

Phytoplasmas are non-culturable, degenerate, Gram-positive prokaryotes, causing hundreds of diseases in plants. They are phloem-limited; as such, they are transmitted by specific phloem-feeding Auchenorrhyncha and Sternorrhyncha. The spread and progress of plant diseases in general are influenced by inoculum density and the presence of a vector. Management of vectors in economically important plants differs for annual versus perennial crops, as losses in annual crops are typically seasonal, while losses in perennial plant systems accumulate over many years. A large body of research that addresses the biology, ecology, vector relationships and epidemiology of crop diseases caused by phytoplasmas has accumulated in the last two decades (see reviews: Christensen *et al.*, 2005; Weintraub and Beanland, 2006; Bertaccini, 2007). This chapter will provide a comprehensive overview of the vector groups and their control.

Vector Taxonomy and Biology

The Hemiptera are a large and diverse order of exopterygote insects which occur in all zoogeographic regions of the world. There are more than 50,000 species in about 100 families. The Hemiptera are now divided into three suborders: Heteroptera (true bugs), Sternorrhyncha (scale insects, aphids, whiteflies, psyllids) and Auchenorrhyncha (leafhoppers, planthoppers, cicadas, treehoppers and spittlebugs). Phylogenetic relationships within the Hemiptera have been summarized by Forero (2008). The single most successful order of phytoplasma vectors is the Hemiptera, and, within the Auchenorrhyncha and Sternorrhyncha, over 200 leafhopper, planthopper and psyllid vectors of phytoplasma, spiroplasma, virus and *Xylella* are already known.

Many more species are to be expected because there are more diseases characterized than there are known disease vectors.

Auchenorrhyncha feeding strategies

Most Auchenorrhyncha feed from phloem tissue, but two superfamilies (Cicadoidea: cicadas; Cercopoidea: spittlebugs) and a subfamily of the Cicadellidae (Cicadellinae) feed from xylem tissue. In addition, the majority of species in the leafhopper subfamily Typhlocybinae feed by removing the cell contents from mesophyll cells.

Since phytoplasmas are phloem-limited, only phloem-feeding insects can potentially acquire and transmit the pathogen. This should assist in the search for any disease vectors. In Weintraub and Beanland's (2006) review of the vectors of phytoplasmas, they point out that this group collectively possesses several characteristics that make its members efficient vectors of phytoplasmas: (i) they are hemimetabolous; thus, nymphs and adults feed similarly and are in the same physical location and often both immatures and adults can transmit phytoplasma; (ii) they feed specifically and selectively on certain plant tissues, which makes them efficient vectors of pathogens residing in those tissues. Furthermore, their feeding is non-destructive, promoting successful inoculation of the plant vascular system without damaging conductive tissues and eliciting defensive responses; (iii) they have a propagative and persistent relationship with phytoplasmas; and (iv) they have obligate symbiotic prokaryotes that are passed to the offspring by transovarial transmission, the same mechanisms that allow the transovarial transmission of phytoplasmas.

Taxonomic groups with phytoplasma vectors

Within the groups of phloem-feeding insects only a small number, primarily in three taxonomic groups, have been confirmed as vectors of phytoplasmas. Weintraub and Beanland (2006, supplemental material) provide a table of all known vector species, which was recently updated (Wilson and Weintraub, 2007). The account by Nielson (1968) of leafhopper vectors, while still useful, is a little difficult; as well as changes in insect taxonomy in the intervening 40 years, perhaps a more significant problem in using Nielson's work is that phytoplasma and virus diseases were poorly understood so 'virus vectors' and 'phytoplasma vectors' are confused. Nielson (1979) discussed and listed the known insect vector species.

The superfamily containing the largest number of vector species is the Membracoidea, within which all known vectors are confined to the family Cicadellidae (leafhoppers). Morphological and molecular evidence indicates that the Membracoidea are a monophyletic superfamily (Dietrich *et al.*, 2001). However, the phylogenetic status and relationships of the families, subfamilies and tribes remain poorly understood, although progress is being made

Table 13.1. Distribution of phytoplasma vector species in tribes of Deltocephalinae and groups of phytoplasmas transmitted.

Tribe	Number of species	Phytoplasma transmitted
Acinophorini	1	Aster yellows, X-disease
Athysanini	12	Aster yellows, X-disease, elm yellows, rice yellow dwarf, stolbur, phyllody, Rhynchosia little leaf
Deltocephalini	8	Aster yellows, X-disease, elm yellows, grassy shoot, maize bushy stunt, phyllody, rice orange leaf
Fieberiellini	2	Aster yellows, X-disease, apple proliferation, Mexican periwinkle virescence
Macrostelini	13	Aster yellows, X-disease, stolbur, Kok-saghyz yellows, Lissers, sugarcane white leaf
Opsiini	11	Aster yellows, peanut witches'-broom, X-disease, elm yellows, clover proliferation, eggplant little leaf, phyllody, purple top
Scaphoideini	3	Aster yellows, X-disease, elm yellows
Scaphytopiini	7	Aster yellows, X-disease, elm yellows, machismo

(Zahniser and Dietrich, 2008). More than 75% of all confirmed phytoplasma vector species are found in the subfamily Deltocephalinae. The feeding habits of species within this subfamily range from monophagous to polyphagous, and members of this group can transmit one or more different phytoplasma taxa (Table 13.1). Most species are found within grassland ecosystems and they may occur abundantly. However, the tribes Opsiini, Macrostelini, Scaphodeini and Scaphtyopiini contain known vector species that are not confined to grass species. Such grass crops as rice, wheat, maize and sugarcane are important to us for food, and therefore diseases and pests have been extensively studied. Little is known about the host relationships of the majority of species, but it is likely that they are narrowly oligotrophic. The subfamily Deltocephalinae is divided into around 23 tribes and over 750 genera with about 7000 described species. The status and relationships of these tribes is under discussion, but it is worthwhile investigating further the distribution of the known vectors within the tribes.

Given that only about 200 vectors are identified and many more phytoplasma diseases have been recognized, the number of species within each tribe will probably increase greatly. It is interesting that some genera within certain tribes seem to be virus vectors (e.g. *Cicadulina* species in Africa), while others seem only to be involved with the transmission of phytoplasma. The subfamily containing the second largest number of confirmed vector species is the Macropsinae. Vector members of the Macropsinae can be monophagous or oligophagous but most feed primarily on woody plants. This subfamily is more highly derived (following the molecular scheme) than all of the remaining subfamilies; the more basal subfamilies of the Deltocephalinae have only one or two vector species. However, in a comparison of the number of competent vectors as a percentage of the total known species for a group, 10% of the Aphrodinae (a less derived subfamily) are phytoplasma vectors, as opposed to the Deltocephalinae, which have only 0.8%.

On the basis of analysis of ribosomal DNA, the morphologically distinct membracids are part of the Cicadellidae; however, to date, no membracids have been confirmed as, or are suspected of, transmitting phytoplasmas. Although membracids are relatively poor transmitters of viruses compared with leafhoppers, it is unknown whether researchers have not considered membracids for use in phytoplasma vector studies because they appear to be a group distinct from the leafhoppers (which are known vectors) or because membracids actually do not transmit phytoplasmas. Because membracids tend to feed on woody hosts, it would not be surprising to find that they transmit phytoplasmas in the groups found primarily in woody plants: western-X (WX), pear decline (PD), apple proliferation (AP) and European stone fruit yellows (ESFY).

Vector species are found in four planthopper families (Fulgoromorpha): Cixiidae, Delphacidae, Derbidae and one species in the Flatidae. The first three families all have at least one species that transmits a phytoplasma in the coconut lethal yellows group. Several species in these families also transmit phytoplasmas from the stolbur (STOL) group. The one flatid vector, *Metcalfa pruinosa* (Say), transmits aster yellows. At present two genera of psyllids include vectors. *Cacopsylla* spp. transmit AP group (16SrX) phytoplasmas to pome and stone fruit trees. The other psyllid genus has one vector species, *Bactericera trigonica* Hodkinson, which transmits a STOL to carrots.

Transmission by vectors

In phytoplasma disease systems that have been characterized, a specific sequence of events is necessary for insects to transmit the pathogen to new hosts. An uninfected insect feeds in the phloem of an infected plant, obtaining nutrition from free amino acids and sugars and ingests phytoplasma particles residing therein. The feeding duration necessary to acquire a sufficient titre of phytoplasma is the acquisition access period (AAP), which can be as short as a few minutes but is generally measured in hours; the longer the AAP, the greater the chance of transmission (Purcell, 1982). The ingested phytoplasma particles must penetrate the insect midgut cells and move into the insect haemocoel, where they are transported throughout the insect body with haemolymph. Phytoplasma particles reach the salivary glands and penetrate the gland cells, where they can multiply. After an incubation period (putatively 10 days to 3 weeks), during which the phytoplasma particles invade the insect tissues and multiply, the insect will release phytoplasma with saliva secreted from the salivary glands when feeding in the phloem of a plant host. The period of time that elapses from initial acquisition to the ability to transmit the phytoplasma is known as the latent period (LP) and is sometimes referred to as the incubation period. The phytoplasma takes up residence in the plant host and begins to multiply. After a latent period (highly variable and host plant dependent), the plant begins to develop symptoms of disease. Once the titre of phytoplasma is sufficiently high, this plant can serve as an acquisition host for any vector species feeding upon it.

Vector–host plant interactions play an important role in limiting or expanding the spread of phytoplasmas. Broadly polyphagous vectors have the potential to inoculate a wider range of plant species, depending on the resistance to infection of each host plant. Several studies have shown that insects that normally do not feed on certain plant species can acquire and transmit phytoplasma to those plants under laboratory conditions. This can also occur under field conditions: the cixiid vector of bois noir (BN) disease of grapevine 'accidentally' transmits the STOL – *Hyalesthes obsoletus* Signoret cannot live on vines. Hence, in many cases, the plant host range of a vector, rather than lack of phytoplasma-specific cell membrane receptors, will limit the spread of phytoplasma by that species (see review, Weintraub and Beanland, 2006).

Vector/Disease Management

Until recently, management of phytoplasma-caused plant diseases focused on spraying insecticides, with or without regard to specifically managing the vector(s). Insect species present at the time a phytoplasma infection is observed may or may not have been present when transmission took place, often weeks or months earlier. Only transmission experiments can provide evidence of the capacity of candidate species actually to transmit phytoplasma to healthy plants.

Conventional insecticides, even when frequently used (e.g. Wally *et al.*, 2004), will not control the appearance of disease, because pathogen transmission occurs faster than insecticides can act and there is often a constant influx of new vectors from surrounding habitats. Mori *et al.* (2008) studied the effects of insecticides applied to the central canopy of grapevines in 18 vineyards on reducing *H. obsoletus* populations and found that there was no significant reduction in vector populations. This was probably due to the fact that *H. obsoletus* prefers other plants and is only incidentally found on grapevine. The use of insecticides might help control vector populations and thus reduce intra-crop transmission. Pilkington *et al.* (2004) have demonstrated that total vector population control can be achieved and disease incidence reduced by just treating crop borders with systemic insecticides. This tactic targets only the vector population, without extensive insecticide application. More recently Saracco *et al.* (2008) have developed a two-pronged strategy vis-à-vis the use of insecticides: to protect plants from infectious migratory vectors use systemics such as neonicotinoids, to suppress vectors use organophosphate insecticides. There are some methods that can be used independently or in conjunction with other methods; many factors need to be considered to arrive at the appropriate management tactics. Some methods are more effective for certain kinds of crops; for example, roguing or destruction of symptomatic plants is more effective in orchards or perennials than in row crops. We will review the effective practices for managing phytoplasma vectors.

Clean propagation material

The first and foremost method of managing vectors is to ensure that they are not transported to new areas. *Scaphoideus titanus* Ball is native to North America and was first found in Europe in 1958 (Bonfils and Schvester, 1960). It is a grapevine specialist, this being the only plant on which it can complete development (Vidano, 1964). As a vector of flavescence dorée in Europe, it causes great economic damage. *S. titanus* presumably entered Europe as eggs under the bark of grapevine canes, and extensive genetics on European and US populations have confirmed its North American origin (Bertin *et al.*, 2007). It is curious that *S. titanus* is only a suspected vector of grapevine yellows in North America (Maixner *et al.*, 1993). A novel and simple method for elimination of phytoplasma and insect eggs from grapevine canes, first used in the mid-1960s (Caudwell, 1966), is by treatment with heat. Today vines are being treated by dipping in hot water (Mannini, 2007). Conversely, it is also important not to transport phytoplasma-infected plants to new areas where potential vectors may be present. Weintraub *et al.* (2004) reported that phytoplasma-infected *Limonium* (Plumbaginaceae) plants were brought into an area where potential vectors were present. Shortly after the seedlings were planted, phytoplasma symptoms were observed in other *Limonium* plantations throughout the area.

Resistant plants

Developing plants that either are resistant to the phytoplasma or deter vector feeding would be a first line of defence. Unfortunately little work has been done on vegetable crops. Carrot (*Daucus carota*) varieties have been tested for resistance to the aster yellows phytoplasma, which occurs in the Great Lakes region of the USA and Canada (Gabelman *et al.*, 1994). For the carrot varieties that they tested, the incidence of phytoplasma disease ranged from 2.5 to 35.3%, indicating that some varieties showed significant resistance.

Before the 1970s, rice pests were kept under natural biological control, but introduction of new cultivars and intensive use of insecticides in the green revolution obliterated that control, causing catastrophic outbreaks of the brown planthopper, *Nilaparvata lugens* (Stål) (Delphacidae), and other pests. Slowly integrated pest management practices with the use of resistant cultivars brought the situation under control (Way and van Emden, 2000). More recently, Padmavathi *et al.* (2005, 2007a, b) have identified genes for resistance to the green leafhopper, *Nephotettix virescens* (Distant); the white-backed planthopper, *Sogatella furcifera* (Horváth); and the brown planthopper. With time, these genes could be incorporated into commercial varieties of rice. In Australia, the pasture legumes, *Stylosanthes* spp., are subject to a phytoplasma-caused little leaf disease (De La Rue *et al.*, 2001). It was found that anthracnose-resistant cultivars of *Stylosanthes* are resistant to phytoplasmas (De La Rue *et al.*, 2003) and are recommended for use for commercial seed production and cattle pasture land. Work on resistance in trees to phytoplasmas and/or vectors has been more intensive and successful (see Seemuller and Harries, Chapter 9, this volume, for a comprehensive discussion).

Roguing

The removal of diseased plants is most effective in orchards where the trees are non-contiguous. Infected plants can be rogued – removed entirely – or ratooned – only symptomatic shoots removed. A classic example of roguing to successfully control X-disease is the removal of the phytoplasma reservoir host, chokecherry (*Prunus virginiana* L.), up to 150 m from peach (*Prunus persica* (L.)) (Parker *et al.*, 1963). There are a number of other successful phytoplasma management examples (Gilmer and Blodgett, 1976; van Steenwyk *et al.*, 1995). Uyemoto *et al.* (1998) developed a new technique of roguing combined with insecticide spraying. To prevent the infected leafhoppers leaving a symptomatic tree as it is being cut down, they sprayed sweet cherry (*Prunus avium*) with diazinon first. They found a significant reduction in disease spread around trees thus removed.

In Australia, papaya (*Carica papaya*) is subject to three phytoplasma diseases: dieback, yellow crinkle and mosaic. Disease incidence varies seasonally: dieback generally occurs in the spring and yellow crinkle in late spring and summer. To reduce phytoplasma inoculum, diseased shoots are ratooned, i.e. the main stem of diseased plants is removed and the presumably pathogen-free lateral shoots allowed to grow, thus maintaining production (Guthrie *et al.*, 1998). Ratooning is effective in reducing the phytoplasma-bearing material, but sometimes lateral shoots test positive for the presence of phytoplasma and eventually show symptoms. Recent work by Esker *et al.* (2006) analysed plant survival post-infection and found that plant age and season in which symptoms first occurred did not affect survival. On the basis of their analyses, they determined that ratooning is not an effective management practice.

Production of phytoplasma-free strawberries is maintained through inspection and roguing of plants with lethal yellows symptoms. Since mother plants produce runners, allowing for movement of phytoplasmas to offspring, this roguing limits the number of diseased plants. However, the appearance of symptoms can take up to 8 weeks, and diseased plants may be unintentionally distributed to growers (Greber and Gowanlock, 1979).

Weed control

While it has long been known that weeds such as field bindweed, *Convolvulus arvensis*, host both vector development and phytoplasmas, it is only recently that work on manipulating habitat to control vectors has elucidated the importance of weeds in the ecology/epidemiology of phytoplasmas. *H. obsoletus* is, to date, the only known vector of the STOL BN, to grapevine in Europe (Maixner *et al.*, 1995). Since grapevine is considered an 'accidental' plant, the planthopper cannot develop on it and prefers not to feed on it. Therefore, acquisition of the phytoplasma from it is unlikely; the level of planthopper infestation with phytoplasma is dependent on the predominant weedy host plants. In Germany, in the presence of *C. arvensis*, up to 80% of the *H. obsoletus* may be infected, whereas if they develop on *Ranunculus* spp. only about 5% are infected (Maixner, 2007). In Italy, *Urtica dioica* is the preferred weed host (Bressan *et al.*, 2007). Stark-Urnau and Kast (2008) showed that use of the systemic glyphosate on weeds reduced vector populations only when the herbicide was applied in the winter. Spring or summer treatments of weeds did not reduce vector populations.

In eastern Africa, Napier grass, *Pennisetum purpureum*, is planted for fodder for cattle, environmental protection (to stabilize soils or to act as windbreaks) and, recently, to manage the most injurious pests of cereals, stem-boring Lepidoptera. Napier grass is planted as a border or trap crop around intercropping maize, *Zea mays*, where the grass is more attractive than maize to stem-borer moths for oviposition but supports only minimal survival of larvae. The system is known as 'push–pull' and has been widely taken up by farmers in recent years. However, two new diseases have become evident – characterized by severe stunting, profuse tillering and lethal yellowing – and have been identified as phytoplasmas (Arocha *et al.*, 2009) and the leafhopper, *Recilia banda Kramer* has been confirmed as a vector (Obura *et al.*, 2009).

These researchers found phytoplasmas in two species of grass in the area and are recommending disease control by weed removal. The use of a 'pull' strategy has been suggested for control of *H. obsoletus* through the use of its preferred host plant, *Vitex agnus castus* (Zahavi *et al.*, 2007). By placing *V. agnus castus* 10, 50 and 70 m from vineyards, it was hoped that *H. obsoletus* would be drawn to those plants and away from the grapevines.

Habitat management

As pointed out by Weintraub and Beanland (2006), vegetation composition surrounding a field/orchard/vineyard has a profound effect on the presence and dispersal of phytoplasma vectors. Vector capture in bidirectional Malaise traps placed in ecotonal regions between vineyards and forests showed that the primary direction of movement was from the wild to the cultivated vegetation. As discussed above, weedy species can greatly influence vector infection levels and distribution, but plant density also affects vector distribution. Lessio and Alma (2004), working with the North American invasive monophagous species *S. titanus*, which transmits flavescence dorée to grapevines, found that the vector would not spread outside the vineyard. Traps were placed at 12 and 24 m beyond the vineyard border and only three individuals were captured over 2 years, whereas within the vineyard 1200 were captured.

Mulching

One means of manipulating the habitat is the application of various organic and synthetic mulches. Synthetic mulches, such as plastic sheeting, can physically prevent the movement of vectors into the soil (*H. obsoletus* lays eggs at or just below soil surface). The use of reflective mulches works in another manner: rather than physical control, the vector is repelled from the plant. Summers and Stapleton (2002) achieved better control of the maize leafhopper vector, *Dalbulus maidis* (DeLong and Wolcott), and higher maize yield with plastic reflective mulch than with insecticide treatments. Similar results were achieved with control of the aster yellows phytoplasma vector *Macrosteles quadrilineatus* (Forbes) (=*fascifrons* Stål) on carrots using aluminium foil mulch. However, Setiawan and Ragsdale (1987) found that the efficacy of the mulch decreased as plant canopy increased. Alternatively, the type of mulching materials used around coconut trees influences the abundance of the planthopper vector of lethal yellows, *Myndus* (=*Haplaxius*) *crudus* Van Duzee (Cixiidae). Fewer nymphs are found around trees mulched with coarse materials such as pine bark nuggets (Howard and Oropeza, 1998).

Physical control covering (barriers)

The most reliable means of controlling phytoplasma vectors is by covering the crop with insect-exclusion screening (IES); however, its applicability is so severely limited due to the logistics of large-scale agriculture in major crops – sugarcane, maize, rice and grapes – that its use cannot even be contemplated. Research showing the positive effects of covering fruit trees is slowly gaining usage with growers worldwide (e.g. of bananas) and we anticipate this trend continuing.

In Australia, there are three different phytoplasma diseases in papaya, the last two being chronic diseases: dieback (causing 68–85% tree death per season in Australia), yellow crinkle (causing 2–27% tree death/season) and mosaic (causing 5–8% tree death/season) (Elder *et al.*, 2002). Elder *et al.* (2002) and Walsh *et al.* (2006) demonstrated that vectors could be 100% controlled by covering the trees with IES. When the effectiveness of IES was compared with systemic insecticide (imidacloprid) treatments and a non-treated control, covered trees had significantly fewer phytoplasma symptoms.

Insect-exclusion screening is the only method to maintain phytoplasma-free vineyards (Mannini, 2007). Production of clonal or mother plants must be covered continuously with IES to maintain the plant free from leafhopper vectors. Tests in tunnels 2.7 m high and 3 m wide showed that plants grew vigorously and had no phytosanitary problems (mildews). Blua *et al.* (2005) studied the effect of a 5-m-high barrier screen to prevent the movement of the sharpshooter *Homalodisca vitripennis* (Germar), a vector of the plant pathogen *Xylella fastidiosa*, to high-value vineyards and nursery stock. They found that the leafhopper behaviour changed, in that they moved away from the barrier to surrounding plants; few actually flew over the barrier. The results of this demonstrated that a barrier could add significantly to vector control.

Barrier sprays

Insecticides kill by a variety of means, both physiologically and physically. Kaolin is a non-abrasive, fine-grained aluminosilicate mineral applied as a particle film. It is a new and improved version of an old type of inorganic chemical control: whitewashing or dusting. Kaolin can act in a couple of ways: killing an insect by suffocating it or coating the plant and obstructing feeding and oviposition sites. Initial work by Puterka *et al.* (2003) demonstrated that kaolin protected grape plants from feeding and oviposition by leafhoppers, by physically coating the plant with a mineral film. Tubajika *et al.* (2007) showed that grapevines treated with kaolin were less likely to become infected with bacteria, and fewer leafhoppers were found in treated fields. The efficacy of kaolin is greatly hindered by water; in dry areas it may be very effective in controlling pests, but in areas with overhead irrigation or heavy rain it is washed away.

Parasitoids/predators

Leafhoppers and planthoppers are attacked by a range of predators. Spiders are very important predators of both adults and nymphs, especially in grassland ecosystems, while Miridae (Hemiptera) may be significant egg predators. There are also specialist parasitoids associated with Auchenorrhyncha (Waloff and Jervis, 1987). Dryinidae (Hymenoptera) deposit eggs into adult and nymphal leafhoppers and planthoppers, and the larvae develop within sacs visible externally on the hopper. The dryinid pupal stage is likely to be in the soil or leaf litter. Mymaridae (Hymenoptera), 'fairy flies', may be very common but rarely seen and attack the egg stage of leafhoppers and planthoppers. The entire development occurs within the egg of the host. Pipunculidae (Diptera) are almost exclusively parasitoids of Auchenorrhyncha. They lay eggs into the adult and nymphal stages and the larvae develop internally, swelling the abdomen in the later stages of growth. Strepsiptera may also be found, often in some species of Delphacidae. Useful summaries of many aspects of predator and parasitoid interactions with delphacid planthoppers may be found in chapters in Denno and Perfect (1994) and especially Perfect and Cook (1994) and Döbel and Denno (1994).

Jiang and Cheng (2004) found that predators could be enhanced by the application of manure to rice paddies to control the whitebacked planthopper. They postulated that biological control of the planthopper was enhanced by providing the predators with alternative food (collembolans) when the planthopper populations were low.

The various species of parasitoids associated with a particular species of Auchenorrhycha may reduce the natural population below an economically important threshold and thus they are not recorded as pests. Some introductions of parasitoids have been made to attempt to control introduced species. The dryinid *Neodryinus typhlocybae* (Ashmead), a specialist parasitoid attacking young *M. pruinosa* instars, was introduced in Italy from 1987 against *M. pruinosa* (Alma *et al.*, 2005). However, parasitoids are all susceptible to reductions in numbers due to pesticide usage, and increases in populations of leafhoppers and planthoppers may result (Heinrichs, 1994). A survey of beet leafhopper, *Circulifer tenellus* (Baker), egg parasitoids found a number of naturally occurring species, but researchers note that low parasitism rates in winter and spring cannot obviate the need for pesticide treatment of uncultivated land (Bayoun *et al.*, 2008). They postulate that, when beet leafhoppers aggregate in the autumn, parasitoids may play a significant role in reducing populations.

Genetic/molecular manipulations

Three forms of manipulations have recently been reviewed (see Weintraub, 2007, and references therein); plant lectins and systemic acquired resistance will be briefly reviewed here. Symbiont control is extensively reviewed in Chapter 15, this volume.

One form of manipulation involves the modification of plant lectins in host plants, which affects various physiological functions of vectors, including blocking the absorption of free amino acids and sugars. Two plant lectins show efficacy in phytoplasma vectors: the snowdrop lectin (*Galanthus nivalis* agglutinin, GNA) and *Allium sativum* leaf lectin (ASAL). These lectins have been shown to cause mortality in phytoplasma vectors in rice.

Systemic acquired resistance (SAR) is a plant defence mechanism that is activated when challenged by either an arthropod or a pathogen. It can be artificially activated by a number of chemicals. The potential for benzothiadiazole (BTH) to elicit a SAR reaction and protect the model plant *Arabidopsis thaliana* Columbia from phytoplasma transmitted by *Colladonus montanus* (Van Duzee) has been demonstrated in the laboratory. The mechanism for this effect is not clear: the plant phloem could have been morphologically modified to prevent phytoplasma from establishing or replicating, but the BTH could also have elicited production of a substance inhibiting vector feeding – fewer leafhoppers survived on BTH-treated plants.

It is important to note that *Bacillus thuringiensis* Berliner transgenic crops have not shown any detrimental effects on leafhopper or planthopper populations. No adverse effects on the brown planthopper, *Nilaparvata lugens* (Stål), were found in laboratory studies on *Bt* rice (Bernal *et al.*, 2002), nor were adverse effects on populations of the phytoplasma vectors *N. lugens, S. furcifera, Nephotettix cincticeps* (Uhler) or *Recilia dorsalis* (Motschulsky) found in the field (Chen *et al.*, 2006).

Antibiotic therapy

Davis *et al.* (1968) demonstrated that, after antibiotic treatment, plants infected with aster yellows went into remission. Tetracycline antibiotics, which inhibit protein synthesis by binding to the 30S ribosomal subunit, are effective against phytoplasmas. Two of the most destructive diseases of stone fruits are caused by X-disease phytoplasmas. Cherry and peach trees were treated with tetracycline antibiotics by high-pressure injection or by gravity infusion, and disease severity was significantly reduced, providing economic control (Lee *et al.*, 1987). Aside from the problems of using antibiotics on food crops, symptom remission is usually temporary, as tetracycline only persists in plants for 1–4 months (McCoy, 1982; Kaminska and Silwa, 2003). Since phytoplasmas are wall-less bacteria, β-lactam antibiotics (such as penicillin) have no bactericidal effects. In an attempt to circumvent these problems, Chen and Chen (1998) attempted to modify maize plants genetically to express a single-chain fragment from antibodies with strong inhibitory activity against a spiroplasma. While the antibody was expressed in transformed cells, it was not able to confer resistance in the whole plant. Malembic-Maher *et al.* (2005) were somewhat more successful when they made genetic constructs of an anti-phytoplasma gene in phloem-specific rice sucrose synthase; symptom appearance was delayed and phytoplasma multiplication was reduced.

Induced resistance

With certain wine grape varietals on various rootstocks, vines often go into spontaneous remission from phytoplasma symptoms (Osler *et al.*, 2003); symptoms may be present for a year or two and then suddenly disappear. Millions of plants of the Prosecco cultivar in Italy recovered within 1–4 years from a phytoplasma infection after the first symptoms appeared, and symptomatic grapes remain at less than 0.1% after 10 years. Maixner (2006) proposed that, if there are no symptoms for 3 consecutive years, the plant should be considered to have recovered from phytoplasmas. Research has focused recently on means of inducing recovery in plants. Stressing plants by severe pruning, pollarding, uprooting or partial uprooting can promote subsequent years of symptom-free plants (Romanazzi and Murolo, 2008). In addition to agrotechnical methods, applications of various chemicals known to induce recovery from various plant pathogens are being tested. Curkovic Perica (2008) has shown that the application of the auxins indole-3-acetic acid (IAA) and indole-3-butyric acid (IBA) to phytoplasma-infected periwinkle shoots induced recovery. Although there were no symptoms, phytoplasma DNA could still be detected in shoots infected with '*Candidatus* Phytoplasma solani'. Field trials in grapevines are currently under way with a number of commercial products based on chitosan, phosetyl-Al, glutathione and oligosaccharides (G. Romanazzi, Italy, 2009, personal communication).

Closing Remarks

As we have shown, there are numerous tactics for managing phytoplasma vectors and diseases. However, the single most effective means of control to date is with physical protection. Floating row covers have not received much attention to date but will probably become more important in the future. These very lightweight nets can be penetrated by water; hence overhead irrigation, fertilizers and herbicides can be used quite freely. At present their drawback is with woody plants, which can easily tear the netting, or in tropical areas, where there is intense solar radiation, which breaks down the netting.

A new and novel means of treating plants involves cryotherapy (Wang and Valkonen, 2008). All phytoplasma-infected sweet potato plants generated from cryo-treated shoot tips were phytoplasma-free. Plants can be treated by this method, from micropropagation or treatment with antibiotics to be phytoplasma-free, but these techniques are not prophylactic; once such plants are exposed to infectious vectors, they too will become infected. Plants treated with chemicals to induce phytoplasma recovery may also be subject to reinfection.

While chemical control of vectors probably will continue for the foreseeable future, we think that vector management will slowly shift to various genetic manipulations of crops to produce truly resistant plants or plants that express some chemical which allows them to be tolerant of phytoplasma infections.

References

Alma, A., Ferracini, C. and Burgio, G. (2005) Development of a sequential plan to evaluate *Neodryinus typhlocybae* (Ashmead) (Hymenoptera: Dryinidae) population associated with *Metcalfa pruinosa* (Say) (Homoptera: Flatidae) infestation in northwestern Italy. *Environmental Entomology* 34, 819–824.

Arocha, Y., Zerfy, T., Abebe, G., Proud, J., Hanson, J., Wilson, M., Jones, P. and Lucas, J. (2009) Identification of potential vectors and alternative plant hosts for the phytoplasma associated with napier grass stunt disease in Ethiopia. *Journal of Phytopathology* 157, 126–132.

Bayoun, I.M., Walker, G.P. and Triapitsyn, S.V. (2008) Parasitization of beet leafhopper eggs, *Circulifer tenellus*, in California. *Journal of Applied Entomology* 132, 412–424.

Bernal, C.C., Aguda, R.M. and Cohen, M.B. (2002) Effect of rice lines transformed with *Bacillus thuringiensis* toxin genes on the brown planthopper and its predator *Cyrtorhinus lividipennis*. *Entomologia Experimentalis et Applicata* 102, 21–28.

Bertaccini, A. (2007) Phytoplasmas: diversity, taxonomy, and epidemiology. *Frontiers in Bioscience* 12, 673–689.

Bertin, S., Guglielmino, C.R., Karam, N., Gomulski, L.M., Malacrida, A.R. and Gasperi, G. (2007) Diffusion of the Nearctic leafhopper *Scaphoideus titanus* Ball in Europe: a consequence of human trading activity. *Genetica* 131, 275–285.

Blua, M.J., Campbell, K., Morgan, D.J.W. and Redak, R.A. (2005) Impact of a screen barrier on dispersion behaviour of *Homalodisca coagulata* (Hemiptera: Cicadellidae). *Journal of Economic Entomology* 98, 1664–1668.

Bonfils, J. and Schvester, D. (1960) Les cicadelles (Homoptera Auchenorrhyncha) dans leurs rapport avec la vigne dans le sud-ouest de la France. *Annales des Epiphyties* 11, 325–336.

Bressan, A., Turata, R., Naixner, M., Spiazzi, S., Boudon-Padieu, E. and Girolami, V. (2007) Vector activity of *Hyalesthes obsoletus* living on nettles and transmitting a stolbur phytoplasma to grapevines: a case study. *Annals of Applied Biology* 150, 331–339.

Caudwell A. (1966) Inhibition of flavescence doree virus *in vivo* by means of heat treatment. *Annales des Epiphyties* 17, 61–66.

Chen, M., Ye, G.Y., Liu, Z.C., Yao, H.W., Chen, X.X., Shen, Z.C., Hu, C. and Datta, S.K. (2006) Field assessment of the effects of transgenic rice expressing a fused gene of *cryAb* and *cry1Ac* from *Bacillus thruingiensis* Berliner on nontarget planthopper and leafhopper populations. *Environmental Entomology* 35, 127–134.

Chen, Y.D. and Chen, T.A. (1998) Expression of engineered antibodies in plants: a possible tool for spiroplasma and phytoplasma disease control. *Phytopathology* 88, 1367–1371.

Christensen, N.M., Axelsen, K.B., Nicolaisen, M. and Schulz, A. (2005) Phytoplasmas and their interactions with hosts. *Trends in Plant Science* 10, 526–535.

Curkovic Perica, M. (2008) Auxin-treatment induces recover of phytoplasma-infected periwinkle. *Journal of Applied Microbiology* 105, 1826–1834.

Davis, R.E., Whitcomb, R.F. and Steere, R.L. (1968) Remission of aster yellows disease by antibiotics. *Science* 161, 793–794.

De La Rue, S.J., Padovan, A. and Gibb, K. (2001) *Stylosanthes* is a host for several phytoplasmas, one of which shows unique 16S–23S intergenic spacer region heterogeneity. *Journal of Phytopathology* 149, 613–619.

De La Rue, S.J., Hopkinson, R., Foster, S. and Gibb, K. (2003) Phytoplasma host range and symptom expression in pasture legume *Stylosanthes*. *Field Crop Research* 84, 327–334.

Denno, R.F. and Perfect, T.J. (1994) *The Planthoppers: their Ecology and Management*. Chapman and Hall, London.

Dietrich, C.H., McKamey, S.H. and Deitz, L.L. (2001) Morphology based phylogeny of the treehopper family Membracidae (Hemiptera: Cicadomorpha: Membracoidea). *Systematic Entomology* 26, 213–239.

Döbel, H.G. and Denno, R.F. (1994) Predator–planthopper interactions. In: Denno, R.F. and Perfect, T.J. (eds) *The Planthoppers: their Ecology and Management*. Chapman and Hall, London, pp. 325–399.

Elder, R.J., Milne, J.R., Reid, D.J., Guthrie, J.N. and Persley, D.M. (2002) Temporal incidence of three phytoplasma-associated diseases of *Carica papaya* and their potential hemipteran vectors in central and south-east Queensland. *Australasian Plant Pathology* 31, 165–176.

Esker, P.D., Gibb, K.S., Padovan, A., Dixon, P.M. and Nutter, F.W. Jr (2006) Use of survival analysis to determine the postincubation time-to-death of papaya due to yellow crinkle disease in Australia. *Plant Disease* 90, 102–107.

Forero, D. (2008) The systematics of the Hemiptera. *Revista Colombiana de Entomología* 34, 1–21.

Gabelman, W.H., Goldman, I.L. and Breitbach, D.N. (1994) Field evaluation and selection for resistance to aster yellows in carrot (*Daucus carota* L.). *Journal of the American Society of Horticultural Science* 119, 1293–1297.

Gilmer, R.M. and Blodgett, E.C. (1976) X-disease. In: *Virus Diseases and Noninfectious Disorders of Stone Fruits in North America*. USDA Agriculture Handbook No. 437. United States Department of Agriculture, Washington D.C., USA, pp. 145–155.

Greber, R.S. and Gowanlock, D.H. (1979) Rickettsia-like and mycoplasma-like organisms associated with two yellow-type diseases of strawberries in Queensland. *Australian Journal of Agricultural Research* 30, 1101–1109.

Guthrie, J.N., White, D.T., Walsh, K.B. and Scott, P.T. (1998) Epidemiology of phytoplasma-associated papaya diseases in Queensland, Australia. *Plant Disease* 82, 1107–1111.

Heinrichs, E.A. (1994) Impact of insecticides on the resistance and resurgence of rice planthoppers. In: Denno, R.F. and Perfect, T.J. (eds) *The Planthoppers: their Ecology and Management*. Chapman and Hall, London, pp. 571–598.

Howard, F.W. and Oropeza, C. (1998) Organic mulch as a factor in the nymphal habitat of *Myndus crudus* (Hemiptera: Auchenorrhyncha: Cixiidae). *Florida Entomologist* 8, 92–97.

Jiang, M.X. and Cheng, J.A. (2004) Effects of manure use on seasonal patterns of arthropods in rice with special reference to modified biological control of white-backed planthopper, *Sogatella furcifera* Horváth (Homoptera: Delphacidae). *Journal of Pest Science* 77, 185–189.

Kaminska, M. and Silwa, H. (2003) Effect of antibiotics on the symptoms of stunting disease of *Magnolia liliiflora* plants. *Journal of Phytopathology* 151, 59–63.

Lee, R.F., Nyland, G. and Lowe, S.K. (1987) Chemotherapy of cherry buckskin and peach yellow leafroll diseases: an evaluation of two tetracycline formulations and methods of application. *Plant Disease* 71, 119–121.

Lessio, F. and Alma, A. (2004) Dispersal patterns and chromatic response of *Scaphoideus titanus* Ball (Homoptera Cicadellidae), vector of the phytoplasma agent of grapevine flavescence dorée. *Agricultural and Forest Entomology* 6, 121–127.

Maixner, M. (2006) Temporal behavior of grapevines infected by type II of Vergilbungskrankheit (bois noir). In: *Extended Abstracts 15th Meeting of the International Council for Study of Virus and Virus-like Diseases of the Grapevine*, Stellenbosch, South Africa, pp. 223–224.

Maixner, M. (2007) Biology of *Hyalesthes obsoletus* and approaches to control this soil-borne vector of bois noir disease. *IOBC/WPRS Bulletin* 30(7), 3–9.

Maixner, M., Pearson, R.C., Boudon-Padieu, E. and Caudwell, A. (1993) *Scaphoideus titanus*, a possible vector of grapevine yellows in New York. *Plant Disease* 77, 408–413.

Maixner, M., Ahrens, U. and Seemuller, E. (1995) Detection of the German grapevine yellows (Vergilbungskrankheit) MLO in grapevine, alternative hosts and a vector by a specific PCR procedure. *European Journal of Plant Pathology* 101, 241–250.

Malembic-Maher, S., Le Gall, F., Danet, J.L., de Borne, F.D., Bove, J.M. and Garnier-Semancik, M. (2005) Transformation of tobacco plants for single-chain antibody expression via apoplastic and symplasmic routes, and analysis of their susceptibility to stolbur phytoplasma infection. *Plant Science* 168, 349–358.

Mannini, F. (2007) Hot water treatment and field coverage of mother plant vineyards to prevent propagation material from phytoplasma infections. *Bulletin of Insectology* 60, 311–312.

McCoy, R.E. (1982) Antibiotic treatment for control of tree diseases associated with mycoplasma-like organism. *Reviews of Infectious Diseases* 4, 157–161.

Mori, N., Pavan, F., Bondavalli, R., Reggiani, N., Paltrinieri, S. and Bertaccini, A. (2008) Factors affecting the spread of 'Bois Noir' disease in north Italy vineyards. *Vitis* 47, 65–72.

Nielson, M.W. (1968) *The Leafhopper Vectors of Phytopathogenic Viruses (Homoptera, Cicadellidae): Taxonomy, Biology, and Virus Transmission*. USDA Technical Bulletin 1382, Washington D.C., USA.

Nielson, M.W. (1979) Taxonomic relationships of leafhopper vectors of plant pathogens. In: Maramorosch, K. and Harris, K. (eds) *Leafhopper Vectors as Plant Disease Agents*. Academic Press, New York, pp. 3–27.

Obura, E., Midega, C.A.O., Masiga, D. Pickett, J.A., Hassan, M., Koji, S. and Khan, Z.R. (2009) *Recilia banda* Kramer (Hemiptera: Cicadellidae) a vector of Napier stunt phytoplasma in Kenya. Naturwissenschafen published online DOI: 10.1007/500114-009-0578-x 4 July 2009.

Osler, R., Carraro, L., Ermacora, P., Ferrini, F., Loi, N. Loschi, A., Martini, M., Mutton, P.B. and Refatti, E. (2003) Roguing: a controversial practice to eradicate grape yellows caused by phytoplasmas. *14th Meeting of the International Council for the Study of Virus and Virus-like Diseases of Grapevine* (ICVG), Locorotondo, Italy, p. 68.

Padmavathi, G., Ram, T., Pasalu, I.C. and Mishra, B. (2005) Gene identification of brown planthopper (*Nilaparvata lugens*) resistance in rice (*Oryza sativa*). *Indian Journal of Agricultural Science* 75, 587–590.

Padmavathi, G., Krishnaiah, N.V., Prasad, G.S.V. and Rao, Y.K. (2007a) Identification of green leafhopper (*Nephotettix virescens* (Distant.)) resistance genes in rice. *Indian Journal of Genetics and Plant Breeding* 67, 118–120.

Padmavathi, G., Ram, T., Ramesh, K., Rao, Y.K., Pasalu, I.C. and Viraktamath, B.C. (2007b) Genetics of whitebacked planthopper, *Sogatella furcifera* (Horvath), resistance in rice. *Sabrao Journal of Breeding and Genetics* 39, 99–105.

Parker, K.G., Palmiter, D.H., Gilmer, R.M. and Hickey, K.D. (1963) X-disease of peach and cherry trees and its control. *Cornell Extension Bulletin* 1100, 1–12.

Perfect, T.J. and Cook, A.G. (1994) Rice planthopper population dynamics: a comparison between temperate and tropical regions. In: Denno, R.F. and Perfect, T.J. (eds) *The Planthoppers: their Ecology and Management*. Chapman and Hall, London, pp. 282–301.

Pilkington, L.J., Gurr, G.M., Fletcher, M.J., Elliott, E., Nikandrow, A. and Nicol, H.I. (2004) Reducing the immigration of suspected leafhopper vectors and severity of Australian lucerne yellows disease. *Australian Journal of Experimental Agriculture* 44, 983–992.

Purcell, A.H. (1982) Insect vector relationships with prokaryotic plant pathogens. *Annual Review of Phytopathology* 20, 397–417.

Puterka, G.J., Reinke, M., Luvisi, D., Ciomperlik, M.A., Bartels, D., Wendel, L. and Glenn, D.M. (2003) Particle film, Surround WP, effects on glassy-winged sharpshooter behavior and its utility as a barrier to sharpshooter infestations in grapes. *Plant Health Progress Online*, doi:10.1094/PHP-2003-0321-RS.

Romanazzi, G. and Murolo, S. (2008) Partial uprooting and pulling to induce recovery in bois noir-infected grapevines. *Journal of Phytopathology* 156, 747–750.

Saracco, P., Marzachì, C. and Bosco, D. (2008) Activity of some insecticides in preventing transmission of chrysanthemum yellows phytoplasma ('*Candidatus* Phytoplasma asteris') by the leafhopper *Macrosteles quadripunctulatus* Kirschbaum. *Crop Protection* 27, 130–136.

Setiawan, D.F. and Ragsdale, D.W. (1987) Use of aluminum-foil and oat-straw mulches for controlling aster leafhopper, *Macrosteles fascifrons* (Homoptera: Cicadellidae), and aster yellows in carrots. *Great Lakes Entomology* 20, 102–109.

Stark-Urnau, M. and Kast, W.K. (2008) Maβnahmen zur Eindämmung des Brennesseltyps der Schwarzholzkrankheit bei Weinreben (*Vitis vinifera*). *Gesunde Pflanzen* 60, 85–89.

Summers, C.G. and Stapleton, J.J. (2002) Management of corn leafhopper (Homoptera: Cicadellidae) and corn stunt disease in sweet corn using reflective mulch. *Journal of Economic Entomology* 95, 325–330.

Tubajika, K.M., Civerolo, E.L., Puterka, G.J., Hashim, J.M. and Luvisi, D.A. (2007) The effects of kaolin, harpin, and imidacloprid on development of Pierce's disease in grape. *Crop Protection* 26, 92–99.

Uyemoto, J.K., Bethell, R.E., Kirkpatrick, B.C., Munkvold, G.P., Marois, J.J. and Brown, K.W. (1998) Eradication as a control measure for X-disease in California cherry orchards. *Acta Horticulturae* 472, 715–721.

van Steenwyk, R.A., Kirkpatrick, B.C., Purcell, A.H., Fouche, C.F., Grant, J.A. and Uyemoto, J.K. (1995) Evaluation of an abatement program for western X-disease in sweet cherry. *Plant Disease* 79, 1025–1028.

Vidano, C. (1964) Scoperta in Italia dello *Scaphoideus littoralis* Ball cicalina americana collegata alla 'Flavescence dorée' della Vite. *L'Italia Agricola* 101, 1031–1049.

Wally, O., Daayf, F., Khadhair, A.H., Adam, L., Elliott, B., Shinners-Carnelley, T., Iranpour, M., Northover, P. and Keyworth, S. (2004) Incidence and molecular detection of yellows-type disease in carrots, associated with leafhoppers in southern Manitoba, Canada. *Canadian Journal of Plant Pathology* 26, 498–505.

Waloff, N. and Jervis, M.A. (1987) Communities of parasitoids associated with leafhoppers and planthoppers in Europe. *Advances in Ecological Research* 17, 281–402.

Walsh, K.B., Guthrie, J.N. and White, D.T. (2006) Control of phytoplasma diseases of papaya in Australia using netting. *Australasian Plant Pathology* 25, 49–54.

Wang, Q.C. and Valkonen, J.P.T. (2008) Efficient elimination of sweetpotato little leaf phytoplasma from sweetpotato by cryotherapy of shoot tips. *Plant Pathology* 57, 338–347.

Way, M.J. and van Emden, H.F. (2000) Integrated pest management in practice – pathways towards successful application. *Crop Protection* 19, 81–103.

Weintraub, P.G. (2007) Insect vectors of phytoplasmas and their control – an update. *Bulletin of Insectology* 60(2), 169–173.

Weintraub, P.G. and Beanland, L. (2006) Insect vectors of phytoplasmas. *Annual Review of Entomology* 51, 91–111.

Weintraub, P.G., Pivonia, S., Rosner, A. and Gera, A. (2004) A new disease in *Limonium latifolium* hybrids. II. Investigating insect vectors. *HortScience* 39, 1060–1061.

Wilson, M.R. and Weintraub, P.G. (2007) An introduction to Hemiptera phytoplasma vectors. *Bulletin of Insectology* 60(2), 177–178.

Zahavi, T., Peles, S., Harari, A.R., Soroker, V. and Sharon, R. (2007) Push and pull strategy to reduce *Hyalesthes obsoletus* population in vineyards by *Vitex agnus castus* as trap plant. *Bulletin of Insectology* 60(2), 297–298.

Zahniser, J.N. and Dietrich, C.H. (2008) Phylogeny of the leafhopper subfamily Deltocephalinae (Insecta: Auchenorrhyncha: Cicadellidae) and related subfamilies based on morphology. *Systematics and Biodiversity* 6, 1–24.

14 Psyllid Vectors and their Control

Barbara Jarausch and Wolfgang Jarausch

RLP AgroScience GmbH, Germany

Introduction

While most phytoplasma vectors belong to the Hemiptera, Auchenorrhyncha (leafhoppers, planthoppers), there is one group of phytoplasmas that is transmitted by psyllid vectors of the superfamily Psylloidea (Hemiptera, Sternorrhyncha). The psyllid-transmitted phytoplasmas cause economically important diseases of fruit trees, such as pear decline (PD), apple proliferation (AP) and European stone fruit yellows (ESFY). The genetically closely related '*Candidatus* (*Ca.*) Phytoplasma pyri', '*Ca.* Phytoplasma mali' and '*Ca.* Phytoplasma prunorum', respectively, are associated with these diseases and together form the 16Sr group X (Seemüller and Schneider, 2004). With one exception their psyllid vectors have been identified rather recently and, because of the economic importance of the diseases, research on these vectors has made important progress in the last decade. Interestingly, the psyllid vector species all belong to the genus *Cacopsylla* and both phytoplasmas and psyllid vectors were geographically limited to Europe and the Palaearctic region. Only PD and peach yellow leaf roll have probably been introduced from Europe to North America along with their vectors. As phytoplasma-infected trees cannot be cured and resistant plant material is not yet available to the growers, preventive control measures such as vector control are of paramount importance to limit the disease spread. For this, knowledge about the biology of the vector species, as well as knowledge about the transmission parameters, is crucial. This chapter will therefore explore the diseases and the phytoplasmas that are transmitted by psyllid vectors, the biology and transmission characteristics of the different psyllid vectors and the possibilities of their control.

Diseases Caused by Psyllid-transmitted Phytoplasmas

The importance of psyllids as phytoplasma vectors has been elucidated only recently, although first reports date from the 1960s. The only recognized psyllid vectors today belong to a single genus of the family Psyllidae (Sternorrhyncha, Psylloidea). They transmit a narrow range of phytoplasmas, all belonging to the 16Sr group X fruit tree phytoplasmas. These diseases, however, are of great and increasing economic importance and thus the impact of the psyllids on fruit production is also high. The diseases transmitted by psyllid vectors are pear decline of *Pyrus*, apple proliferation of *Malus* and European stone fruit yellows and peach yellow leaf roll of *Prunus*.

There are two other reports about psyllids involved in the transmission of aster yellows (AY) or stolbur-type (STOL) phytoplasmas to carrot. Leclant *et al.* (1974) found *Trioza nigricornis* (Förster) as a vector; Font *et al.* (1999) obtained transmission with *Bactericera trigonica* Hodk. As these reports were not confirmed by other groups, these putative psyllid vectors will not be within the scope of the present chapter.

Psyllid-transmitted Phytoplasmas of 16Sr Group X

All four phytoplasma diseases of *Pyrus*, *Malus* and *Prunus* are caused by distinct but closely related phytoplasmas. They all belong to the 16Sr group X, the apple proliferation cluster (Seemüller *et al.*, 1998b). This phylogenetic clade comprises the AP agent, '*Ca.* Phytoplasma mali'; the PD agent, '*Ca.* Phytoplasma pyri'; and the ESFY agent, '*Ca.* Phytoplasma prunorum', as well as the peach yellow leaf roll (PYLR) agent and the newly identified PD (PDTW) agent from Taiwan (Seemüller *et al.*, 1998b; Seemüller and Schneider, 2004 ; Liu *et al.*, 2007). All these phytoplasmas are phylogenetically closely related to each other and share 16S rDNA sequence similarities between 98.6 and 99.6%. Further members of this subclade are '*Ca.* Phytoplasma spartii' (associated with Spartium witches'-broom), '*Ca.* Phytoplasma rhamni' (associated with Buckthorn witches'-broom) and '*Ca.* Phytoplasma allocasuarinae' (associated with Allocasuarina yellows), which exhibit between 94 and 97.2% 16S rDNA sequence similarity with '*Ca.* Phytoplasma pyri' (Marcone *et al.*, 2004).

On a non-ribosomal fragment analysed for '*Ca.* Phytoplasma mali' and '*Ca.* Phytoplasma prunorum' an overall sequence heterogeneity of 10.3% was found (Jarausch *et al.*, 2000a). Within '*Ca.* Phytoplasma mali', three subtypes have been defined, which are currently used in epidemiological studies to monitor their geographic distribution (Jarausch *et al.*, 2000b). Recently, higher dissimilarities between the 16Sr group X phytoplasmas were found by analysing new molecular markers. Danet *et al.* (2007) reported sequence dissimilarities in the *Imp* gene of 36% between '*Ca.* Phytoplasma prunorum' and '*Ca.* Phytoplasma mali', 29% between '*Ca.* Phytoplasma prunorum' and '*Ca.* Phytoplasma pyri' and 28% between '*Ca.* Phytoplasma mali' and '*Ca.* Phytoplasma pyri'. Respective values for the *ace*F marker were 11, 12 and 10%, and for the *pnp* marker 7, 6 and 5%. By analysing the *sec*Y gene, 8, 7 and 10% dissimilarities were found, respectively.

Regarding 16S rDNA sequence similarity, PYLR phytoplasma is most closely related to '*Ca.* Phytoplasma pyri' (99.6% identity) and was therefore regarded as subtype of '*Ca.* Phytoplasma pyri' by Seemüller and Schneider (2004). However, the data reported by Morton *et al.* (2003) for the sequence identities of the putative hydrophilic domain of the immunodominant membrane protein indicate that PYLR phytoplasma is distinct from the European isolates of '*Ca.* Phytoplasma pyri'.

Various phytoplasma strains have been transmitted from diseased apricot, cherry and Japanese plum trees to the experimental host *Catharanthus roseus* (periwinkle). These strains proved to be members of either the AY or the STOL or X-disease phytoplasma group and are not related to '*Ca.* Phytoplasma prunorum' (Lorenz *et al.*, 1994; Seemüller *et al.*, 1998b).

Recently, the nucleotide sequence of the entire chromosome of '*Ca.* Phytoplasma mali' strain AT was determined (Kube *et al.*, 2008). This work and comparative studies by pulsed-field gel electrophoresis showed that the genome size of '*Ca.* Phytoplasma mali' strains varies, ranging between 600 and 640 kb. Furthermore, restriction digestion of entire chromosomes with endonuclease *Ceu*I revealed that '*Ca.* Phytoplasma mali' strains have a linear chromosome, unlike the previously sequenced phytoplasmas, which are characterized by circular chromosomes.

Table 14.1 gives an introductory overview of the most important phytoplasma diseases and their agents that are transmitted by psyllid vectors. So far, all important fruit crop diseases caused by phytoplasmas of the 16Sr X group are vectored by psyllids. Plate 4 shows images of psyllid vector species.

Table 14.1. Phytoplasma diseases of fruit crops, their agents, their psyllid vectors and the vector's host plant.

Psyllid species	Phytoplasma	Disease	Reproduction host plant
Cacopsylla picta	'*Ca.* Phytoplasma mali'	Apple proliferation	*Malus*
Cacopsylla melanoneura	'*Ca.* Phytoplasma mali'	Apple proliferation	*Crataegus, Malus*
Cacopsylla pruni	'*Ca.* Phytoplasma prunorum'	European stone fruit yellows	*Prunus* species
Cacopsylla pyri	'*Ca.* Phytoplasma pyri'	Pear decline	*Pyrus*
Cacopsylla pyricola	'*Ca.* Phytoplasma pyri'	Pear decline	*Pyrus*
*Cacopsylla pyrisuga**	'*Ca.* Phytoplasma pyri'	Pear decline	*Pyrus*
*Cacopsylla qianli**	PDTW phytoplasma	Pear decline – Taiwan	*Pyrus*
*Cacopsylla chinensis**	PDTW phytoplasma	Pear decline – Taiwan	*Pyrus*
Cacopsylla pyricola	PYLR phytoplasma	Peach yellow leaf roll	*Prunus persica*

*Presumed vectors.

Pear decline

Pear decline (PD) is one of the most important diseases of pear and induces a more or less rapid decline of the tree. A decline-like disorder called 'moria del pero' was already reported in Italy at the beginning of the 20th century (Mader, 1908; Refatti, 1964). In North America, PD was first described in British Columbia (McLarty, 1948) and then spread along the Pacific coast (Woodbridge *et al.*, 1957; Nichols *et al.*, 1960). The disease presumably originates from Europe and has been introduced into North America as well as its vector (Seemüller, 1989). Devastating epidemics of PD were observed in the 1950s and 1960s along the Pacific coast of North America and in Italy (Refatti, 1964). Today, PD probably occurs wherever pear is grown in Europe and North America. Recently, PD has also been reported from the Asian part of Turkey (Sertkaya *et al.*, 2008) and from Iran (Salehi *et al.*, 2008), and a similar decline of pear has been found in Taiwan (Liu *et al.*, 2007). New outbreaks of PD are currently observed across Europe.

Disease development and symptom expression of PD are significantly influenced by the rootstock and the stage of the disease. Three different forms can be distinguished: quick decline, slow decline and reddening of the foliage with leaf curl (Seemüller, 1989). Quick decline is the sudden wilt and death of the trees in summer or autumn, which is favoured by abiotic stress of the trees, such as heat and drought. Slow decline is a progressive weakening of the tree grown on oriental and less susceptible *Py. communis* rootstocks. Reddening of the foliage in late summer or autumn is a mild form of slow decline and occurs in trees on more tolerant rootstocks. Affected trees may exhibit reduced vigour, yield and fruit size. There exists no specific symptom of PD that is reliable in visual diagnosis.

The PD agent can infect most or all *Pyrus* spp. (Seemüller *et al.*, 1998a). Natural infections have been found in commercial scion and rootstock cultivars of *Py. communis* and *Py. pyricola* and in rootstocks or own-rooted trees of *Py. ussuriensis*, *Py. calleryana*, *Py. elaeagrifolia* and quince (Blodgett *et al.*, 1962; Schneider, 1970). It could be further experimentally transmitted by graft inoculation to various other *Pyrus* spp. (Seemüller *et al.*, 1998a).

Apple proliferation

Apple proliferation (AP) is one of the economically most important phytoplasma diseases in Europe. AP was first reported in northern Italy (Rui *et al.*, 1950). Since then, AP has been detected in most European countries where apple is grown commercially. The highest incidences of AP are found in temperate climatic zones of middle, western, southern and eastern Europe (Kunze, 1989). It is not known whether this restriction is due to a missing vector or unsuitable temperatures northwards or southwards of this zone. The geographic distribution of AP is restricted to Europe and neighbouring regions. It has only recently been detected outside Europe in the Asian part of Turkey (Canik and Ertunc, 2007).

AP-infected trees are not usually killed by the phytoplasma infection. However, AP seriously impairs fruit size, fruit quality, yield and vigour of the trees (Kunze, 1989). Fruits of infected trees are poor in taste, and fruit size and weight are reduced by 30–60%, rendering them unmarketable. AP induces specific symptoms, which allow a reliable diagnosis (Seemüller, 1990). These are enlarged stipules and witches'-brooms, which are the effect of suppression of apical dominance, resulting in growth of dormant axillary buds on the upper parts of vigorous shoots. These typical symptoms can be best seen in late summer and autumn. Early break of leaf buds in spring is also related to AP. Foliar reddening in late summer is often a first indication of the disease.

All commercially grown cultivars and rootstocks of the domestic apple *Malus* × *domestica* are susceptible to the disease. In graft inoculation experiments, 58 ornamental and wild *Malus* spp. and subspecies, as well as 40 hybrids of different *Malus* spp., which were used as rootstocks, could be infected with the AP agent (Kartte and Seemüller, 1991). Whereas most of the wild *Malus* spp. were moderately to highly susceptible to infection, a group of *M. sieboldii*-derived hybrids showed a good level of resistance to AP (Bisognin *et al.*, 2008).

European stone fruit yellows

European stone fruit yellows (ESFY) is the common name of several economically important decline diseases of stone fruits (*Prunus* spp.) in Europe. A decline by apoplexy was first reported for apricot in France (Chabrolin, 1924) and for Japanese plum in Italy (Goidànich, 1933). Since then several diseases have been described, such as apricot chlorotic leaf roll of apricot (*Pr. armeniaca*), leptonecrosis and decline of Japanese plum (*Pr. salicina*), peach yellows and decline of peach (*Pr. persica*). Similar disorders have been found on European plum (*Pr. domestica*), almond (*Pr. amygdalus*) and flowering cherry (*Pr. serrulata*) (Lorenz *et al.*, 1994). Molecular characterization of the pathogen revealed that all these diseases are caused by a similar organism, the ESFY phytoplasma (Lorenz *et al.*, 1994; Jarausch *et al.*, 1998). So far, ESFY is restricted to Europe and neighbouring regions and is distinct from stone fruit phytoplasma diseases found in North America. It has been reported from most southern and central European countries, with its northern limit in south-east England. Outside Europe, ESFY has been found in Turkey and most recently in Azerbaijan (Jarausch *et al.*, 2000a; Danet *et al.*, 2007).

ESFY is the limiting factor in the production of apricot and Japanese plum in several major stone-fruit-growing areas of Europe, where it causes considerable economic losses due to the high mortality of infected trees. Its economic incidence is less important in peach, almond and European plum. Disorders in cherry are mostly not related to the ESFY agent but caused by other types of phytoplasmas. Typical symptoms of ESFY useful for diagnostics are off-season growth and premature break of leaf buds before flowering in late winter, and leaf yellowing or reddening in combination with leaf roll in summer. The ESFY agent infects all scion and rootstock cultivars belonging to a wide range of *Prunus* spp., including *Pr. armeniaca, Pr. salicina, Pr. persica, Pr. amygdalus, Pr. domestica, Pr. cerasifera, Pr. insistitia, Pr. spinosa, Pr. marianna, Pr. avium* and *Pr. cerasus*, as well as various interspecific hybrids used as rootstocks (Jarausch *et al.*, 1998, 2000a).

Peach yellow leaf roll disease

Whereas ESFY disease of *Prunus* is limited to Europe, peach yellow leaf roll (PYLR) is a disease of stone fruits in North America, caused by a similar phytoplasma. It is distinct from western X (WX) disease, also occurring on *Prunus* spp. in North America. It differs from WX-disease by symptomatology, speed of disease spread and by the insect vector (Blomquist and Kirkpatrick, 2002b). Whilst WX-disease is transmitted exclusively by leafhoppers, the main vector of PYLR is a pear psyllid. PYLR was first described in the middle of the 20th century and caused major losses in peach in the late 1970s (Purcell *et al.*, 1981). Symptoms of PYLR are yellow, downward-rolled leaves in late summer, which often have swollen midribs. PYLR induces a rapid decline of peach and thus resembles an ESFY infection of peach. PYLR was probably introduced to North America and is predominantly found in peach orchards in the neighbourhood of pear orchards (Purcell *et al.*, 1981). Its highest incidences are found in northern California and it seems to be restricted to this area.

Taxonomy of Psyllid Vectors

The Hemiptera (= Rhynchota) are divided into three suborders: Sternorrhyncha, Auchenorrhyncha and Heteroptera. Among the Sternorrhyncha, the jumping plant-lice or psyllids form the well-defined superfamily Psylloidea. About 3000 species of this superfamily are described worldwide, including about 400 species in Europe (Burckhardt, 1994). Currently the superfamily Psylloidea is subdivided into six families: Psyllidae (including Aphalaridae and Spondyliaspididae), Calophyidae, Phacopteronidae, Carsidaridae, Homotomidae and Triozidae (Burckhardt, 1994). All recognized phytoplasma vectors are found in one subfamily of Psyllidae. In this subfamily Psyllinae, the genus *Cacopsylla* includes all important vector species for fruit tree phytoplasmas.

For the determination of the most important *Cacopsylla* spp. found on *Rosaceae* in Europe, a determination key is available on the Internet at www.psyllidkey.info (Burckhardt *et al.*, 2008). Currently, this key only exists in the German language but an English version will be available soon.

Biology of Psyllid Vectors

Most psyllids are phloem feeders, and nymphs as well as adults feed on plant sap. They are particularly interesting for their highly specialized host requirements. Most psyllids are monophagous, having only one host plant; others are oligophagous, with a few, closely related host plants. Only a few are polyphagous (Ossiannilsson, 1992; Burckhardt, 1994). The adults are often able to exploit other food plants temporarily. The vast majority of psylloid species are bisexual, and only a few cases of facultative parthenogenesis are known (Hodkinson, 1974). The egg is oblong with a basal pedicel, which is inserted in the plant tissue. Depending on the species, the eggs are laid on the new buds, in crevices of the bark or on leaves, where they can produce pit-like deformations on the leaf blade. The larval development passes through five instars, which are more or less strongly flattened dorsoventrally (Burckhardt, 1994). Many tropical and southern temperate species are polyvoltine, with overlapping generations. In contrast, northern temperate species tend to be univoltine or bivoltine. Species overwinter as eggs, young nymphal instars or adults. These often migrate to shelter plants, such as conifers, and return to the hosts in spring (Burckhardt, 1994). Psylloids occur in all biogeographical regions, from sea to alpine level, and often have geographically restricted ranges.

Important univoltine vector species overwinter as adults and have an obligate alternation of host plants for reproduction and overwintering. As far as is known, this applies for *C. pruni*, *C. melanoneura* and *C. picta* (Cermák and Lauterer, 2007; Mayer and Gross, 2007; Thébaud *et al.*, 2009). In these cases, overwintering was observed only on conifers at higher altitudes. Migration to the respective reproduction or overwintering plant seems to be direct and may take place even over long distances, e.g. 27 km in the case of *C. pruni* (Thébaud *et al.*, 2009).

Recent studies indicate that chemical components of the host plants have an impact on the migration behaviour of psyllids. Mayer and Gross (2007) showed that the migration of *C. melanoneura* between reproduction, overwintering and transitional hosts corresponds with changing host plant preferences during the life cycle of this psyllid. Gross and Mekonen (2005) found that plant odours influence the host-finding behaviour of the monophagous *C. picta* as well as the oligophagous *C. melanoneura*. During recent studies, Mayer *et al.* (2008a, b) revealed that volatiles emitted from phytoplasma-infected apple plants were more attractive to its vector *C. picta* than those from uninfected ones, showing the meaning of chemo-ecological aspects for the analysis of vector–plant–pathogen interaction.

When feeding, psyllid nymphs may cause considerable damage to their host plants, since their salivary injections may contain phytotoxins and thus produce serious necroses and malformations. Certain phytoplasma vector species may also occur as direct plant pests, e.g. pear psyllids, which make them economically particularly important (Ossiannilsson, 1992).

Vectors of pear decline

In Europe, three recognized or presumed vectors of PD live on pear: *C. pyri* (Linnaeus), *C. pyricola* (Foerster) and *C. pyrisuga* (Foerster). *C. pyri* is reported from Europe, the Caucasus, Central Asia, the Russian Far East and China; *C. pyricola* occurs naturally in the Western Palaearctic and was introduced into the USA and Canada in the early 19th century. The two species are oligophagous on *Pyrus* spp. such as *P. communis*, *P. eleagrifolia*, *Py. pyraster*, *Py. amygdaliformis* and *Py. salicifolia* (Burckhardt, 1994). The biology of *C. pyri* and *C. pyricola* is similar: both are polyvoltine. Thus, *C. pyri* can produce four to five generations in central Europe and up to eight generations in southern France. Two morphologically distinct forms can be distinguished: a darkish winter form (*C. pyri* f. *pyri*) and a light summer form (*C. pyri* f. *pyrarboris*). *C. pyricola* has four to five generations in France and three to four in the USA. The darker winter form (*C. pyricola* f. *simulans*) appears as one and the lighter summer form (*C. pyricola* f. *pyricola*) as three to four generations per year, respectively. The oviposition of the winter form, on leaf buds and midribs of the leaves, coincides with rising temperatures in early spring (Burckhardt, 1994). In contrast, *C. pyrisuga* is univoltine; the adults overwinter on conifers and re-migrate to *Pyrus* by the middle of March to April. Egg deposition takes place in two different steps: first at the beginning of April and secondly in the middle of May, followed by a 6-week-long larval development and the emergence of new adults in June. All three pear psyllids can cause direct damage on pear trees: the larvae affect plant growth by withdrawal of plant sap, and the secreted honeydew burns plant tissue and favours the growth of sooty mould; adults damage the plants by injection of salivary phytotoxins.

First reports of pear psyllids as vectors for phytoplasmas came from the Pacific coast of North America. Jensen *et al.* (1964) identified *C. pyricola* as the vector of '*Ca.* Phytoplasma pyri' at a time when the disease was thought to be virus-borne. Since then no further vector has been described for the USA. However, the distribution of the putative vectors of '*Ca.* Phytoplasma pyri' in Europe and the whole Palaearctic biogeographic region is diverse: while for Great Britain only *C. pyricola* has been described as vector (Davies *et al.*, 1992), *C. pyri* was identified as the main vector in France (Lemoine, 1984), Italy (Carraro *et al.*, 1998a) and Spain (Garcia-Chapa *et al.*, 2005). Recently, Kucerova *et al.* (2007) presented preliminary data from the Czech Republic, where they found naturally infected individuals of *C. pyri* as well as of *C. pyrisuga*. The vector capability of *C. pyrisuga* is up to now not confirmed.

Vectors of apple proliferation

Two psyllids, *C. picta* (Foerster) (syn. *C. costalis*) and *C. melanoneura* (Foerster), are recognized vectors of '*Ca.* Phytoplasma mali' (Frisinghelli *et al.*, 2000; Tedeschi *et al.*, 2002; Jarausch *et al.*, 2003). *C. picta* is distributed only in Europe and is monophagous on *Malus* spp. The insect completes one generation per year and overwinters as an adult on overwintering plants (conifers). At the end of winter (March/April), *C. picta* re-migrants move from the overwintering sites to apple trees for oviposition. The insects of the new generation feed on the primary host until the beginning of July, when they leave the apple trees as adults (Mattedi *et al.*, 2008; Tedeschi *et al.*, 2009).

Cacopsylla melanoneura has a Palaearctic distribution and is oligophagous on Rosaceae such as *Crataegus*, *Malus* and *Pyrus* spp. The life cycle is similar to that of *C. picta* but the overwintering adults appear earlier in the year on *Crataegus* or apple trees and the new generation abandons the host plant earlier than *C. picta* to migrate to the overwintering plants (Mattedi *et al.*, 2008). In most of the studied areas, both species are present (Carraro *et al.*, 2001a; Jarausch *et al.*, 2003; Delić *et al.*, 2005; Mattedi *et al.*, 2008), in others only *C. melanoneura* has been found (Tedeschi *et al.*, 2002). Several studies on the vector capacity of *C. picta* and *C. melanoneura* and on the role of hawthorn as source of '*Ca.* Phytoplasma mali' in different European regions led to contradictory results. Detailed transmission trials identified *C. picta* in Germany (Jarausch *et al.*, 2003) and northern Italy (Frisinghelli *et al.*, 2000; Carraro *et al.*, 2008) as the main vector of '*Ca.* Phytoplasma mali'. In contrast, *C. melanoneura* was repeatedly identified as the main vector in the Aosta Valley (Tedeschi *et al.*, 2002), whereas the German population of *C. melanoneura* hardly acquired '*Ca.* Phytoplasma mali' from infected apple and was not able to transmit the phytoplasma (Mayer *et al.*, 2009). Furthermore, the German population preferred hawthorn as a host plant, which, however, was not found to be infected with the phytoplasma, whereas the north-western Italian population seems to be able to move between apple and hawthorn (Tedeschi *et al.*, 2009). Accordingly, hawthorn has been found to be infected with '*Ca.* Phytoplasma mali' and thus may play a role in the epidemiology of AP in this region.

'*Ca.* Phytoplasma mali' is the only phytoplasma that has been reported to be transmitted by psyllids and a leafhopper. Early findings of Krczal *et al.* (1988) have been confirmed by Tedeschi and Alma (2006) that *Fieberiella florii* (Stål) is able – at least under experimental conditions – to transmit the phytoplasma. However, the importance of *F. florii* for the spread of AP remains questionable, as *F. florii* was not or almost not found in the regular insect captures carried out in apple-growing regions of northern Italy and south-west Germany, where AP is actually spreading (Mattedi and Jarausch, unpublished data). '*Ca.* Phytoplasma mali' was also detected by molecular means in different aphid species captured on infected trees (Cainelli *et al.*, 2007). Unsuccessful transmission trials and low concentrations of the phytoplasma measured by quantitative PCR in the aphids indicate that aphids are not able to transmit AP.

Vectors of European stone fruit yellows

Cacopsylla pruni (Scopoli) has been identified as the only vector of '*Ca.* Phytoplasma prunorum' in various European countries, such as Italy, France and Germany (Carraro *et al.*, 1998b; Jarausch *et al.*, 2001, 2007a, 2008). *C. pruni* is a European and central Asian species that is known from almost all of Europe (Lauterer, 1999). This psyllid is strictly oligophagous on *Prunus* spp., completes one generation per year and overwinters as an adult on shelter plants, usually conifers. At the end of winter/early spring, *C. pruni* re-migrants move from the overwintering plants back to *Prunus* for oviposition. The insects of the new generation feed on the reproduction hosts until the beginning of July, when they leave the stone fruits as adults to move to overwintering hosts (Carraro *et al.*, 2001b, 2004; Thébaud *et al.*, 2009).

Acquisition and Transmission Characteristics of Psyllid Vectors

The great majority of plant pathogens are transmitted by insects of the hemipteran assemblage. The complex and specific interactions between hemipteran vectors and the pathogens they transmit has been studied in detail, mostly for plant viruses (Ng and Falk, 2006; Hogenhout *et al.*, 2008a). Currently four mechanisms are described for the transmission of viruses by insects of the hemipteroid assemblage: non-persistent, semi-persistent, persistent circulative and persistent propagative (Hogenhout *et al.*, 2008a). Besides viruses, phytoplasmas are one of the most important insect-transmitted plant pathogens. As phytoplasmas are phloem-limited, only phloem-feeding insects can potentially acquire and transmit the pathogen (Weintraub and Beanland, 2006).

Phloem-feeding insects acquire the phytoplasma passively during feeding in the phloem of infected plants. The following process of phytoplasma passage and multiplication in the insect body comprises, similarly for all vector species, the latent or incubation phase and the infectivity period, where the insect can transmit the pathogen. Detailed descriptions of the cellular processes of transport and multiplication in the insect body have been reported for leafhoppers (Weintraub and Beanland, 2006; Hogenhout *et al.*, 2008b). So far, these mechanisms of phytoplasma transmission remain to be demonstrated for psyllid vectors as well. Several publications show that psyllids transmit the pathogen in a persistent (circulative or propagative) manner (Carraro *et al.*, 2001b, c; Thébaud *et al.*, 2009).

Acquisition of a phytoplasma by an insect does not imply that the insect is a vector, since phytoplasmas may be acquired but not re-injected by feeding. Therefore, the natural infection rate of a psyllid species is not necessarily correlated with its transmission capacity, which has to be proven by transmission trials. The length of time needed for an individual to become infectious is of paramount importance for the disease spread as well as the control strategies. Research with univoltine vector species, for example, focuses on the question whether emerging new adults are able to acquire the phytoplasma on infected trees and to transmit it before migration. The consequence would be a spread of the disease inside an orchard, whereas the transmission by overwintered adults would lead to a transmission within a region.

Vectors of pear decline

After the identification of *C. pyricola* as the vector for '*Ca.* Phytoplasma pyri' in California (Jensen *et al.*, 1964), many investigations in the USA and Europe followed, in order to determine the infection rate of the psyllids and to analyse the transmission parameters. In England, transmission trials carried out with field-collected *C. pyricola* yielded transmission rates between 3 and 61%, depending on the collection site of the psyllids (Davies *et al.*, 1992). Acquisition of '*Ca.* Phytoplasma pyri' by *C. pyricola* from experimentally infected pear seedlings was best in August and lowest in winter. In California, Blomquist and Kirkpatrick (2002a) detected the pathogen in both winter form and summer form of *C. pyricola* but without a clear seasonal trend. The number of phytoplasmas per psyllid was estimated to range from 1×10^6 to 8.2×10^7, with higher titre in the winter form. They concluded that psyllid-mediated spring infections could happen well before '*Ca.* Phytoplasma pyri' would normally recolonize the upper part of the tree from the roots. In Italy, Carraro *et al.* (1998a, 2001c) detected '*Ca.* Phytoplasma pyri' in 55% of groups of *C. pyri* collected from March to October in the orchards, and 30% of the inoculated test plants became infected. They could furthermore show that *C. pyri* retained infectivity during winter but could not transmit PD to dormant plants. Garcia-Chapa *et al.* (2005) found that the percentage of infected individuals is similar from June to August but reaches a rate of almost 100% in September, coinciding with the maximum phytoplasma titre in the aerial plant parts. The highest transmission rate to an artificial sucrose medium was obtained in August and also in October. Although the percentage of infected psyllids was similar for both genders, '*Ca.* Phytoplasma pyri' transmission by females was significantly higher than by males.

PD has also been found in Taiwan (PDTW), where the European species *C. pyri* and *C. pyricola* are not present. Liu *et al.* (2007) found two other *Cacopsylla* species, *C. qianli* and *C. chinensis*, infected with the PDTW phytoplasma. Their role in transmission of PDTW in Taiwan remains to be clarified.

Vectors of apple proliferation

Both psyllid vectors of AP transmit the pathogen in a persistent manner; the presence of phytoplasma-infected and infective psyllids among the first remigrants collected in apple orchards suggests winter-retention of the pathogen in both species: *C. picta* (Jarausch *et al.*, 2004; Mattedi *et al.*, 2008) and *C. melanoneura* (Tedeschi *et al.*, 2003). In north-west Italy, Tedeschi *et al.* (2003) estimated that 3.5% of *C. melanoneura* re-migrants are infected and 0.8% of the new generation; in north-east Italy, the natural infection rate of *C. picta* was found to be 9 and 13%, respectively, for overwintering and offspring adults (Carraro *et al.*, 2008). In Germany, about 10% of overwintered *C. picta* and 0.2% of overwintered *C. melanoneura* were naturally infected with the pathogen (Jarausch *et al.*, 2004, 2007b). In transmission trials conducted between 2002 and 2006 with overwintered adults of *C. picta*, between 8 and 45% transmission was obtained, depending on the year (Jarausch *et al.*, 2007b). The phytoplasma concentration in the infected individuals of *C. picta* was extremely high and ranged between 10^6 and 10^8, as measured by quantitative

PCR (Jarausch *et al.*, 2007b). During experimental transmission trials in different laboratories in Germany and Italy, it was confirmed that both generations of *C. picta* can transmit the agent efficiently (Jarausch *et al.*, 2004; Carraro *et al.*, 2008). Thus, *C. picta* appears to be the main vector of '*Ca.* Phytoplasma mali' in these regions. In contrast, transmission by *C. melanoneura* has been found in only one case in Trentino (Mattedi *et al.*, 2008) but has been described repeatedly for the population captured in the Aosta Valley (Tedeschi *et al.*, 2002; Tedeschi and Alma, 2004). The overwintered as well as the new adults of *C. melanoneura* transmitted the pathogen. The natural transmission period in the orchards investigated by bait plant trials in Trentino was found to be during the migration period of the new generation of *C. picta* (Mattedi *et al.*, 2008). Consequently, *C. picta* and *C. melanoneura* are able to transmit '*Ca.* Phytoplasma mali' during the entire period when they are present on apple trees.

Vectors of European stone fruit yellows

The transmission of '*Ca.* Phytoplasma prunorum' by different developmental stages of *C. pruni* was studied in detail under controlled conditions (Carraro *et al.*, 1998b, 2001b, 2004). The overwintered as well as the new adults of *C. pruni* were able to transmit the agent to healthy test plants. The overwintered adults were already infected and infectious when they reached their primary hosts in early spring. They concluded that *C. pruni* transmits the winter-retained phytoplasma that had been acquired the previous year. The overwintered psyllids continued to transmit the pathogen in a persistent manner until their death. The adults of the new generation were already highly infectious when they abandoned the stone fruits in summer, and therefore the natural transmission period lasts as long as the vector is present on *Prunus*. In Germany, Jarausch *et al.* (2007a, 2008) found 2–3% of the field-collected overwintered adults naturally infected by '*Ca.* Phytoplasma prunorum'. During transmission trials under controlled conditions, the vector capacity of overwintered and new-generation adults of *C. pruni* was consistently lower than that described by Carraro *et al.* (2001b, 2004). Similar low infectivity and transmission rates of only 0.6% were confirmed in France by Jarausch *et al.* (2001) and Thébaud *et al.* (2008). Thébaud *et al.* (2009) demonstrated that the population of *C. pruni* in southern France has an extremely long 'effective latency' period, which lasts the overwintering period. During this time, the phytoplasma concentration within the insects continuously rises, reaching a maximum of 10^7 phytoplasmas per insect at re-migration. They concluded that only overwintered adults can efficiently transmit the disease and thus the disease spread is monocyclic. A possible vertical (= transovarial) transmission was not observed by Carraro *et al.* (1998b) and Thébaud *et al.* (2009), whilst Tedeschi *et al.* (2006) found indications for the existence of this passage in *C. pruni*.

Naturally infected individuals of *C. pruni* were found in several European countries, such as Italy (Carraro *et al.*, 1998b), France (Yvon *et al.*, 2004),

Spain (Laviña *et al.*, 2004), Czech Republic (Fialová *et al.*, 2004), Germany (Jarausch *et al.*, 2007a) and Bosnia–Herzogovina (Delić *et al.*, 2008). These investigations showed that wild *Prunus* spp. play an important role in the epidemiology of '*Ca.* Phytoplasma prunorum'. Whereas low populations of *C. pruni* were found on cultivated *Prunus* spp., such as *Pr. armeniaca*, *Pr. persica*, *Pr. amygdalus* and *Pr. domestica*, much higher vector densities were reported from different wild *Prunus* spp. such as *Pr. spinosa*, *Pr. cerasifera*, *Pr. domestica* and *Pr. salicina*. Interestingly, the wild *Pr. spinosa* and *Pr. cerasifera*, which presented reservoirs for the pathogen and the vector, rarely showed any typical symptoms (Carraro *et al.*, 2002; Jarausch *et al.*, 2008). In conclusion, some wild *Prunus* spp. play an important role in the epidemiology of ESFY disease, as the cycle of '*Ca.* Phytoplasma prunorum', as well as of its vector *C. pruni*, can be completed independently from the presence of infected cultivated stone fruit trees.

Vectors of peach yellow leaf roll

Insect vectors for PYLR phytoplasma were looked for in California in the 1980s and 1990s. As two similar diseases, western X and PYLR, caused by two genetically distinct phytoplasmas, exist in the same region, only the application of molecular methods enabled the proof that a psyllid is the main vector of PYLR. Experimental transmission of PYLR phytoplasma to peach seedlings was achieved with field-collected *C. pyricola* from naturally infected peach trees (Guerra, 1997). In field surveys for leafhoppers and psyllids in diseased peach orchards, only *C. pyricola* proved to be infected with PYLR phytoplasma, as confirmed by molecular means (Blomquist and Kirkpatrick, 2002b). Ten to 25% of groups of ten individuals tested positive, indicating a high infection rate. Infected psyllids were captured from peach as well as from pear grown in the neighbourhood. The population dynamics of *C. pyricola* were similar in peach and pear, with low densities in summer and an increase in autumn. Thus, the spread of PYLR is dependent on adjacent pear orchards, where presumably the vector reproduces (Purcell *et al.*, 1981).

Control of Psyllid Vectors

As there is no applicable means to cure a phytoplasma-infected fruit tree, insecticide treatments were the first measures to control the spread of fruit tree phytoplasma diseases whenever a vector species was identified. However, classical pest management is hampered for various reasons: insecticide resistance of polyvoltine psyllid species, low abundance of univoltine vectors, missing or missing homologation of appropriate products and not least environmental protection considerations. Therefore, alternative strategies of pest control, as well as long-term solutions based on natural resistance of the plant, are currently under investigation.

Vectors of pear decline

For years, the control of pear psyllids relied upon a system of integrated pest management (Solomon *et al.*, 1989). Chemical control is difficult, as pear psyllids have a pronounced capability to produce resistance against insecticides due to their host specificity and the high reproductivity. A survey initiated by the Insecticide Resistance Action Committee (IRAC) showed that *C. pyri* developed resistance against several organophosphates and pyrethroids in many European and American countries (Sterck and Highwood, 1992). In France, Buès *et al.* (1999) found a high tolerance against organophorous insecticides in *C. pyri*, whereas in the USA *C. pyricola* mainly developed resistance against pyrethroids (van de Baan and Croft, 1991). Concordantly, both pear psyllids showed a geographic and seasonal variability in the susceptibility to insecticides in European and American studies. Thus, the summer form of *C. pyri*, as well as of *C. pyricola*, was, in each case, more sensitive than the winter form (van de Baan and Croft, 1991; Buès *et al.*, 1999). These results led to the conclusion that traditional control strategies are not sufficient for the management of insecticide-resistant pear psyllids. Thus, alternative control mechanisms, such as the release of predators and the use of repellents, were evaluated for several years. In particular, the predatory bug *Anthocoris nemoralis* was integrated into selective insecticide programmes in Italy and the USA (Civolani and Pasqualini, 2003; Daugherty *et al.*, 2007). However, despite a positive effect due to the release of *A. nemoralis*, additional chemical treatment was still needed. Recently, a promising new approach has risen, in the form of processed-kaolin particle film technology (Puterka *et al.*, 2000; Pasqualini *et al.*, 2007). First results showed a good efficacy against *C. pyri* compared with mineral oil and untreated plots. Since kaolin repels insects, effects on beneficials and phytotoxic effects are low, and it therefore might be an alternative control strategy for *C. pyri* in organic and IPM orchards.

Vectors of apple proliferation

Considering the phenology of the two species, it appears that their control is possible only when the insects are present on cultivated plants. In Trentino, field trials have been carried out since 1999, in order to find efficient insecticides to control *C. picta* and *C. melanoneura* and to determine the timing of the treatments. Ethofenprox was found to be the most efficient product to control the overwintering adults of both species before blossom (Mattedi *et al.*, 2007). *C. melanoneura* was also efficiently depleted with organophosphates. The control strategy was aimed at preventing the reproduction of both species on apple. A particular problem arose for the control of overwintered adults of *C. picta* in years when oviposition coincided with the period of blossom, when insecticides cannot be applied. In this case, the strategy can be focused on the control of the development of the new generation. Organophosphates as well as neonicotinoids (thiametoxan, thiacloprid) were found to be appropriate products to control the larval development of *C. picta* (Mattedi *et al.*, 2007). The results of the transmission trials showed that both generations of *C. picta* can transmit the phytoplasma. Consequently, in areas where the disease is present, both the re-migrants and the new generation must be controlled. Therefore the precise prediction of the migration phase and the larval development is indispensable for an efficient control of the vectors of '*Ca.* Phytoplasma mali'.

Vectors of European stone fruit yellows

Only very few attempts have been undertaken to control ESFY by classical means of spraying. Poggi Pollini *et al.* (2007) conducted a trial for vector control after a severe outbreak of the disease in the province of Trentino (Italy). They treated four different experimental orchards with diverse pesticides to control the vector *C. pruni*. The monitoring of ESFY-like symptoms during the following seasons demonstrated, however, that most of the applications had no efficacy in controlling the disease. In areas where the disease is endemic (present on wild plants) and the populations of *C. pruni* are abundant, new strategies for vector control have to be developed.

Conclusions

For many years vectors of important European fruit tree diseases have been searched for, though without success, as they are univoltine psyllids with a particular biological cycle. The vectors of AP and ESFY colonize their host plant early in the vegetative season, when phytoplasma symptoms are less pronounced. Their population densities – at least on the cultivated host plants – are often low, which renders a classical control by insecticides difficult. However, they can be highly efficient vectors of the phytoplasmas, and the natural infection rate of vectors such as *C. picta* is very high. Experimental transmission trials demonstrated that both generations of the univoltine vectors, the overwintered adults as well as the offspring larvae and adults, can transmit '*Ca.* Phytoplasma mali' and '*Ca.* Phytoplasma prunorum'. Bait plant trials conducted in apple orchards indicated that the migrating new generation of *C. picta* can already be an efficient vector. Thus, the disease can be spread inside an orchard if the psyllids acquire the phytoplasma on infected trees, and it can be spread over larger distances inside a region by the remigrating adults. In contrast, the spread of ESFY seems to be predominantly monocyclic, signifying that infectious re-migrants transmit the phytoplasma on a regional scale.

Polyvoltine vector species of PD can be present in the orchards in much higher population densities. However, acquisition and transmission efficiencies of the different generations seem to vary considerably. The highest risk for transmission was found with the August generation and the winter generation (September/October). It is noteworthy that the winter form acquires the phytoplasma when its concentration is highest in the tree and that an infected psyllid can retain infectivity until the next season.

The observed transmission efficiencies of the different vectors varied – at least under the experimental conditions – considerably between the region and the year of trial, and according to the origin of field-collected insects. There is a strong indication that these differences are not purely technical but reflect genetic variation of phytoplasma strains as well as of vector populations. This would narrow further the relationship between pathogen and vector, which is already very strict. Although AP, PD and ESFY are often present in the same region and the 16Sr X group phytoplasmas are closely related to each other and are transmitted by vectors belonging to the same genus, there is no indication for an interference in the spread of the different diseases.

The control of the fruit tree phytoplasma diseases is based on prevention. This has to start with the use of healthy planting material. Then the spread of the disease by the psyllid vector has to be controlled according to a risk assessment based on knowledge about the presence of vector populations and about the infection pressure. Control of psyllids was always successful if the vector population was high enough and appropriate insecticides were available. However, there is no guarantee of limiting the disease spread if highly efficient vectors are present in low abundance. Therefore, these measures have to be accompanied by the uprooting of diseased trees.

Outlook

Further studies are currently under way to characterize the genetic and biological differences of phytoplasma strains of '*Ca.* Phytoplasma pyri', '*Ca.* Phytoplasma mali' and '*Ca.* Phytoplasma prunorum' (Danet *et al.*, 2007). Research has started to characterize, in parallel, different populations within the same vector species. Using microsatellite markers, Sauvion *et al.* (2007) found indications for the existence of at least two different populations of *C. pruni* in France. There is further indication that the contradictory transmission results obtained for *C. melanoneura* from different regions is also linked to different populations. Quantitative real-time PCR is a powerful tool to study the multiplication of the phytoplasma inside an individual insect (Pedrazzoli *et al.*, 2007) and may be applied to distinguish between non-vectors, which just acquired the phytoplasma by sucking, and real vectors, which are characterized by multiplication of the phytoplasma in the salivary glands (Cainelli *et al.*, 2007). These tools can be applied to study the pathogen–vector relationship in more detail and to better define the risks for the disease spread of different phytoplasma strains and vector populations.

A new approach for a possible biological vector control has been elucidated very recently. Mayer *et al.* (2008a, b) discovered that phytoplasma-infected trees heightened their attractiveness to *C. picta* by the release of a sesquiterpene. A synthetic ß-caryophellene was highly attractive to newly emerged adults. They proposed that this new type of phytopathogen-induced plant allomone may represent a promising compound for mass trapping of *C. picta*.

References

Bisognin, C., Schneider, B., Salm, H., Grando, M.S., Jarausch, W., Moll, E. and Seemüller, E. (2008) Apple proliferation resistance in apomictic rootstocks and its relationship to phytoplasma concentration and simple sequence repeat genotypes. *Phytopathology* 98, 153–158.

Blodgett, E.C., Schneider, H. and Aichele, M.D. (1962) Behavior of pear decline disease on different stock–scion combinations. *Phytopathology* 52, 679–684.

Blomquist, C.L. and Kirkpatrick, B.C. (2002a) Frequency and seasonal distribution of pear psylla infected with the pear decline phytoplasma in California pear orchards. *Phytopathology* 92, 1218–1226.

Blomquist, C.L. and Kirkpatrick, B.C. (2002b) Identification of phytoplasma taxa and insect vectors of peach yellow leaf roll disease in California. *Plant Disease* 86, 759–763.

Buès, R., Boudhinhon, J.F., Toubon, J.F. and Faivre d'Arcier, F. (1999) Geographic and seasonal variability of resistance to insecticides in *Cacopsylla pyri* L. (Hom., Psyllidae). *Journal of Applied Entomology* 123, 289–297.

Burckhardt, D. (1994) Psylloid pests of temperate and subtropical crop and ornamental plants (Hemiptera, Psylloidea): a review. *Entomology (Trends in Agricultural Science)* 2, 173–186.

Burckhardt, D., Jarausch, B. and Jarausch, W. (2008) www.psyllidkey.info: ein elektronischer Schlüssel zur Bestimmung der Psylliden auf Rosaceen in Mitteleuropa. *Mitteilungen aus dem Julius-Kühn-Institut* 417, 177.

Cainelli, C., Forno, F., Mattedi, L. and Grando, M.S. (2007) Can apple aphids be vectors of *Candidatus* Phytoplasma mali? *IOBC/WPRS Bulletin* 30(4), 261–266.

Canik, D. and Ertunc, F. (2007) Distribution and molecular characterization of apple proliferation phytoplasma in Turkey. *Bulletin of Insectology* 60(2), 335–336.

Carraro, L., Loi, N., Ermacora, P., Gregoris, A. and Osler, R. (1998a) Transmission of pear decline by using naturally infected *Cacopsylla pyri*. *Acta Horticulturae* 472, 665–668.

Carraro, L., Osler, R., Loi, N., Ermacora, P. and Refatti, E. (1998b) Transmission of European stone fruit yellows phytoplasma by *Cacopsylla pruni*. *Journal of Plant Pathology* 80, 233–239.

Carraro, L., Osler, R., Loi, N., Ermacora, P. and Refatti, E. (2001a) Fruit tree phytoplasma diseases diffused in nature by psyllids. *Acta Horticulturae* 550, 345–350.

Carraro, L., Loi, N. and Ermacora, P. (2001b) Transmission characteristics of the European stone fruit yellows phytoplasma and its vector *Cacopsylla pruni*. *European Journal of Plant Pathology* 107, 695–700.

Carraro, L., Loi, N. and Ermacora, P. (2001c) The life cycle of pear decline phytoplasma in the vector *Cacopsylla pyri*. *Journal of Plant Pathology* 83, 87–90.

Carraro, L., Ferrini, F., Ermacora, P. and Loi, N. (2002) Role of wild *Prunus* species in the epidemiology of European stone fruit yellows. *Plant Pathology* 51, 513–517.

Carraro, L., Ferrini, F., Labonne, G., Ermacora, P. and Loi, N. (2004) Seasonal infectivity of *Cacopsylla pruni*, the vector of European stone fruit yellows phytoplasma. *Annals of Applied Biology* 144, 191–195.

Carraro, L., Ferrini, F., Labonne, G., Ermacora, P. and Loi, N. (2008) Infectivity of *Cacopsylla picta* (syn. *Cacopsylla costalis*), vector of '*Candidatus* Phytoplasma mali' in north east Italy. *Acta Horticulturae* 781, 403–407.

Cermák, V. and Lauterer, P. (2007) Overwintering of psyllids (Hemiptera, Psylloidea) in South Moravia (Czech Republic) with respect to the vectors of the apple proliferation cluster phytoplasmas. 4th European Hemiptera Congress, Ivrea, Italy, 10–14 September 2007.

Chabrolin, C. (1924) Quelques maladies des arbres fruitiers de la Vallée du Rhône. *Annales Epiphytes* 10, 265–333.

Civolani, S. and Pasqualini, E. (2003) *Cacopsylla pyri* L. (Hom., Psyllidae) and its predators' relationship in Italy's Emilia-Romagna region. *Journal of Applied Entomology* 127, 214–220.

Danet, J.-L., Bonnet, P., Jarausch, W., Carraro, L., Skoric, D., Labonne, G. and Foissac, X. (2007) *Imp* and *sec*Y, two new markers for MLST (multilocus sequence typing) in the 16SrX phytoplasma taxonomic group. *Bulletin of Insectology* 60(2), 339–340.

Daugherty, M.P., Briggs, C.J. and Welter, S.C. (2007) Bottom-up and top-down control of pear psylla (*Cacopsylla pyricola*): fertilization, plant quality, and the efficacy of the predator *Anthocoris nemoralis*. *Biological Control* 43, 257–264.

Davies, D.L., Guise, C.M. and Adams, A.N. (1992) Parry's disease is similar to pear decline and is associated with mycoplasma-like organisms transmitted by *Cacopsylla pyricola*. *Plant Pathology* 41, 195–203.

Delić, D., Martini, M., Ermacora, P., Carraro, L. and Myrta, A. (2005) First report of fruit tree phytoplasmas and their psyllid vectors in Bosnia and Herrzegovina. *Journal of Plant Pathology* 87, 150.

Delić, D., Martini, M., Ermacora, P., Carraro, L. and Myrta, A. (2008) Identification of fruit tree phytoplasmas and their vectors in Bosnia and Herzegovina. *Acta Horticulturae* 781, 429–434.

Fialová, R., Navrátil, M., Válová, P., Lauterer, P., Kocourek, F. and Poncarová-Voráčková, Z. (2004) Epidemiology of European stone fruit yellows phytoplasma in the Czech Republic. *Acta Horticulturae* 657, 483–487.

Font, I., Abad, P., Albinana, M., Espino, A.I., Dally, E.L., Davies, R.E. and Jorda, C. (1999) Amarilleos y enrojecimientos en zanahoria: una enfermedad a diagnostico. *Boletín de Sanidad Vegetal Plagas* 25, 405–415.

Frisinghelli, C., Delaiti, L., Grando, M.S., Forti, D. and Vindimian, M.E. (2000) *Cacopsylla costalis* (Flor, 1861), as a vector of apple proliferation in Trentino. *Journal of Phytopathology* 148, 425–431.

Garcia-Chapa, M., Sabaté, J., Lavina, A. and Batlle, A. (2005) Role of *Cacopsylla pyri* in the epidemiology of pear decline in Spain. *European Journal of Plant Pathology* 111, 9–17.

Goidànich, G. (1933). Un deperimento dei susini. *Bolletino Stazione Patologia Vegetale Roma* 13, 160–173.

Gross, J. and Mekonen, N. (2005) Plant odours influence the host finding behaviour of apple psyllids (*Cacopsylla picta; C. melanoneura*). *IOBC WPRS Bulletin* 28(7), 351–355.

Guerra, L.J. (1997) Biological and molecular characterization of phytoplasmas infecting fruit and nut trees in California. PhD thesis, University of California, Davies.

Hodkinson, I.D. (1974) The biology of the Psylloidea (Homoptera): a review. *Bulletin of Entomological Research* 64, 325–339.

Hogenhout, S.A., El-Desouky, A., Whitfield, A.E. and Redinbaugh, M.G. (2008a) Insect vector interactions with persistently transmitted viruses. *Annual Review of Phytopathology* 46, 327–359.

Hogenhout, S.A., Oshima, K., El-Desouky, A., Kakizawa, S., Kingdom, H.N. and Namba, S. (2008b) Phytoplasmas: bacteria that manipulate plants and insects. *Molecular Plant Pathology* 9, 327–359.

Jarausch, B., Schwind, N., Jarausch, W., Krczal, G., Seemüller, E. and Dickler E. (2003) First report of *Cacopsylla picta* as a vector for apple proliferation phytoplasma in Germany. *Plant Disease* 87, 101.

Jarausch, B., Schwind, N., Jarausch, W. and Krczal, G. (2004) Overwintering adults and springtime generation of *Cacopsylla picta* (synonym *C. costalis*) can transmit apple proliferation phytoplasmas. *Acta Horticulturae* 657, 409–413.

Jarausch, B., Fuchs, A., Mühlenz, I., Lampe, I., Harzer, U. and Jarausch, W. (2007a) Research on European stone fruit yellows (ESFY) in Germany. *Bulletin of Insectology* 60(2), 389–390.

Jarausch, B., Fuchs, A., Schwind N., Krczal, G. and Jarausch, W. (2007b) *Cacopsylla picta* as most important vector for '*Candidatus* Phytoplasma mali' in Germany and neighbouring regions. *Bulletin of Insectology* 60(2), 189–190.

Jarausch, B., Mühlenz, I., Beck, A., Lampe, I., Harzer, U. and Jarausch, W. (2008) Epidemiology of European stone fruit yellows in Germany. *Acta Horticulturae* 781, 417–422.

Jarausch, W., Lansac, M., Saillard, C., Broquaire, J.M. and Dosba, F. (1998) PCR assay for specific detection of European stone fruit yellows phytoplasmas and its use for epidemiological studies in France. *European Journal of Plant Pathology* 104, 17–27.

Jarausch, W., Saillard, C., Broquaire, J.M., Garnier, M. and Dosba, F. (2000a) PCR-RFLP and sequence analysis of a non-ribosomal fragment for genetic characterization of European stone fruit yellows phytoplasmas infecting various *Prunus* species. *Molecular and Cellular Probes* 14, 171–179.

Jarausch, W., Saillard, C., Helliot, B., Garnier, M. and Dosba, F. (2000b) Genetic variability of apple proliferation phytoplasmas as determined by PCR-RFLP and sequencing of a non-ribosomal fragment. *Molecular and Cellular Probes* 14, 17–24.

Jarausch, W., Danet, J.L., Labonne, G., Dosba, F., Broquaire, J.M., Saillard, C. and Garnier, M. (2001) Mapping the spread of apricot chlorotic leaf roll (ACLR) in southern France and implication of *Cacopsylla pruni* as a vector of European stone fruit yellows (ESFY) phytoplasmas. *Plant Pathology* 50, 782–790.

Jensen, D.D., Griggs, W.H., Gonzales, C.Q. and Schneider, H. (1964) Pear decline virus transmission by pear psylla. *Phytopathology* 54, 1346–1351.

Kartte, S. and Seemüller, E. (1991) Susceptibility of grafted *Malus* taxa and hybrids to apple proliferation disease. *Journal of Phytopathology* 131, 137–148.

Krczal, G., Krczal, H. and Kunze, L. (1988) *Fieberiella florii* (Stål), a vector of apple proliferation agent. *Acta Horticulturae* 235, 99–106.

Kube, M., Schneider, B., Kuhl, H., Dandekar, T., Heitmann, K., Migdoll, A.M., Reinhardt, R. and Seemüller, E. (2008) The linear chromosome of the plant-pathogenic mycoplasma '*Candidatus* Phytoplasma mali'. *BMC Genomics* 9, 306.

Kucerova, J., Talacko, L., Lauterer, P., Navratil, M. and Fialova, R. (2007) Molecular tests to determine *Candidatus* Phytoplasma pyri presence in psyllid vectors from a pear tree orchard in the Czech Republic – a preliminary report. *Bulletin of Insectology* 60(2), 191–192.

Kunze, L. (1989) Apple proliferation. In: Fridlund, P.R. (ed.) *Virus and Viruslike Diseases of Pome Fruits and Simulating Noninfectious Disorders*. Cooperative Extension College of Agriculture and Home Economics, Washington State University, Pullmann, Washington, pp. 99–113.

Lauterer, P. (1999) Results of investigations on Hemiptera in Moravia, made by Moravian Museum (Psylloidea 2). *Acta Musei Moraviae, Scientae Biologicae* (Brno) 84, 71–151.

Laviña, A., Sabaté, J., García-Chapa, M., Batlle, A. and Torres, E. (2004) Occurrence and epidemiology of European stone fruit yellows phytoplasma in Spain. *Acta Horticulturae* 657, 489–494.

Leclant, F., Marchoux, G. and Gianotti, J. (1974) Demonstration of the vector role of the psyllid *Trioza nigricornis* Foerst. (Homopter. insect) in the transmission of a proliferation disease of *Daucus carota* L. *Comptes Rendus des Séances de l'Académie de Sciences D* 278, 57–59.

Lemoine, J. (1984) Is *Psylla pyri* a vector for pear decline in France? *IOBC/WPRS Bulletin* 7(5), 245–251.

Liu, H.-L., Chen, C.-C. and Lin, C.-P. (2007) Detection and identification of the phytoplasma associated with pear decline in Taiwan. *European Journal of Plant Pathology* 117, 281–291.

Lorenz, K.-H., Dosba, F., Poggi Pollini, C., Llácer, G. and Seemüller, E. (1994) Phytoplasma diseases of *Prunus* species in Europe are caused by genetically similar organisms. *Zeitschrift für Pflanzenkrankheiten und Pflanzenschutz* 101, 567–575.

Mader, C. (1908) La mortalità dei peri nella plaga di Bolzano-Gries. *Almanacco Agrario, Sezione Di Trento Del Consiglio Provinciale D'Agricoltura Pel Tirolo* 26, 347–350.

Marcone, C., Gibb, K.S., Streten, C. and Schneider, B. (2004) '*Candidatus* Phytoplasma spartii', '*Candidatus* Phytoplasma rhamni' and *Candidatus* Phytoplasma allocasuarinae', respectively associated with spartium witches'-broom, buckthorn witches'-broom and allocasuarina yellows diseases. *International Journal of Systematic and Evolutionary Microbiology* 54, 1025–1029.

Mattedi, L., Forno, F., Cainelli, C., Grando, S. and Jarausch, W. (2007) Transmission of *Candidatus* Phytoplasma mali by psyllid vectors in Trentino. *IOBC/WPRS Bulletin* 30(4), 39.

Mattedi, L., Forno, F., Cainelli, C., Grando, M.S. and Jarausch, W. (2008) Research on *Candidatus* Phytoplasma mali transmission by insect vectors in Trentino. *Acta Horticulturae* 781, 369–374.

Mayer, C.J. and Gross, J. (2007) Different host plant odours influence migration behaviour of *Cacopsylla melanoneura* (Förster), an insect vector of the apple proliferation phytoplasma. *IOBC WPRS Bulletin* 30(4), 177–184.

Mayer, C.J., Vilcinskas, A. and Gross, J. (2008a) Phytopathogen lures its insect vector by altering host plant odor. *Journal of Chemical Ecology* 34, 1045–1049.

Mayer, C.J., Vilcinskas, A. and Gross, J. (2008b) Pathogen-induced release of plant allomone manipulates vector insect behaviour. *Journal of Chemical Ecology* 34, 1518–1522.

Mayer, C.J., Jarausch, B., Jarausch, W., Jelkmann, W., Vilcinskas, A. and Gross, J. (2009) *Cacopsylla melanoneura* has no relevance as vector of apple proliferation in Germany. *Phytopathology* 99, 729–738.

McLarty, H.R. (1948) Killing of pear trees. *Annual Report of the Canadian Plant Disease Survey* 28, 77.

Morton, A., Davies, D.L., Blomquist, C.L. and Barbara, D.J. (2003) Characterization of homologues of the apple proliferation immunodominant membrane protein gene from three related phytoplasmas. *Molecular Plant Pathology* 4, 109–114.

Ng, J.C.K. and Falk, B.W. (2006) Virus–vector interactions mediating nonpersistent and semipersistent transmission of plant viruses. *Annual Review of Phytopathology* 44, 183–212.

Nichols, C.W., Schneider, H., O'Reilly, H.J., Shalla, T.A. and Griggs, W.H. (1960) Pear decline in California. *Bulletin of the California Department of Agriculture* 49, 186–192.

Ossiannilsson, F. (1992) The Psylloidea (Homoptera) of Fennoscandia and Denmark. *Fauna Entomologica Scandinavica* 26, 1–347.

Pasqualini, E., Civolani, S. and Grappadelli, L.E. (2007) Particle film technology: approach for a biorational control of *Cacopsylla pyri* (Rhynchota Psyllidae) in northern Italy. *Bulletin of Insectology* 55, 39–42.

Pedrazzoli, F., Gualandri, V., Forno, F., Mattedi, L., Malagnini, V., Salvadori, A., Stoppa, G. and Ioriatti, C. (2007) Acquisition capacities of the overwintering adults of the psyllid vectors of '*Candidatus* Phytoplasma mali'. *Bulletin of Insectology* 60(2), 195–196.

Poggi Pollini, C., Bianchi L., Forno, F., Franchini,. S., Giunchedi, L., Gobber, M., Mattedi, L., Miorelli, P., Pignatta, D., Profaizer, D., Ratti, C. and Reggiani, N. (2007) Investigation on European stone fruit yellows in experimental apricot orchards in the province of Trentino (Italy). *Bulletin of Insectology* 60(2), 323–324.

Purcell, A.H., Nyland, G., Raju, B.C. and Heringer, M.R. (1981) Peach yellow leaf roll epidemic in northern California: effects of peach cultivar, tree age, and proximity to pear orchards. *Plant Disease* 65, 365–368.

Puterka, G.J., Glenn, D.M., Sekutowski, D.G., Unruh, T.R. and Jones, S.K. (2000) Progress toward liquid formulations of particle film for insect and disease control in pear. *Environmental Entomology* 29, 329–339.

Refatti, E. (1964) Pear moria in Italy. *FAO Plant Protection Bulletin* 12, 1–7.

Rui, D., Ciferri, R. and Refatti, E. (1950) La virosi degli 'scopazzi del melo' nel Veronese. *Notiziario sulle malattie delle piante* 13, 7–11.

Salehi, M., Izadpanah, K., Taghavi, S.M. and Rahimian, H. (2008) Characterization of a phytoplasma associated with pear decline in Iran. *Journal of Phytopathology* 156, 493–495.

Sauvion, N., Lachenaud, O., Genson, G., Rasplus, J.-Y. and Labonne, G. (2007) Are there several biotypes of *Cacopsylla pruni*? *Bulletin of Insectology* 60(2), 185–186.

Schneider, H. (1970) Graft transmission and host range of the pear decline causal agent. *Phytopathology* 60, 204–207.

Seemüller, E. (1989) Pear decline. In: Fridlund, P.R. (ed.) *Virus and Viruslike Diseases of Pome Fruits and Simulating Noninfectious Disorders*. Cooperative Extension College of Agriculture and Home Economics. Washington State University, Pullman, Washington, pp. 188–201.

Seemüller, E. (1990) Apple proliferation. In: Jones, A.L. and Aldwinkle, H.S. (eds) *Compendium of Apple and Pear Diseases*. APS Press, St Paul, Minnesota, pp. 67–68.

Seemüller, E. and Schneider, B. (2004) '*Candidatus* Phytoplasma mali', '*Candidatus* Phytoplasma pyri' and '*Candidatus* Phytoplasma prunorum', the causal agents of apple proliferation, pear decline and European stone fruit yellows, respectively. *International Journal of Systematic and Evolutionary Microbiology* 54, 1217–1226.

Seemüller, E., Lorenz, K.-H. and Lauer, U. (1998a) Pear decline resistance in *Pyrus communis* rootstocks and progenies of wild and ornamental *Pyrus* taxa. *Acta Horticulturae* 472, 681–691.

Seemüller, E., Marcone, C., Lauer, U., Ragozzino, A. and Göschl, M. (1998b) Current status of molecular classification of the phytoplasmas. *Journal of Plant Pathology* 80, 3–26.

Sertkaya, G., Sertkaya, E. and Kaya, K. (2008) Detection of pear decline disease in pear and quince in the eastern Mediterranean region of Turkey. *Acta Horticulturae* 781, 511–515.

Solomon, M.G., Cranham, J.E., Easterbrook, M.A. and Fitzgerald, J.D. (1989) Control of the pear psyllid *Cacopsylla pyricola* in south east England by predators and pesticides. *Crop Protection* 8, 197–205.

Sterck, G. and Highwood, D.P. (1992) Implementation of the IRAC anti-resistance guidelines with IPM programs for Belgian apple and pear orchards. *Proceedings of the British Crop Protection Conference on Pests and Disease* 2, 517–526.

Tedeschi, R. and Alma, A. (2004) Transmission of apple proliferation phytoplasma by *Cacopsylla melanoneura* (Homoptera: Psyllidae). *Journal of Economical Entomology* 97, 8–13.

Tedeschi, R. and Alma, A. (2006) *Fieberiella florii* (Stål) (Homoptera Auchenorrhynca) as a vector of '*Candidatus* Phytoplasma mali'. *Plant Disease* 90, 284–290.

Tedeschi, R., Bosco, D. and Alma, A. (2002) Population dynamics of *Cacopsylla melanoneura* (Homoptera: Psyllidae), a vector of apple proliferation phytoplasma in northwestern Italy. *Journal of Economical Entomology* 95, 544–551.

Tedeschi, R., Visentin, C., Alma, A. and Bosco, D. (2003) Epidemiology of apple proliferation (AP) in northwestern Italy: evaluation of the frequency of AP-positive psyllids in naturally infected populations of *Cacopsylla melanoneura* (Homoptera: Psyllidae). *Annals of Applied Biology* 142, 285–290.

Tedeschi, R., Ferrato, V., Rossi, J. and Alma, A. (2006) Possible phytoplasma transovarial transmission in the psyllids *Cacopsylla melanoneura* and *Cacopsylla pruni. Plant Pathology* 55, 18–24.

Tedeschi, R., Lauterer, P., Brusetti, L., Tota, F. and Alma, A. (2009) Composition, abundance and phytoplasma infection in the hawthorn psyllid fauna of northwestern Italy. *European Journal of Plant Pathology* 123, 301–310.

Thébaud, G., Yvon, M., Labonne, G. and Alary, R. (2008) European stone fruit yellows: consequences of the life cycle of the vector and of the multiplication of the phytoplasma in the insect on the epidemiology of the disease. *Acta Horticulturae* 781, 423–428.

Thébaud, G., Yvon, M., Alary, R., Sauvion, N. and Labonne, G. (2009) Efficient transmission of '*Candidatus* Phytoplasma prunorum' is delayed by eight months due to a long latency in its host-alternating vector. *Phytopathology* 99(3), 265–273.

van de Baan, H.E. and Croft, B.A. (1991) Resistance to insecticides in winter and summer form of pear psylla, *Psylla pyricola*. *Pesticide Science* 32, 225–233.

Weintraub, P.G. and Beanland, L.A. (2006) Insect vectors of phytoplasmas. *Annual Review of Entomology* 51, 91–111.

Woodbridge, C.G., Blodgett, E.C. and Diener, T.O. (1957) Pear decline in the Pacific northwest. *Plant Disease Reporter* 41, 569–572.

Yvon, M., Labonne, G. and Thébaud, G. (2004) Survival of European stone fruit yellows phytoplasma outside fruit crop production areas: a case study in southeastern France. *Acta Horticulturae* 657, 477–481.

15 Microbial Symbionts of Auchenorrhyncha Transmitting Phytoplasmas: a Resource for Symbiotic Control of Phytoplasmoses

Alberto Alma,[1] Daniele Daffonchio,[2] Elena Gonella[1] and Noura Raddadi[1]

[1]*Università degli Studi di Torino, Italy;* [2]*Università degli Studi di Milano, Italy*

Introduction

Symbiotic associations between microorganisms and insect hosts are known to have an important role in the evolution of both partners, making an important contribution to the success of the insect lifestyle. Arthropod-associated microbes have several relationships with their hosts, with positive, neutral or negative effects on them. Mutualistic symbionts may perform an essential function required for the survival of the host, as in the case of primary endosymbiotic bacteria. Primary symbionts (P-symbionts) are often located within specialized organs of the host, the bacteriomes, which consist of aggregates of cells called bacteriocytes. Bacteriomes are transmitted maternally to the progeny through eggs, permitting the vertical transfer of bacteria. The associations between P-symbionts and their insect hosts are ancient, dating back hundreds of millions of years. Co-cladogenesis, i.e. the observation that the phylogenetic trees based on the P-symbiont genes are similar to the ones based on the host genes, indicating a co-evolution, has been reported for several insects, including mealybugs, aphids, whiteflies and psyllids (Baumann, 2006; Takiya *et al.*, 2006). A common feature of P-symbionts is that they undergo degenerative evolution involving irreversible loss of genes and regulatory capacities, making their survival dependent on their host, and hence are not culturable on commercial media. Other symbionts, commonly called secondary or guest symbionts, are not necessary for the host's survival but have positive effects that improve their host's fitness, such as rescue from heat stress, resistance to natural enemies, host plant specialization and reproduction, and compensation for the loss of the essential symbiotic microbes. Several secondary symbionts can be transmitted horizontally through host populations by co-feeding and are usually inherited by the offspring. Among the microorganisms that have no effect or negative action on the host fitness, several heritable bacteria are of particular interest, as they exploit the insect by manipulating its sexuality or reproduction.

This chapter will deal with the current understanding about microbial symbionts associated with insect vectors of phytoplasmas, with a particular regard to the achievements towards the development of innovative strategies to prevent pathogen diffusion, as it was proposed for controlling other bacterial phytopathogenic agents. The first attainments with reference to the study of the microbiota associated with phytoplasma vectors are those concerning insects transmitting the aetiological agents of grapevine yellows (GY). Indeed, among the phytoplasma-caused diseases with major economic effects in Europe, GY, which cause severe losses in many wine-producing regions, are of increasing relevance. These diseases are spread in several grapevine-growing areas in the world, such as central and southern Europe, the Middle East, northern and southern Africa, North and South America and Australia (Boudon-Padieu, 2003). The spread of such pathogens is mainly due, in nature, to insect vectors belonging to the Auchenorrhyncha of the families Cicadellidae and Cixiidae (leafhoppers and planthoppers).

The most important grapevine yellows undermining European grapevine production is flavescence dorée (FD), a quarantine disease caused by a phytoplasma of the elm yellows (EY) group. The vector of this bacterium is the Nearctic leafhopper *Scaphoideus titanus* Ball, which feeds strictly on grapevine phloem, completing its whole life cycle on this species and transmitting the phytoplasma to healthy grapes. Different American wild grape species were reported to host the phytoplasma, besides European grapevine (Beanland *et al.*, 2006). Another important GY is bois noir (French name, BN), also called Vergilbungskrankheit (VK) (German name) and legno nero (LN) (Italian name), which afflicts viticulture in different European countries and in the Middle East. The phytoplasma responsible for BN, belonging to the stolbur (STOL) group, is transmitted by the polyphagous planthopper *Hyalesthes obsoletus* Signoret (Hemiptera: Cixiidae), which lives on several wild and cultivated herbaceous dicotyledons, often growing near the vineyards. These plants can be infected by STOL, sometimes without symptoms, and work as a reservoir for the phytopathogenic agent (Lessio *et al.*, 2007).

As no direct cure for such diseases is available at present, the control is focused on the prevention of disease transmission by the vectors, by using insecticides and removing diseased plants. In some cases, the application of control strategies is mandatory, as it happens in several European countries for FD. However, insect resistance to insecticides is undermining the outcome of vector control, together with the considerable impact of chemical insecticides on the non-target invertebrate fauna and on the general environment. Alternative approaches are required to substitute the highly impacting use of insecticides in reducing disease transmission. An emerging alternative for long-term control is symbiotic control, i.e. the exploitation of microorganisms living in symbiosis with the insect vectors to control the transmission of the microbial pathogen (Miller *et al.*, 2006). Such a strategy is based on the evidence that a strict association occurs between insects feeding on restricted diets, such as blood or plant sap, and their microbial symbionts. Symbiotic control strategies could take advantage of such associations by exploiting the competition with the pathogenic agent in vector colonization, by reducing vector competence, producing antagonistic molecules or unbalancing the vector populations.

Symbiotic control was first proposed for controlling the spread of a human disease, Chagas' disease, which is caused by an insect-transmitted protozoon, by using a transformed endosymbiont of the insect vector *Rhodnius prolixus* Stål. The endosymbiont was transformed to prevent the insect vector competence for the trypanosome, *Trypanosoma cruzi*, responsible for the disease (Beard *et al.*, 2001). Also the use of symbiotic bacteria of tsetse flies, *Wigglesworthia glossinidia* and *Sodalis glossinidius*, was proposed to control the transmission of sleeping sickness through the expression of antagonistic factors (Aksoy and Rio, 2005).

A model for the exploitation of symbiotic microorganisms for symbiotic control strategies against the spread of phytoplasmas is provided by the results obtained concerning one of the most important bacterial diseases of grapevine affecting American wine production: Pierce's disease (PD). PD is caused by the γ-Proteobacterium *Xylella fastidiosa* attacking the xylem of the plant. It has several host species besides grapevine and it is transmitted by sharpshooter leafhoppers (Cicadellidae) and spittlebugs (Cercopidae). However, the most troubling vector is the glassy-winged sharpshooter (GWSS) *H. vitripennis* (Germar) (Hemiptera: Cicadellidae) (formerly *Homalodisca coagulata* (Say)), whose recent accidental introduction is responsible for the explosion of PD in California, currently one of the most affected areas.

The possibility of cultivating *X. fastidiosa* in artificial media increases the feasibility of a symbiotic control approach. Indeed, different studies were able to indicate several molecules with an antimicrobial activity against this bacterium (Li and Gray, 2003; Kuzina *et al.*, 2006). Moreover, recently the role of cell to cell signalling in the virulence of *X. fastidiosa* has been demonstrated. This mechanism involves the production of a diffusible signalling factor (DSF), mediated by the *rpfC* gene, which modulates the virulence and the effectiveness of the colonization of the insect vector (Chatterjee *et al.*, 2008). Knowledge of the processes ruling the host–microbe interaction may provide an efficient target for symbiotic control strategies.

A Model for Exploiting Symbionts to Control Plant Diseases Transmitted by Insects

Homalodisca vitripennis, vector of Pierce's disease to grapevine

PD is a lethal disease, found in several regions of the American continent, especially in regions with mild winters and a longer growing season. It is distributed from the southern wine-producing states of North America, such as Florida, Texas and California, through Mexico to some parts of Central America and north-western South America, such as Costa Rica and Venezuela (Hopkins and Purcell, 2002). PD has not been reported outside America; nevertheless commercial exchanges of plant material with Europe could support its introduction. The causal agent *X. fastidiosa* has been isolated in Taiwan and in the Kosovo region in Europe (Berisha *et al.*, 1998).

Xylella fastidiosa is transmitted among the host plants by different insect vectors. The species *X. fastidiosa* contains several strains differing in the host range, pathogenicity, DNA similarity and nutritional requirements, which determine the grade of fastidiousness for isolation and cultivation outside the hosts. *X. fastidiosa* is responsible for several diseases in different fruit trees and other species. Symptoms of *X. fastidiosa*'s infection include leaf chlorosis, leaf scorch and crop loss, and can determine plant death. *X. fastidiosa* also colonizes several plant species without causing symptoms. Such hosts are not without relevance, since they can be a reservoir for the transmission of the pathogen to sensitive plants by insect vectors, leading to severe damage to agricultural production.

Different species of froghoppers and leafhoppers are known to be vectors of *X. fastidiosa* strains that determine PD and other diseases across the southern United States, together with the glassy-winged sharpshooter *H. vitripennis* (Blua *et al.*, 2001). Even though they have variable transmission efficiencies, these insect vectors do not necessitate any measurable latent period to transmit *X. fastidiosa* efficiently. Multiplication and circulation of bacterial cells within the insect body are not required and, once infective, vectors are able to transmit the pathogen indefinitely (Hopkins and Purcell, 2002). As far as vertical transmission through the vector is concerned, no evidence of transovarial transmission of *X. fastidiosa* has been reported yet. Vector nymphs stop transmitting after moulting, but they recover the capacity for transmission after feeding on an infected plant, suggesting that bacteria are transmitted from the external surface of the vector's foregut, which is replaced when moulting occurs (Redak *et al.*, 2004).

In the past decade, the disease has increased in prominence, causing serious production losses, especially in southern California, because of the accidental introduction of the GWSS, first detected in the state at the end of the 1980s. *H. vitripennis* is a big insect and is able to feed on a large host plant range, including 73 plant species in 35 families. In California and the south-eastern United States, the GWSS is documented to produce at least two generations per year. Eggs are laid inside the epidermis of the lower leaf blade of host plants. Nymphs feed on leaf petioles or small stems while they progress through four moults, before becoming winged adults in 10–12 weeks. Adults of *H. vitripennis*, in contrast to other sharpshooters of the tribe Cicadellini, are able to feed on mature woody tissues of plants. For this reason, overwintering adults can feed on dormant vines, although other hosts such as citrus plants can represent a reservoir for these insects when other species are quiescent (Hopkins and Purcell, 2002; Almeida and Purcell, 2003). Infected adults, remaining infective for their life cycle, can inoculate 2-year-old woody tissues of grapevines and may transmit the pathogen to vines at the base of canes even during winter, suggesting a polycyclic spread of the pathogen throughout the year (Almeida and Purcell, 2003).

The transmission efficiency appears to be lower and more variable than the efficiency of the other important vector, *Graphocephla atropunctata* (Signoret) (Almeida and Purcell, 2003); nevertheless *H. vitripennis* is an effective vector in the field, due to its dispersal ability and its inclination to aggregate in high numbers along crop borders and to feed on woody tissues. This leads to the establishment of a high incidence of chronic infections during summer and autumn, and implies that the control of *H. vitripennis* populations in vineyards may need to be maintained all year long.

The current management of PD includes removal of diseased grapevines, use of insecticides and biological control agents to reduce GWSS populations in citrus groves, urban areas and vineyards.

Co-primary symbionts of *Homalodisca vitripennis*

Due to its relevant role in the exponential increase of PD damage to Californian viticulture, *H. vitripennis* was studied for other biological aspects that could give a perspective for effective disease control. Hence the study of the occurrence of microbial symbionts and of the nature of symbiotic relationships with this devastating insect vector was begun a few years ago. A first molecular screening of the total bacterial community associated with the GWSS and other sharpshooters showed the occurrence of dominant γ-Proteobacteria forming a separate clade, with a divergence of the 16S rRNA gene from the closest relatives exceeding 10%. Furthermore, these bacteria showed a substantial divergence (~7%) between symbionts of the different insect species, suggesting the ancient origin of the clade they belong to. Such symbionts, initially reported as the 't-symbionts' of several leafhoppers (Buchner, 1965), were described as organisms belonging to the Enterobacteriales group in the γ-Proteobacteria, characterized by a small genome size and a biased nucleotide composition favouring adenine and thymine (A + T), and were named '*Candidatus* Baumannia cicadellinicola' (Moran *et al.*, 2003). The irregularly spherical cells of these organisms were typically within bacteriomes, common in sap-feeding insects, which often host symbionts that provide nutrients for the insects feeding on nutritionally unbalanced diets. *H. vitripennis* has bilaterally paired bacteriomes composed of two parts – red and yellow pigmented – both hosting *Baumannia*. Together with this γ-Proteobacterium, further bacterial symbionts were discovered to be associated with the GWSS. Besides secondary symbionts of the genus *Wolbachia*, frequently detected in the haemolymph of the insect (Takiya *et al.*, 2006; Curley *et al.*, 2007), bacteriome-associated bacteria other than *Baumannia,* belonging to the phylum Bacteroidetes, were found only in the yellow portion of the specialized organs in the GWSS and other Auchenorrhyncha. According to phylogenetic studies of the distinctive clade formed by these bacteria, they were proposed to be part of a very ancient association with an ancestor of all Auchenorrhyncha. These vertically transmitted symbionts, characterized by a distinctive strap-like shape, have now been designated as '*Ca.* Sulcia muelleri' (Moran *et al.*, 2005).

Sequencing of the whole genome of these symbionts was useful for the investigation of traits of intracellular symbionts (Wu *et al.*, 2006; McCutcheon and Moran, 2007). *Baumannia* exhibited intermediate features between endosymbionts and free-living bacteria, representing a model for studies of the evolutionary dynamics of intracellular symbionts (Wu *et al.*, 2006). *Sulcia* revealed one of the smallest reported genomes, second only to *Carsonella rudii*, the primary endosymbiont of certain psyllids, with traits previously known only for endosymbionts of the γ-Proteobacteria group (McCutcheon and Moran, 2007).

Attempts at predicting the metabolic capabilities of these two endosymbionts from their genome sequence indicated that both have, as expected from their small genome size, a relatively limited repertoire of synthetic capabilities. Some pathways, such as those concerning vitamin and cofactor synthesis, were complete in *Baumannia*, suggesting that it provides the host with these compounds, which have very low concentrations in the xylem sap. Nevertheless, many expected pathways were missing in *Baumannia*, especially those involving the production of essential nutrients lacking in the xylem sap, as is the case for amino acids. This implies that both the host and *Baumannia* must obtain amino acids from other external sources (Wu *et al.*, 2006). Conversely, the genome of *Sulcia* exhibited the whole pathways for the synthesis of amino acids essential for the host, although the origin of the necessary nitrogen is unknown. These findings indicated that the two symbionts work in concert, and possibly even share metabolites, to produce all of the nutrients needed by the host to survive on xylem sap (Wu *et al.*, 2006; McCutcheon and Moran, 2007). The complementarity between the host and each symbiont is extended to the mutual dependence between the symbionts, which are dependent on each other for many essential metabolic intermediates, providing a simple model of genomic co-evolution, a central process in the evolution of most organisms living in stable associations (Wu *et al.*, 2006). Microscopy studies using specific probes for the two endosymbionts supported such a strict association between the two symbionts, which were observed to live in close proximity within the host bacteriomes and sometimes with a single *Sulcia* cell surrounded by closely adjacent *Baumannia* cells (Moran *et al.*, 2005).

The occurrence of this long-term co-inheritance of multiple symbionts during the diversification of a eukaryotic host was confirmed by investigations on the distribution of *Baumannia* and *Sulcia* among sharpshooters and related leafhoppers. Combined phylogenetic studies of host species and symbionts supported a congruent evolutionary history between sharpshooters, *Sulcia* and *Baumannia*. It was suggested that *Sulcia* was ancestrally present in a host lineage that acquired *Baumannia* at the same approximate time as the switch to a feeding mode on xylem sap. This is consistent with the view that the symbiont's metabolic capabilities were a requirement for the new lifestyle of the host. After the acquisition of *Baumannia*, both *Sulcia* and *Baumannia* diversified in parallel with their sharpshooter hosts, through strict maternal transmission (Takiya *et al.*, 2006). In view of their ancient associations with hosts, together with their nutritional contributions to host metabolism, the two symbionts of *H. vitripennis* have been defined as 'co-primary' symbionts (Takiya *et al.*, 2006).

The discovery of such a tripartite association raises the likelihood that bacterial symbiosis has been a major element governing the ecological diversification of Auchenorrhyncha, involving not only benefits for the host, such as nutrient supply, but also some constraining factors, since gene losses and genome degradation in the symbionts may limit the ecological capabilities of the host. These constraints also occur in the association between insects and single primary symbionts, and may be even more complex in the case of multiple bacterial partners, as the evolution of each symbiont is likely to have had major consequences for the others, as well as for the hosts (Takiya *et al.*, 2006). For this reason, the genomes of these two symbionts provide new findings on the ecological interactions between *H. vitripennis* and the bacteria living in its body, and supply potential targets for controlling the spread of diseases such as PD.

Use of *Alcaligenes* endophytes for symbiotic control and paratransgenesis

Among the strategies for the control of PD diffusion, a symbiotic control approach has been proposed as a promising method for long-term control of *X. fastidiosa*, by rendering *H. vitripennis* incompetent for pathogen transmission through the use of paratransgenic insects. Paratransgenesis can be defined as the genetic alteration of the symbiotic biota associated with an insect vector to produce anti-pathogen factors that disrupt pathogen transmission. With paratransgenesis, the genetically modified organism is not the insect itself but rather the microbial symbionts it carries. Paratransgenesis was first proposed to control the transmission of Chagas' disease (Beard *et al.*, 2001). In the case of PD, specific attention was focused on the possible role of grape endophytic bacteria that can also be associated with the insect vector in delivering anti-*X. fastidiosa* factors. Among the cultivable bacteria isolated from the GWSS, the γ-Proteobacterium *Alcaligenes xylosoxidans denitrificans*, also associated with grape phloem and sharing the same niche as *X. fastidiosa*, has been selected as a candidate for paratransgenesis (Bextine *et al.*, 2004). Bacteria classified as endophytic organisms identify those bacteria that colonize tissues and internal structures of plants without any negative effect on plant physiology or growth. The exploitation of non-virulent bacteria residing in the plant for control purposes has been proposed in several models, as in the case of the use of benign strains of *X. fastidiosa* limiting the development of Pierce's disease (Hopkins, 2005). An advantage in using *A. xylosoxidans denitrificans* is that it can easily be maintained in culture and manipulated for the study and the application of the symbiotic control approach.

The capability of *A. xylosoxidans denitrificans* to colonize the cibarial region of *H. vitripennis*, a necessary condition for an effective anti-pathogen activity, was tested by introducing bacterial cells in an artificial feeding system.

To monitor the movement of *A. xylosoxidans denitrificans*, the dsRed gene, encoding for a red fluorescent protein, was inserted in the bacterial genome with the *Himar*1, a mariner-family transposable element that was originally identified in *Haematobia irritans*. Successful delivery to and colonization of transformed *A. xylosoxidans denitrificans* in the foregut regions of GWSS were then observed by fluorescent microscopy, supporting the possibility of managing *A. xylosoxidans denitrificans* as a paratransgenic symbiotic control agent (Bextine *et al.*, 2004). Subsequently, in order to assess the feasibility of a plant-based delivery system for symbiotic control strategies, the ability of a genetically marked strain of *A. xylosoxidans denitrificans* to colonize several host plants was evaluated. A strain of *A. xylosoxidans denitrificans*, originally isolated from the cibarial region of the foregut of the GWSS, was transformed, using the same *Himar*1 transposition system, to express an EGfp protein. Seedlings of different potential host plants were inoculated with transformed bacteria and analysed by quantitative real-time PCR. The host plants that were less hospitable to the bacterium were also less desirable host plants for the insect, indicating that the entities in the tritrophic interaction composed of plant, insect and microbe are closely associated and have developed a beneficial relationship for each of them (Bextine *et al.*, 2005).

The ability of *A. xylosoxidans denitrificans* to colonize both the potential host plants of *X. fastidiosa* and the insect vectors allows the potential employment of this bacterium for driving anti-pathogen factors. In theory, according to the microbial ecological data provided, such factors could operate within the insect or directly inside the plant, in order to cure infected grapevines by neutralizing or eliminating existing *X. fastidiosa* colonies, making the removal and replacement of diseased grapevines unnecessary and reducing costs associated with lost yield in subsequent seasons (Bextine *et al.*, 2005).

In order to make the paratransgenic agent a useful tool for PD biocontrol, the impact related to its release in soil, water and plant environments was evaluated for a risk assessment, underlying how transformed *A. xylosoxidans denitrificans* colonizes plants in preference to other environments. Additional studies are under way concerning possible gene transfer events or virulence factors associated with the paratransgenic agent when it is in the presence of other bacteria, as well as the occurrence of changes in plant physiology or in the xylem bacterial community after the introduction of modified *A. xylosoxidans denitrificans*. Furthermore, the possibility of the transformed endophyte colonizing grape berries or surviving the winemaking process intact was investigated (Miller *et al.*, 2006).

Prerequisites for a Successful Symbiotic Control Approach

A successful symbiotic control strategy, involving both a paratransgenic approach and the use of non-recombinant symbionts, should engage microbial agents that satisfy several requirements (Fig. 15.1).

REQUIRED FEATURE		RELATED ADVANTAGES
Stable association with the disease-transmitting vector	⇒	Wide distribution of the control factor
Dominance within the microbial community	⇒	Increased chance of success in antagonistic activity
Co-localization with the microorganism responsible for the disease	⇒	Control factor activity in site of pathogen development
Cultivability and predisposition to be genetically manipulated *in vitro*	⇒	Possibility to apply paratransgenic strategies
Efficient spread within field populations of the vector	⇒	Extensive dispersal of the transmission-blocking factor

Fig. 15.1. Schematic presentation of the main requirements of a microbial symbiont for use in symbiotic control strategies.

A potential symbiotic control agent should first be stably associated with the disease-transmitting vector, possibly playing an actual role in permitting the host's survival or improving the insect's fitness. Such a strict symbiotic relationship will guarantee a high prevalence of the microbe in different host populations, enabling the control factor to be widely distributed within insects. Besides being prevalent within vector populations, the control agent should be dominant within the microbial community of single individuals, as a high density in the insect body results in increased chance of a successful antagonistic action.

Another important necessary feature of a microbe able to act as a vehicle for anti-disease strategies is a co-localization with the microorganism responsible for the disease, allowing the potential transmission-blocking factor to be located in appropriate sites of pathogen development. Such localization should occur equally in the juvenile and in the adult stages, especially in those cases where both nymphs and adults are able to acquire and host the pathogenic agent.

In addition to the previous characteristics, a further requisite needed by an effective symbiotic control agent is the capability to be cultivated and genetically manipulated *in vitro*. Transformants should be stable and competitive with wild-type strains, and should maintain their symbiotic function without threatening the insect fitness or becoming pathogenic to other coexisting organisms. The possibility of easily isolating and transforming the microbial symbiont will be necessary for the development of paratransgenic approaches; moreover, it is also required when the exploitation of naturally occurring antagonistic factors is considered. Indeed, the genetic manipulation of these potential agents is necessary for an exhaustive study of the activity and the role played in the host's biology, i.e. by creating knockout mutants, or through the use of strains genetically marked with fluorescent proteins to investigate the microbial distribution patterns in the insect's body.

Finally, the symbiotic control agent should be efficiently delivered and spread within field populations of the vector. Thus, a transmission pathway is required to guarantee the symbiont's dispersal. Microorganisms may be vertically transmitted to the offspring or horizontally transferred between individuals exploiting their feeding or reproductive behaviour (co-feeding and venereal transmission, respectively).

As several features should be possessed by a candidate for symbiotic control, the development of this kind of approach should follow a detailed study of interactions between the vector and its naturally associated microorganisms. A first step in exploring the symbiotic relationships between host and symbiont is to characterize the whole microbial diversity associated with the considered vector. Microbial community fingerprinting methods may provide an estimation of the composition of bacterial species residing in the insect's body. These methods target conserved genes of molecular chronometers such as the rRNA genes. By using universal primers for bacteria, rRNA genes can be amplified by PCR, and the different amplified products from the different bacteria can be separated on the basis of their length (LH-PCR, length heterogeneity-PCR; T-RFLP, terminal-restriction fragment length polymorphism) or their sequence polymorphisms (SSCP, single strand conformation polymorphisms) (Alma *et al.*, 2008).

Once the microorganisms composing the microbial community of the disease-transmitting vector have been identified, further molecular-biology-based analyses need to be performed, in order to evaluate prevalence and dominance features of the most promising candidates, i.e. by means of qualitative and quantitative PCR screenings. In addition, *in situ* hybridization and electron microscopy may provide information on the natural distribution of microorganisms in the insect's organs. Evidence of the presence in key organs may indicate a co-localization with the pathogenic agents and even suggest possible transmission pathways for the spread among individuals; nevertheless, to demonstrate the symbiont's transfer, experiments involving the analysis of insect biology are needed. Parallel to these studies, the isolation on artificial media of the potential symbiotic control agents should be attempted. If possible, genetic manipulation of the candidate control agent will be the following step towards the development of an innovative control strategy, in order to explore the symbiotic effects on the host's fitness and behaviour, or to investigate potential antagonistic or competitive interactions with the pathogen.

A Case Study: Auchenorrhyncha Vectors of the Phytoplasma Agents of Grapevine Yellows

Phytoplasmas are responsible for grapevine diseases that are among the main problems of European viticulture. Grapevine yellows currently present in Europe are associated with phytoplasmas belonging to the EY group 16SrV and to the STOL group 16SrXII. Different Auchenorrhyncha are efficient vectors of GY agents, as well as of other phytoplasmas, as they possess several features necessary for an effective transmission: nymphs and adults feed similarly on phloem cells; phytoplasmas are propagative and persistent in them; and both immature stages and adults can transmit them (Weintraub, 2007). The species belonging to the families Cixiidae and Cicadellidae, which are known to transmit the grape yellows phytoplasmas, are *H. obsoletus*, *S. titanus* and *Oncopsis alni* (Schrank). The last one transmits the agent of Palatinate grape yellows (PGY), which is currently the least disturbing disease among grapevine yellows owing to a limited localization, restricted to the grapevine-growing region of Palatinate in Germany.

Among other plant pests in the Auchenorrhyncha is the maize leafhopper *Dalbulus maidis* (DeLong & Wolcott) (Cicadellidae), which is the major vector of two maize-stunting pathogens: the corn stunt spiroplasma (CSS, *Spiroplasma kunkelii*) and the maize bushy stunt phytoplasma (MBSP). These pathogens are responsible for major crop losses in North America (California and Gulf Coast states), Mexico and Central and South America (Ebbert and Nault, 2001). A beneficial mutual interaction between *D. maidis* and *S. kunkelii* has been reported in several studies. Indeed, the infection of this vector by *S. kunkelii* does not have negative effects on its longevity and lifespan but improves its survival in the absence of the almost unique feeding source (which is maize) and/or in both cool and warm conditions in the laboratory and under field conditions. On the other hand, *S. kunkelii* overwinters in the body of its insect vector, as no host plant other than maize is known (Ebbert and Nault, 2001; Ammar and Hogenhout, 2006, and references therein).

For a first insight into the microbial diversity associated with insects transmitting phytoplasmas, the most important vectors of GY agents, *S. titanus* and *H. obsoletus*, were taken into account. These two vectors have different lifestyles – the former strictly ampelophagous and the latter polyphagous, only occasionally feeding on grapevine – and dissimilar biological features, which may reflect diversely affiliated microbial communities.

The vector of FD is *S. titanus*, a Nearctic leafhopper accidentally introduced into Europe, currently diffused, with a non-uniform pattern, in the main wine-growing areas. *S. titanus* is strictly associated with grapevine, where it accomplishes one generation per year and overwinters in the egg stage, laid in 2-year-old bark. In the second half of May, eggs begin hatching and continue until after the first 10 days of July. After hatching, the development of five juvenile instars occurs. Nymphs feed on foliar veins on the lower side of the basal leaves, close to the canes and to the trunk in which the overwintering eggs were laid. BN is caused by a phytoplasma non-specific for grapevine, transmitted by not strictly ampelophagous vector(s). Such an epidemiological situation neatly differs from FD, reflecting the life cycle of the aetiological agent of BN: involving different host plants besides grapevine, the final host plant, and presumably different vectors besides *H. obsoletus*, which is presently the only confirmed vector. The only reported insect vector of BN is *H. obsoletus*, a polyphagous planthopper widespread in Europe, the Middle East, Asia Minor and Afghanistan, which lives on dicotyledonous weeds. In Europe, there is one generation per year, overwintering as juveniles on the roots of different wild herbaceous plants. Among the host plants, the most common ones are nettle (*Urtica dioica*) and bindweed (*Convolvulus arvensis*); nevertheless *H. obsoletus* can occasionally be found on grapevine. As STOL infects a large number of wild and cultivated plants, many of which are commonly found in the vineyard agroecosystem, this planthopper may inoculate the phytoplasma from such host plants, diffusing BN throughout the vineyards (Alma *et al.*, 2008).

Microbial symbionts inhabiting the body of *Scaphoideus titanus* and *Hyalesthes obsoletus*

Because of the economic impact of GY and the important role of *S. titanus* and *H. obsoletus* in spreading these diseases, and in the light of the promising results obtained by studying the microbiota residing in *H. vitripennis*, the possibilities of the exploitation of symbiotic microorganisms as symbiotic control agents were explored. The first studies concerning the microbial symbionts associated with phytoplasma vectors were recently initiated for the FD vector *S. titanus*. Microbial community fingerprinting methods were employed for diversity screening (Marzorati *et al.*, 2006), showing the presence of a complex microbial community. The length heterogeneity PCR technique (LH-PCR), discriminating different bacteria by sequence length differences in portions of the 16S rRNA gene that include two variable sequence regions, was employed for a survey of the bacterial microbiota associated with *S. titanus* (Marzorati *et al.*, 2006). The resulting fragments with different lengths, recognized in the electropherogram by different peaks, showed different bacterial species repeatedly occurring in field-recovered leafhopper individuals. Parallel to LH-PCR, PCR-denaturing gradient gel electrophoresis (DGGE), separating bacterial species by sequence polymorphisms on a portion of the 16S rRNA gene, was performed, and single fragments obtained by polyacrylamide gel electrophoresis were sequenced, showing that the leafhopper is inhabited by Bacteroidetes of the genera '*Ca.* Cardinium' and *Chryseobacterium*; α-Proteobacteria of the genus *Asaia*; and γ-Proteobacteria of the genus *Stenotrophomonas* (Marzorati *et al.*, 2006). Besides the use of molecular tools, further ultrastructural investigations involving transmission electron microscopy (TEM) confirmed a fairly heterogeneous microbial community. Several bacterial morphologies with different ultrastructural characteristics were observed in tissues of the leafhopper, in agreement with the molecular fingerprinting results.

Some bacteria found in *S. titanus* are co-localized in organs such as the insect gut and salivary glands, which are key points for the multiplication and the completion of the phytoplasma's life cycle within the insect host. Since the occurrence in the same organs as the pathogen is one of the necessary features for a potential symbiotic control agent, these observations open the perspective that a paratransgenic approach exploiting those symbionts to control phytoplasma proliferation is theoretically possible. Recently, bacteria of the genus *Asaia* were reported to be the dominant symbionts in the malaria vector *Anopheles stephensi* Liston and were proposed as potential symbiotic control agents, in order to control the transmission of the malaria parasite (Favia *et al.*, 2007). Recent studies underlined that *Asaia* is dominant in the microflora of wild *S. titanus* populations, and a strain of this bacterium tagged with a Gfp was observed to colonize several organs, including salivary glands and reproductive organs (Crotti *et al.*, 2008). In-depth knowledge of the role of *Asaia* symbionts in the host would be a further step towards the design of a symbiotic control approach for FD.

Very few studies have been performed on *H. obsoletus*, the vector of BN. Recently, Moran and colleagues (2005) screened several families of Auchenorrhyncha for the presence of bacterial symbionts and discovered that most of them are inhabited by '*Ca.* Sulcia muelleri'. This study did not include any cixiid, the family of *H. obsoletus*. However, all the families close to Cixiidae, such as Dictyopharidae or Fulgoridae, that were tested were positive to these symbionts, leaving open the question of whether *Sulcia* can also be a primary symbiont of *H. obsoletus*. Such a possibility would be consistent with the hypotheses raised by Moran and colleagues, suggesting that this symbiont has been present throughout the diversification of this major insect group and is one of the oldest (Moran *et al.*, 2005). According to these assumptions, *Sulcia* probably resides in more host species than are currently known (Moran *et al.*, 2005). Molecular ecology-based screening of the microbial community of *H. obsoletus* actually showed the occurrence of a bacterium related to '*Ca.* Sulcia muelleri', with a high prevalence in insect populations and a distribution in several host organs, suggesting that this symbiont could be stably associated to the cixiid (Gonella *et al.*, 2008).

Inherited microorganisms associated with *Scaphoideus titanus*: bacterial *Cardinium* endosymbionts and yeast-like symbionts (YLS)

According to the classical community fingerprinting approach based on the application of LH-PCR, PCR-DGGE and sequencing, the major symbionts associated with *S. titanus* were found to be bacteria in the genus *Cardinium*, showing, when tested by PCR with specific primers, a minimal field infection rate of more than 94% (Marzorati *et al.*, 2006). On the basis of the 16S rRNA gene sequence, the closest relative to *Cardinium* associated with *S. titanus* is a symbiont of the tick *Ixodes scapularis* Say. Its phylogenetic branch incorporated endosymbionts of several mites, such as *Metaseiulus*, *Oppiella*, *Petrobia* and *Brevipalpus*, for example the feminizing symbiont of *Brevipalpus phoenicis* (Geijskes) (Weeks *et al.*, 2003), while a separated branch was composed of '*Ca.* Cardinium hertigii' endosymbionts of the parasitoid wasp *Encarsia pergandiella* Howard (Zchori-Fein *et al.*, 2004), *Aspidiotus paranerii* Gerson (Weeks *et al.*, 2003) and *Plagiomerus diaspidis* Crawford (Zchori-Fein and Perlman, 2004).

Bacteria of the genus *Cardinium* can be easily recognized in the insect's tissues when observed by electron microscopy, as they show a peculiar morphological motif (Bigliardi *et al.*, 2006; Sacchi *et al.*, 2008). This is a brush-like structure resembling the parallel roads of ancient Roman campsites (Zchori-Fein *et al.*, 2004), called *cardi* in Latin, from which the generic name derives. Such a brush-like array of microtubule-like structures, residing within a cell typically showing a two-layered envelope composed of an outer cell wall and an inner plasma membrane, is considered a morphological signature of the genus. The microtubule-like complex consists of a system of parallel microtubule elements, a fibrous electron-dense plaque and a set of electron-dense structures adhering to the outer leaflet of the bacterial plasma membrane (Sacchi *et al.*, 2008). The metabolic and physiological meaning of this complex tubular structure is unknown; it might perhaps represent a membrane system where enzymatic activities occur.

Cardinium intracellular symbionts are fairly widespread within arthropods. Besides insects, mites and ticks, spiders and nematodes were reported to host endosymbionts phylogenetically related to *Cardinium*. Pekár and Šobotník (2007) found structures with the same micromorphology as *Cardinium* cells in the femoral organs of spiders. Similarly, bacterial cells presenting the typical *Cardinium* morphological markers were identified in several tissues of the plant-parasitic nematode *Heterodera glycines* Ichinohe, although a phylogenetic analysis of the endosymbionts presenting these structures showed they are sufficiently distant from *Cardinium* to be attributed to a new genus named *Paenicardinium* (Noel and Atibalentja, 2006). In some of these cases, the relationship between *Cardinium* endosymbionts and their hosts was reported to be associated with different effects on the reproductive behaviour or reproductive alterations, such as parthenogenesis, feminization of genetic males and cytoplasmic incompatibility. Cytoplasmic incompatibility occurs when crosses between symbiont-infected males and uninfected females fail to produce progeny, whereas both crosses between uninfected males and infected females and those between infected males and infected females are fertile (Ishikawa, 2003; Zchori-Fein *et al.*, 2004). *Cardinium*-induced cytoplasmic incompatibility is known to occur with the same dynamics governing *Wolbachia*-related cytoplasmic incompatibility (Perlman *et al.*, 2008).

Cardinium endosymbionts are actually the sole model of manipulators of the host's reproduction, together with the α-Proteobacterium *Wolbachia*. These microorganisms, able to spread within insect host populations by controlling the host's reproduction, were recently revealed to be of major interest among arthropods' symbionts. Thanks to their capacity to enter female germline cells, they can be vertically transmitted to the progeny. *Wolbachia*, the most studied sexual manipulator of arthropods, was reported to manipulate the reproduction of different hosts by cytoplasmic incompatibility, feminization of genetic males, male killing and induction of parthenogenesis (Bandi *et al.*, 2001). Reproductive manipulators appear to be promising for use in symbiotic control strategies, through the development of insect-vector interference strategies. In particular, cytoplasmic incompatibility was proposed as a method to suppress or modify natural populations of arthropod pests in a way analogous to the sterile insect technique, by means of the release of incompatible male insects to control wild populations of disease vectors (Zabalou *et al.*, 2004).

The *Cardinium* living in *S. titanus* is currently the only case of a sexual endosymbiont stably associated with insect vectors of phytoplasmas, although a preliminary overview of the microbiota related to *H. obsoletus* underlined some individuals affected by *Wolbachia* (Gonella *et al.*, 2008). However, nothing is known concerning biases in the sex ratio of this leafhopper, and both males and females tested by specific PCR showed similar infection rates; hence no obvious indication of an interference with the sex ratio can be predicted. Such a high prevalence of *Cardinium* in both sexes of *S. titanus* could be explained by the selection of infected individuals caused by cytoplasmic incompatibility or by a mutualistic interaction with the host.

Although the ability of *Cardinium* to manipulate *S. titanus*'s reproduction was not proven, TEM examinations of several tissues of the insect showed that this bacterium colonizes different organs, including the ovaries of females, indicating that this bacterium is vertically transmitted to the offspring (Marzorati *et al.*, 2006; Sacchi *et al.*, 2008). TEM examination of the apical region of the ovary revealed the presence of a particular cell morphotype, with the cytoplasm filled with *Cardinium* bacteria. Such structures are similar to the cells harbouring symbiotic bacteria described in a variety of insects, including cockroaches and aphids (Nardon and Nardon, 1998). These observations, together with the detection of *Cardinium* in the initial phases of embryonic development and during the nymphal stages, suggest that these bacteriocyte-like cells might play an active role in the transmission of the symbionts to the progeny, similar to other previously described models. In addition to the ovaries, *Cardinium* was observed in different tissues and organs of *S. titanus*, such as fat bodies, a proper localization for nutritional provisions with the host, or salivary glands. The presence of *Cardinium* cells in the salivary glands seemed to indicate a complex life cycle of the endosymbiont, involving both insect and plant hosts. Such a hypothesis is supported by the proof of a release of this bacterium in artificial media and in grapevine leaves (Pajoro *et al.*, 2008). The intriguing possibility of a symbiont transfer through the plant could guarantee multiple means of transmission of the potential biocontrol agent within natural vector populations.

Together with bacterial symbionts, eukaryotic yeast-like symbionts (YLS) are harboured by numerous Hemiptera, such as aphids and planthoppers (Ishikawa, 2003), as well as by some cockroaches and parasitic wasps (Gibson and Hunter, 2008). Molecular phylogenetic analyses showed that YLS belong to the class Sordariomycetes in the subphyulm Ascomycotina (Ishikawa, 2003). They were reported to be transmitted to the progeny directly from mothers by transovarial infection and to be present at every developmental stage of the hosts.

The association between YLS and their hosts was investigated in the insect model *Nilaparvata lugens* (Stål), the Asian rice brown planthopper (Sasaki *et al.*, 1996). A nutritional role played by these eukaryotic symbionts was studied, and it was proposed that they are essential for the normal development of their hosts. As an example, YLS were suggested to be responsible for recycling nitrogen contained in the uric acid wastes produced by the host, by way of uricase enzymes (Sasaki *et al.*, 1996). A nutritional function is also provided by fungal symbionts colonizing the gut of the tobacco beetle *Lasioderma serricorne* (Fabricius): these microorganisms are able to detoxify plant material ingested by the beetle (Dowd, 1989). The role of the YLS associated with Cerataphidini aphids has not yet been assigned; in spite of this, it was proposed that they are functionally homologous with *Buchnera*, the primary endosymbiotic bacterium of aphids. Such a possibility is connected with the phylogenetic clade formed by YLS-containing aphid species, nested within the large cluster of *Buchnera*-harbouring species. This evidence suggests that YLS-infected species form a clade descendent from the common ancestor of aphids, in which the eukaryotic symbiont replaced *Buchnera* (Ishikawa, 2003).

A YLS was recently found in *S. titanus* (Sacchi *et al.*, 2008), within specialized cells of fat bodies, resembling mycetocytes. Gene sequence analysis and *in situ* hybridization led to the identification of these YLS as members of the class Sordariomycetes, with the fungus *Bionectria pityrodes* as the closest relative. At ultrastructural level, these microorganisms appear rod-shaped with a two-layered cell wall: an outer, electron-dense layer, 25 nm in thickness, and an inner, electron-clear layer, 100 nm thick. Some of the yeast cells show a protuberance, indicating their reproduction by means of budding. These organisms were observed with a high prevalence and in a high concentration within the tissues of *S. titanus*, also suggesting, in this case, a metabolic involvement of the YLS necessary for the host's development, possibly connected with nitrogen recycling, as reported for *N. lugens*. Such a hypothesis is supported by evidence that adult individuals of *S. titanus* reared in the laboratory on a diet based on a sucrose solution without any nitrogen source were able to live in those conditions for periods equivalent to the typical adult lifespan in the field (Pajoro *et al.*, 2008). Based on these observations, further investigation of the possible role of the YLS in nitrogen metabolism would be worthwhile, as well as exploring the possibility of an involvement of this microorganism in developing symbiotic control strategies.

As described in the *N. lugens* model, as well as in different leafhoppers and planthoppers, the YLS of *S. titanus* were observed to be transovarially transmitted to offspring following a route from the thin layer of the fat body to the extracellular space and then to the follicle cells (Sacchi *et al.*, 2008). Nevertheless, the vertical transmission of the YLS seems to be in some way limited. Indeed, few yeast cells colonize ovaries and embryos of *S. titanus*, compared with the dense populations of yeasts described in structures, named symbiote balls, harboured by *N. lugens*, suggesting a lower rate of vertical transmission (Sacchi *et al.*, 2008). What kind of constraint rules such a limitation for the spread of YLS still has to be clarified.

Conclusions

The use of biocontrol agents is attracting increasing interest in pest management owing to their harmless effects on the environment in comparison with chemical compounds. Biocontrol microorganisms can impair the insect life cycle in many cases by producing factors that are toxic to the target insect species, for example the production of delta-endotoxins by *Bacillus thuringiensis* that are active against larvae of different insect orders. Another important approach for pest biocontrol, which has had a growing interest in recent years, consists of the use of bacterial symbionts of insect vectors of diseases as agents for blocking the transmission of pathogens and is termed symbiotic control (Beard *et al.*, 2001; Rio *et al.*, 2004). This strategy, involving the use of symbionts able to interfere with pathogen vectoring capacity, often exploits the microbial ecology of the host's body. Hence, a careful and extensive investigation of the microbiota associated with insect vectors is required as a step towards the development of symbiotic control approaches.

In recent years, symbiont research was started for grapevine diseases caused by phytopathogenic bacteria. Intensive studies concerning the molecular characterization of bacteriocyte-associated symbionts of the main Pierce's disease vector, *H. vitripennis*, were carried out (Moran *et al.*, 2005; Takiya *et al.*, 2006; Wu *et al.*, 2006; McCutcheon and Moran, 2007). The promising results obtained concerning the microbial symbionts of *H. vitripennis*, their role in the insect's biology and their capability to be spread through vector and plant populations represent an example of how the microbiota associated with insect vectors of plant-pathogenic bacteria can be exploited with the purpose of exploring new approaches for limiting the pathogen's spread. This could constitute a model for the development of a symbiotic control strategy for phytoplasmosis control also. In the light of such a potential, efforts were started on symbiont research concerning phytoplasma vectors of grapevine, *S. titanus* and *H. obsoletus* (Bigliardi *et al.*, 2006; Marzorati *et al.*, 2006; Crotti *et al.*, 2008; Gonella *et al.*, 2008; Sacchi *et al.*, 2008).

As a whole, these studies have shown that the microbial communities associated with the insect vectors of phytopathogenic bacteria are made up of a diversity of symbiotic associations, often composed of multiple symbiont species, as indicated in the main studies on the symbiosis of Auchenorrhyncha (Moran *et al.*, 2005). Although the intricate interaction patterns occurring both with the host and among microbes are still to be clarified, some of them are likely to play a role in host nutrition, while others, transmitted to the offspring, could be responsible for alterations of the host's reproduction. The activity and influence of the microbial communities colonizing these insects should be considered in explaining unclear aspects of insect biology and evolution. Moreover, the knowledge of interactions and functions played by microorganisms could be useful to take advantage of identified bacteria for symbiotic control purposes.

Such a control method, potentially capable of managing successfully and sustainably the phytoplasma agroeconomic problem, must seriously take into consideration a risk assessment of the impact after delivery of the symbiotic control agent in the environment. This evaluation should consider toxicity to the insect host and to the plant, and possible effects on the ecology of their naturally associated microbial communities. If a paratransgenic agent is to be employed, fitness alteration of both the microbial strain and the host, horizontal gene transfer to wild-type strains of the symbiotic control agent and transgene instability must be assessed. Finally, specific concerns expressed by any stakeholder organization involved in the winemaking processes ought to be considered.

Acknowledgements

We sincerely thank all the technical staff of the laboratories at the Universities of Turin and Milan for experimental support and stimulating discussion and comments on the work. C. Bandi (University of Milan, Italy), A. Cherif (University of Tunis, Tunisia), G. Favia (University of Camerino, Italy), M. Mandrioli (University of Modena and Reggio Emilia, Italy) and L. Sacchi (University of Pavia, Italy) are acknowledged for helpful discussion. Support comes from the Italian Ministry for Research (MIUR), in the ambit of projects PRIN 2007 'Caratterizzazione del microbiota associato a *Scaphoideus titanus* e *Hyalesthes obsoletus*, cicaline vettrici di fitoplasmi nella vite ed isolamento e studio della localizzazione di batteri acetici simbionti'. AA and DD acknowledge support of the Cost Action FA0701: 'Arthropod Symbiosis: from Fundamental Studies to Pest and Disease Management'.

References

Aksoy, S. and Rio, R.V.M. (2005) Interactions among multiple genomes: tsetse, its symbionts and trypanosomes. *Insect Biochemistry and Molecular Biology* 35, 691–698.

Alma, A., Sacchi, L., Favia, G., Bandi, C. and Daffonchio, D. (2008) Symbiotic microorganisms in leafhopper and planthopper vectors of phytoplasmas in grapevine. In: Bourtzis, K. and Miller, T.A. (eds) *Insect Symbiosis*, Vol. 3. CRC Press, Taylor and Francis Group, Boca Raton, Florida, pp. 339–359.

Almeida, R.P.P. and Purcell, A.H. (2003) Transmission of *Xylella fastidiosa* to grapevines by *Homalodisca coagulata* (Hemiptera: Cicadellidae). *Journal of Economic Entomology* 96, 264–271.

Ammar, E.D. and Hogenhout, S.A. (2006) Mollicutes associated with arthropods and plants. In: Bourtzis, K. and Miller, T.A. (eds) *Insect Symbiosis*, Vol. 2. CRC Press, Taylor and Francis Group, Boca Raton, Florida, pp. 97–118.

Bandi, C., Dunn, A.M., Hurst, G.D. and Rigaud, T. (2001) Inherited microorganisms, sex-specific virulence and reproductive parasitism. *Trends in Parasitology* 17, 88–94.

Baumann, P. (2006) Diversity of prokaryote–insect associations within the Sternorrhyncha (psyllids, whiteflies, aphids, mealybugs). In: Bourtzis, K. and Miller, T.A. (eds) *Insect Symbiosis*, Vol. 2. CRC Press, Taylor and Francis Group, Boca Raton, Florida, pp. 1–24

Beanland, L., Noble, R. and Wolf, T.K. (2006) Spatial and temporal distribution of North American grapevine yellows disease and of potential vectors of the causal phytoplasmas in Virginia. *Environmental Entomology* 35, 332–344.

Beard, C.B., Dotson, E.M., Pennington, P.M., Eichler, S., Cordon-Rosales, C. and Durvasula, R.V. (2001) Bacterial symbiosis and paratransgenic control of vector-borne Chagas disease. *International Journal of Parasitology* 31, 621–627.

Berisha, B., Chen, Y.D., Zhang, G.Y., Xu, B.Y. and Chen, T.A. (1998) Isolation of Pierce's disease bacteria from grapevines in Europe. *European Journal of Plant Pathology* 104, 427–433.

Bextine, B.R., Lauzon, C., Potter, S.E., Lampe, D. and Miller, T.A (2004) Delivery of a genetically marked *Alcaligenes* sp. to the glassy winged sharpshooter for use in a paratransgenic control strategy. *Current Microbiology* 48, 327–331.

Bextine, B.R., Lampe, D., Lauzon, C., Jackson, B. and Miller, T.A. (2005) Establishment of a genetically marked insect-derived symbiont in multiple host plants. *Current Microbiology* 50, 1–7.

Bigliardi, E., Sacchi, L., Genchi, M., Alma, A., Pajoro, M., Daffonchio, D., Marzorati, M. and Avanzati, A.M. (2006) Ultrastructure of a novel *Cardinium* sp. symbiont in *Scaphoideus titanus* (*Hemiptera*: *Cicadellidae*). *Tissue and Cell* 38, 257–261.

Blua, M.J., Redak, R.A., Morgan, D.J.W. and Costa, H.S. (2001) Seasonal flight activity of two *Homalodisca* species (Homoptera: Cicadellidae) that spread *Xylella fastidiosa* in Southern California. *Journal of Economic Entomology* 94(6), 1506–1510.

Boudon-Padieu, E. (2003) The situation of grapevine yellows and current research directions: distribution, diversity, vectors, diffusion and control. In: *Proceedings of the 14th Conference ICVG*. 12–17 September, Locorotondo, Italy, Dipartimento di Protezione delle Piante e Microbiologia Applicata, Italy, pp. 47–53.

Buchner, P. (1965) *Endosymbiosis of Animals with Plant Microorganisms*. John Wiley and Sons, New York.

Chatterjee, S., Wistrom, C. and Lindow, S.E. (2008) A cell–cell signaling sensor is required for virulence and insect transmission of *Xylella fastidiosa*. *Proceedings of the National Academy of Sciences of the USA* 105, 2670–2675.

Crotti, E., Pajoro, M., Damiani, C., Ricci, I., Negri, I., Rizzi, A., Clementi, E., Raddadi, N., Scuppa, P., Marzorati, M., Pasqualini, L., Bandi, C., Sacchi, L., Favia, G., Alma, A. and Daffonchio, D. (2008) *Asaia*, a transformable bacterium, associated with *Scaphoideus titanus*, the vector of 'flavescence dorée'. *Bulletin of Insectology* 61(1), 219–220.

Curley, C.M., Brodie, E.L., Lechner, M.G. and Purcell, A.H. (2007) Exploration for facultative endosymbionts of glassy-winged sharpshooter (Hemiptera: Cicadellidae). *Annals of the Entomological Society of America* 100, 345–349.

Dowd, P.F. (1989) *In situ* production of hydrolytic detoxifying enzymes by symbiotic yeasts of cigarette beetles (Coleoptera: Anobiidae). *Journal of Economic Entomology* 82, 396–400.

Ebbert, M.A. and Nault, L.R. (2001) Survival in *Dalbulus* leafhopper vectors improves after exposure to maize stunting pathogens. *Entomologia Experimentalis et Applicata* 100, 311–324.

Favia, G., Ricci, I., Damiani, C., Raddadi, N., Crotti, E., Marzorati, M., Rizzi, A., Urso, R., Brusetti, L., Borin, S., Mora, D., Scuppa, P., Pasqualini, L., Clementi, E., Genchi, M., Corona, S., Negri, I., Grandi, G., Alma, A., Kramer, L., Esposito, F., Bandi, C., Sacchi, L. and Daffonchio, D. (2007) Bacteria of the genus *Asaia* stably associate with *Anopheles stephensi*, an Asian malarial mosquito vector. *Proceedings of the National Academy of Sciences of the USA* 104, 9047–9051.

Gibson, C.M. and Hunter, M.S. (2008) Inherited fungal and bacterial endosymbionts of a parasitic wasp and its cockroach host. *Microbial Ecology* 57, 542–549.

Gonella, E., Negri, I., Marzorati, M., Brusetti, L., Pajoro, M., Mandrioli, M., Tedeschi, R., Daffonchio, D. and Alma, A. (2008) Study of the bacterial community affiliated to *Hyalesthes obsoletus*, the insect vector of 'bois noir' phytoplasma of grape. *Bulletin of Insectology* 61(1), 221–222.

Hopkins, D.L. (2005) Biological control of Pierce's disease in the vineyard with strains of *Xylella fastidiosa* benign to grapevine. *Plant Disease* 89, 1348–1352.

Hopkins, D.L. and Purcell, A.H. (2002) *Xylella fastidiosa*: cause of Pierce's Disease of grapevine and other emergent diseases. *Plant Disease* 86, 1056–1066.

Ishikawa, H. (2003) Insect symbiosis: an introduction. In: Bourtzis, K. and Miller, T.A. (eds) *Insect Symbiosis*. CRC Press, Taylor and Francis Group, Boca Raton, Florida, pp. 1–21.

Kuzina, L.V., Miller, T.A. and Cooksey, D.A. (2006) *In vitro* activities of antibiotics and antimicrobial peptides against the plant pathogenic bacterium *Xylella fastidiosa*. *Letters in Applied Microbiology* 42, 514–520.

Lessio, F., Tedeschi, R. and Alma, A. (2007) Population dynamics, host plants and infection rate with stolbur phytoplasma of *Hyalesthes obsoletus* Signoret in north-western Italy. *Journal of Plant Pathology* 89, 97–102.

Li, Z.T. and Gray, D.J. (2003) Effect of five antimicrobial peptides on the growth of *Agrobacterium tumefaciens, Escherichia coli* and *Xylella fastidiosa*. *Vitis* 2, 95–97.

Marzorati, M., Alma, A., Sacchi, L., Pajoro, M., Palermo, S., Brusetti, L., Raddadi, N., Balloi, A., Tedeschi, R., Clementi, E., Corona, S., Quaglino, F., Bianco, P.A., Beninati, T., Bandi, C. and Daffonchio, D. (2006) A novel *Bacteroidetes* symbiont is localized in *Scaphoideus titanus*, the insect vector of flavescence dorée in *Vitis vinifera*. *Applied and Environmental Microbiology* 72, 1467–1475.

McCutcheon, J.P. and Moran, N.A. (2007) Parallel genomic evolution and metabolic interdependence in an ancient symbiosis. *Proceedings of the National Academy of Sciences of the USA* 104, 19392–19397.

Miller, T.A., Lauzon, C., Lampe, D., Durvasula, R. and Matthews, S. (2006) Paratransgenesis applied to control insect-transmitted plant pathogens: the Pierce's disease case. In: Bourtzis, K. and Miller, T.A. (eds) *Insect Symbiosis*, Vol. 2. CRC Press, Taylor and Francis Group, Boca Raton, Florida, pp. 247–263.

Moran, N.A., Dale, C., Dunbar, H., Smith, W.A. and Ochman, H. (2003) Intracellular symbionts of sharpshooters (Insecta: Hemiptera: Cicadellinae) form a distinct clade with a small genome. *Environmental Microbiology* 5, 116–126.

Moran, N.A., Tran, P. and Gerardo, N.M. (2005) Symbiosis and insect diversification: an ancient symbiont of sap-feeding insects from the bacterial phylum Bacteroidetes. *Applied and Environmental Microbiology* 71, 8802–8810.

Nardon, P. and Nardon, C. (1998) Morphology and cytology of symbiosis in insects. *Annales de la Societé Entomologique de France* 34, 105–134.

Noel, G.R. and Atibalentja, D. (2006) '*Candidatus* Paenicardinium endonii', an endosymbiont of the plant-parasitic nematode *Heterodera glycines* (Nemata: Tylenchida), affiliated to the phylum Bacteroidetes. *International Journal of Systematic and Evolutionary Microbiology* 56, 1697–1702.

Pajoro, M., Marzorati, M., Negri, I., Sacchi, L., Daffonchio, D. and Alma, A. (2008) Investigation over the life cycle of ST1-C the endosymbiont of *Scaphoideus titanus*. *Bulletin of Insectology* 61(1), 217–218.

Pekár, S. and Šobotník, J. (2007) Comparative study of the femoral organ in *Zodarion* spiders (Araneae, Zodariidae). *Arthropod Structure and Development* 36, 105–112.

Perlman, S.J., Kelly, S.E. and Hunter, M.S. (2008) Population biology of cytoplasmic incompatibility: maintenance and spread of *Cardinium* symbionts in a parasitic wasp. *Genetics* 178, 1003–1011.

Redak, R.A., Purcell, A.H., Lopes, J.R.S., Blua, M.J., Mizell, R.F. and Andersen, P.C. (2004) The biology of xylem fluid-feeding insect vectors of *Xylella fastidiosa* and their relation to disease epidemiology. *Annual Review of Entomology* 49, 243–270.

Rio, R.V.M., Hu, Y. and Aksoy, S. (2004) Strategies for the home team: symbioses exploited for vector-borne disease control. *Trends in Microbiology* 12, 325–336.

Sacchi, L., Genchi, M., Clementi, E., Bigliardi, E., Avanzati, A.M., Pajoro, M., Negri, I., Marzorati, M., Gonella, E., Alma, A., Daffonchio, D. and Bandi, C. (2008) Multiple symbiosis in the leafhopper *Scaphoideus titanus* (Hemiptera: Cicadellidae): details of transovarial transmission of *Cardinium* sp. and yeast-like endosymbionts. *Tissue and Cell* 40, 231–242.

Sasaki, T., Kawamura, M. and Ishikawa, H. (1996) Nitrogen recycling in the brown planthopper, *Nilaparvata lugens*: involvement of yeast-like endosymbionts in uric acid metabolism. *Journal of Insect Physiology* 42, 125–129.

Takiya, D.M., Tran, P.L., Dietrich, C.H. and Moran, N.A. (2006) Co-cladogenesis spanning three phyla: leafhoppers (Insecta: Hemiptera: Cicadellidae) and their dual bacterial symbionts. *Molecular Ecology* 15, 4175–4191.

Weeks, A.R., Velten, R. and Stouthamer, R. (2003) Incidence of a new sex-ratio-distorting endosymbiotic bacterium among arthropods. *Proceedings of the Royal Society of London, Series B Biology* 270, 1857–1865.

Weintraub, P. (2007) Insect vectors of phytoplasmas and their control – an update. *Bulletin of Insectology* 60(2), 169–173.

Wu, D., Daugherty, S.C., Van Aken, S.E., Pai, G.H., Watkins, K.L., Khouri, H., Tallon, L.J., Zaborsky, J.M., Dunbar, H.E., Tran, P.L., Moran, N.A. and Eisen, J.A. (2006) Metabolic complementarity and genomics of the dual bacterial symbiosis of sharpshooters. *PLoS Biology* 4, 1079–1092.

Zabalou, S., Reigler, M., Theodorakopoulou, M., Stauffer, C., Savakis, C. and Bourtzis, K. (2004) *Wolbachia*-induced cytoplasmic incompatibility as a means for insect pest population control. *Proceedings of the National Academy of Sciences of the USA* 101, 15042–15045.

Zchori-Fein, E. and Perlman, S.J. (2004) Distribution of the bacterial symbiont *Cardinium* in arthropods. *Molecular Ecology* 13, 2009–2016.

Zchori-Fein, E., Perlman, S.J., Kelly, S.E., Katzir, N. and Hunter, M.S. (2004) Characterization of a 'Bacteroidetes' symbiont in *Encarsia* wasps (*Hymenoptera: Aphelinidae*): proposal of '*Candidatus* Cardinium hertigii'. *International Journal of Systematic and Evolutionary Microbiology* 54, 961–968.

16 Transmission Specificity and Competition of Multiple Phytoplasmas in the Insect Vector

Domenico Bosco and Romina D'Amelio

Università degli Studi di Torino, Italy

Introduction

Phytoplasmas are transmitted by insects in the order Hemiptera. However, vector species are restricted in only a few families of the suborder Auchenorrhyncha: namely, Cercopidae, Cixiidae, Derbidae, Delphacidae, Cicadellidae and Psyllidae (Weintraub and Beanland, 2006). Within a family, some species are known to be phytoplasma vectors, while others are not. 'Transmission specificity' defines the fact that only selected species can act as vectors of a pathogen. Transmission of phytoplasmas by insects involves, at several levels, elements of host–pathogen specificity. Insect vectors can acquire more than one phytoplasma species/strain, either by feeding on multiple-infected source plants or by feeding sequentially on different plants infected by different phytoplasmas. The acquisition of multiple phytoplasmas leads to their interaction in the vector insects. This chapter analyses the concepts of phytoplasma transmission specificity and competition in the insect vectors.

Transmission Specificity

Host range

The host range of both phytoplasmas and insects greatly influences the chances that a phytoplasma and a potential vector will come into contact. For example, *Cacopsylla* spp., which are monophagous on pome or stone fruits, are only vectors of '*Candidatus* (*Ca.*) Phytoplasma mali', '*Ca.* Phytoplasma pyri' and '*Ca.* Phytoplasma prunorum', which infect pome or stone fruits (Weintraub and Beanland, 2006). *Orosius argentatus* Evans, a very polyphagous leafhopper species on herbaceous hosts (Larsen and Walter, 2007), transmits different phytoplasma species/strains to several plant species.

Sometimes insects can transmit a given phytoplasma but under natural conditions they are not vectors because they do not feed on the natural host plant of the phytoplasma. As an example *Euscelidius variegatus* Kirschbaum, known as a laboratory vector of flavescence dorée on broad bean, is not the natural vector to grapevine (Boudon-Padieu *et al.*, 1989). Therefore, the host plant is sometimes crucial for successful transmission, and a phytoplasma vector can fail to transmit to or acquire from certain host plants. It is worth noting that most phytoplasma strain collections are maintained in periwinkle, which is an excellent host for a huge variety of phytoplasmas and can easily be grafted, so that phytoplasmas can be maintained continuously (Favali *et al.*, 2008). Unfortunately this plant is a poor host for many vectors, which hampers the possibility of obtaining experimentally infected insects, even with efficient vectors.

Insect feeding preference

Feeding preferences have a major role in transmission specificity. Vector insects can be polyphagous, oligophagous or strictly monophagous, according to their ability to feed and reproduce on many, few or one host plant, respectively. Similarly, phytoplasmas may be generalists, infecting several different plant species, or specialists, infecting one or a few related plant species. A generalist phytoplasma can be transmitted by several vector species. Examples of plant-generalist and vector-generalist associations are aster yellows, which are transmitted by tens of vector species to hundreds of host plants, and X-disease ('*Ca.* Phytoplasma pruni') phytoplasma, which is also transmitted by several vector species to many host plants (Lee *et al.*, 2000). For some generalist phytoplasmas, e.g. beet-leafhopper-transmitted virescence (BLTV), only one vector species, *Circulifer tenellus* (Baker), has been identified. So far, we can define these associations as plant-generalists and vector-specialists. Plant-specialist phytoplasmas can be transmitted by a narrow range of vector species, but some plant-specialists are known to be transmitted by different vector species. Among the latter are maize bushy stunt vectored by *Dalbulus maidis* (DeLong & Wolcott), *D. elimatus* (DeLong & Wolcott) and *Graminella nigrifrons* (Forbes), and apple proliferation, which is transmitted by *Cacopsylla costalis* (Forster), *Cacopsylla melanoneura* (Forster) and *Fieberiella florii* (Stål) (Weintraub and Beanland, 2006). These associations can be defined as plant-specialists and vector-generalists. Finally, some plant-specialist phytoplasmas are transmitted by a specific vector, e.g. flavescence dorée (FDP), which is transmitted by *Scaphoideus titanus* Ball (Schvester *et al.*, 1963). We can define these associations as plant-specialists and vector-specialists.

However, it is likely that a large number of new vector species, as well as new host plants, have not yet been discovered, and in the coming years new data on phytoplasma–vector–plant associations will become available, and may conflict with our present knowledge in terms of insect and host-plant specificity.

Insect feeding behaviour

Phytoplasmas are known to be transmitted by planthoppers, leafhoppers and psyllids in a persistent, propagative manner (Marzachì *et al.*, 2004). Clearly, the transmission of phloem-restricted pathogens like phytoplasmas is correlated with the mode of phloem-feeding behaviour, but we know that Hemiptera show plasticity in their feeding sites. Phloem-, xylem- and parenchyma-feeding guilds (Tonkyn and Whitcomb, 1987) are not strict categories and, especially among vascular-feeder leafhoppers, the distinction between the phloem-feeding and xylem-feeding guilds is blurred (Wayadande, 1994). Therefore, although phloem-feeding behaviour is a prerequisite for transmission of phytoplasmas and other phloem-limited plant pathogens, we cannot exclude the possibility that species feeding primarily on xylem transmit phytoplasmas. Many aphids, whiteflies and mealybugs are phloem-feeders on plant species infected with phytoplasmas, but so far none of them has been found to be a vector of phytoplasmas. Recently, apple aphids were found to be positive in PCR assays for AP phytoplasmas, and were suspected to be vectors, but the results of transmission experiments seem to exclude this possibility (Cainelli *et al.*, 2007). A phloem-feeding habit is thus necessary but insufficient for phytoplasma transmission.

Geographic range

The geographic range of the insect vector, pathogen and host plants plays a role in transmission specificity. A number of insect species are not vectors because they are restricted to areas where a given phytoplasma is absent. When a vector is introduced in a new geographic area, it comes into contact with new phytoplasmas, and the association can result in dramatic spread of a disease. This explains the epidemic spread of flavescence dorée in the vineyards of southern Europe following the introduction of *S. titanus* in the 1950s (Bonfils and Schvester, 1960; Vidano, 1964). The leafhopper *E. variegatus* is Palaearctic in origin and was introduced into North America, where it became a vector of X-disease and American aster yellows (Jensen, 1969). These examples demonstrate that species other than known vectors have the potential for transmission, and transmission can also occur even when the phytoplasma and insect have never interacted previously during their evolution.

Phytoplasma–vector recognition and colonization

Phytoplasmas must overcome the gut barrier in order to colonize the insect body and multiply (Marzachì *et al.*, 2004). To do so, they probably adhere to the midgut epithelial cell membrane and enter the midgut intra- or intercellularly (by endo- or diacytosis). The mechanisms of phytoplasma–vector recognition, adhesion and transport through insect cells are mostly unknown. Some information on possible mechanisms of mollicute movement through vector barriers can be obtained from cultivable spiroplasmas, whose interactions with the insect host have been studied in some detail (Liu *et al.*, 1983; Fletcher *et al.*, 1998; Kwon *et al.*, 1999; Özbek *et al.*, 2003). Transmission of spiroplasmas by leafhoppers is believed to be mediated by recognition of specific spiroplasma membrane proteins, in a process of receptor-mediated endocytosis. Surface proteins of *Spiroplasma citri* were found to be involved in its adherence to vector insect cells. A significant reduction in adherence to insect cultured-cell monolayers in micro-titre plate adhesion assays after treatment of the spiroplasma cells with proteases suggested that surface proteins of spiroplasmas are involved in the spiroplasma–vector cell interactions (Fletcher *et al.*, 1998). Further evidence of the role of adhesion-related protein in the interaction with the vector was provided by Berho *et al.* (2006a), who found that non-insect-transmissible strains of *S. citri* lacked plasmids encoding adhesion-related proteins. Tranformation of an *S. citri* non-insect-transmissible strain with a plasmid encoding for the P32 adhesion-related protein restored transmissibility (Berho *et al.*, 2006b). Glycoproteins from the vector leafhopper *Circulifer haematoceps* (Mulsant & Rey) were found to interact with spiralin, the most abundant membrane protein of *S. citri* (Killiny *et al.*, 2005).

A schematic model of spiroplasma movement through two barriers within the leafhopper vector has been proposed by Fletcher *et al.* (1998). In the midgut, spiroplasmas adhere to receptors on the apical plasmalemma and are taken into the cytoplasm by endocytosis. After migrating through the cell, they are released by exocytosis into the space between the basal plasmalemma and the basal lamina, before crossing into the haemolymph, which transports them to the salivary glands. In the salivary glands, spiroplasmas pass through the basal lamina and adhere to receptors on the plasmalemma outer surface, taken up by endocytosis and released by exocytosis into the salivary ducts.

Phytoplasmas apparently attach to cells of their vectors: FD phytoplasmas were shown to adhere to nitrocellulose-bound extracts of the salivary glands, haemolymph, gut and fat bodies of several insects, including non-vector species (Lefol *et al.*, 1993). Many investigations have been carried out on the most abundant phytoplasma membrane protein, Amp, and its role in mediating the interaction with the vector has been proposed (Barbara *et al.*, 2002; Kakizawa *et al.*, 2006). A specific interaction between Amp of the onion yellows phytoplasma and the insect microfilament complex (actin and myosin) of the vector, but not of a non-vector, species has been reported (Suzuki *et al.*, 2006), and this interaction has been proposed as a determinant of insect vector specificity. Recently, Amp from chrysanthemum yellows phytoplasma (CYP, '*Ca.* Phytoplasma asteris') has been cloned and sequenced (Galetto *et al.*, 2008b); the CYP fusion protein has been found to interact with three major vector proteins, besides actin and myosin (Galetto *et al.*, 2008a). All these studies suggest that a key determinant in transmission specificity is recognition and adhesion of phytoplasmas to insect membranes and that this recognition triggers phytoplasma entrance into and colonization of the vector body.

Phytoplasma multiplication

Phytoplasmas are transmitted in a persistent, propagative manner by insect vectors in the order Hemiptera (Marzachì *et al.*, 2004). Once in the haemocoel of a vector species, they circulate and multiply in the body cavity, and pass through the salivary glands before being excreted together with homopteran saliva during successive nutrition. A vector insect must be permissive to phytoplasma colonization and must sustain phytoplasma multiplication. An insect that is unable to sustain multiplication will not be a vector. However, multiplication of phytoplasmas in insects does not always result in infective vector(s) (Purcell *et al.*, 1981; Vega *et al.*, 1993). This 'permissivity' can be considered necessary but insufficient by itself for phytoplasma transmission.

We can speculate that the innate immune system of insect vectors plays a major role in enabling phytoplasma multiplication in and colonization of the insect body. Unfortunately this topic has not yet been investigated for phytoplasma vectors, while considerable efforts have been made in the field of mosquito vectors (Baton *et al.*, 2008). It is possible that in some species phytoplasmas are recognized by specific binding of insect-recognition receptors, following which a variety of defence reactions are activated, leading to phytoplasma inactivation, while in other species phytoplasmas escape the insect immune response and can multiply and colonize the body, including the salivary glands. So far clear relationships between phytoplasma titre in the insect body and transmission capability have not been demonstrated, but there is some evidence that the level of phytoplasma multiplication is positively correlated with transmission efficiency both among and within species. For CYP, the species sustaining the most active phytoplasma multiplication, *Macrosteles quadripunctulatus* Kirschbaum, is also a more efficient vector compared with *E. variegatus* and *Euscelis incisus* (Bosco *et al.*, 2007). A comparison between the fate of CYP phytoplasmas in *E. variegatus* Kirschbaum individuals which failed to transmit and those which transmitted (Galetto *et al.*, 2009) revealed that the majority of non-transmitters acquired the phytoplasmas but sustained multiplication at a significantly lower titre than transmitters. For *E. variegatus* infected with CYP there seemed to be intraspecific variation in phytoplasma permissivity.

Salivary gland barrier

The resistance of the salivary glands to phytoplasma infection may explain the fact that some species acquire phytoplasmas but are not vectors. After investigating variation in vector competency of *E. variegatus*, Galetto *et al.* (2009) reported that 700 CYP cells per ng of insect DNA were found in the head of non-transmitter leafhoppers, versus 4000–8000 cells in the head of transmitters. Therefore, phytoplasma titre in salivary glands can be an indication of transmission efficiency.

Bressan *et al.* (2006) found that all 15 hopper species microinjected with a flavescence dorée phytoplasma (FDP) suspension were positive in PCR assays for FDP, but only three of them successfully transmitted the phytoplasmas to the feeding medium. The other tested species, belonging to Cixiidae, Delphacidae, Membracidae, Flatidae and Aphrophoridae, failed to transmit. The most likely explanation for these non-vector species is 'resistance' of the salivary glands. The three vector species were Cicadellidae, belonging to the Deltocephalinae subfamily (*Anoplotettix fuscovenosus* (Ferrari), *E. incisus* and *E. variegatus*). Since the natural vector of FDP, *S. titanus*, also belongs to the Deltocephalinae, it can be concluded that FDP specificity can act at the subfamily level. 'Resistance' of the salivary glands may also explain the failure of transmission by apple aphids that acquired a high titre of AP phytoplasmas (Cainelli *et al.*, 2007). The main elements of transmission specificity are summarized in Fig. 16.1.

Epidemiological Cycles of Phytoplasmas

Often phytoplasma infection is due to a single phytoplasma strain/species, and vector insects can acquire this phytoplasma and transmit it to other plants of the same species or other susceptible species. Therefore, the epidemiological cycle is simple, since a single phytoplasma is vectored among susceptible plants of one or more botanical species. Among susceptible plant species, some are unable to sustain acquisition by insect vectors and therefore can be considered dead-end hosts, at least for a given vector. For example, potato and cyclamen are dead-end hosts for aster yellows, and peach is a dead-end host for X-disease (Purcell, 1982; Alma *et al.*, 2000). The observation that a plant can be infected by a phytoplasma but is not suitable for vector acquisition leads to the conclusion that the feeding times required for inoculation and acquisition are probably different, with inoculation time being shorter. It is possible that the vector feeds with some difficulty in the phloem of dead-end plant species. It has been found that shorter feeding times are generally required for inoculation versus acquisition, and very efficient inoculation can take place in a few hours (Bressan *et al.*, 2007; Saracco *et al.*, 2008). Purcell (1982) proposed that this is probably because: (i) more inoculum can be delivered than can be acquired per unit time by a given vector; and (ii) a given amount of inoculum has greater impact on infection of plants compared with insects; phytoplasmas are injected directly into the sieve tubes of plants, where they multiply, whereas they have to overcome anatomical barriers in the insects in order to reach the salivary glands. While dead-end plants terminate the phytoplasma cycle, other susceptible plants may become a source of inoculum for different vector species with different host ranges. This can result in the spread of a phytoplasma to other plants and possibly to other crops (Lee *et al.*, 1998).

It appears that vectors can act in 'closed' or 'open' epidemiological cycles. A closed cycle is represented by a phytoplasma that circulates between a main, if not exclusive, host plant and a main, if not exclusive, vector species. The epidemiological cycle is therefore restricted to a single plant (crop), as in the case of FD, which is transmitted by *S. titanus* to grapevine. An open cycle

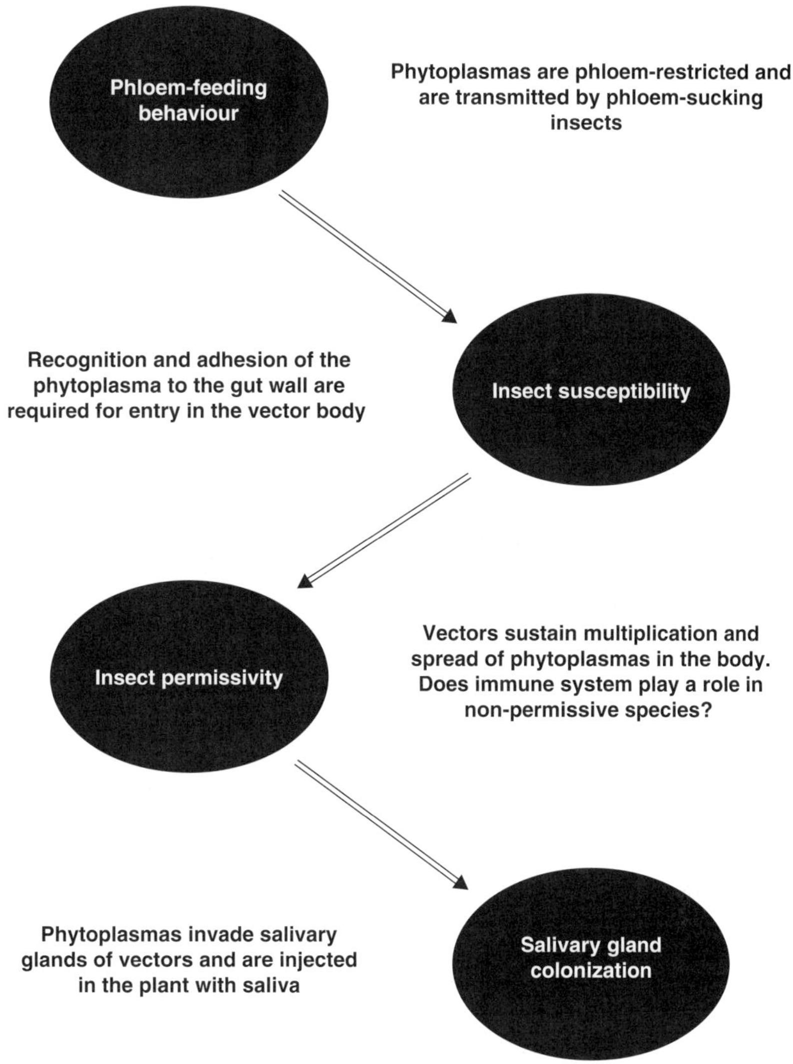

Fig. 16.1. Elements of insect transmission specificity.

is represented by a phytoplasma that circulates among different host plants because its vector(s) can, regularly or accidentally, feed on different plants (crops). Bois noir (BN), transmitted by *Hyalesthes obsoletus* Signoret from weeds to grapevine (Maixner, 1994; Maixner *et al.*, 1995), and peach X-disease, transmitted by *Paraphlepsius irroratus* (Say) and other species from chokecherry to peach (Rosenberger and Jones, 1978), are examples of such open cycles.

Sometimes, in nature, multiple phytoplasmas (or mollicutes) infect the same plant and interact/compete for the same host (Lee *et al.*, 2000). Multiple infections of genetically related or unrelated phytoplasmas are common in nature, and the availability of molecular diagnostic tools has greatly increased findings of simultaneous infections of multiple phytoplasmas in plants. Even if multiple phytoplasma infections are relatively common, the pattern of competition of different strains/species in the same plant has seldom been investigated. It appears that, in some cases, the competition results in interaction and mutual suppression among different strains (Freitag, 1964), but in other cases the presence of different phytoplasma species is repeatedly detected at different times in the same perennial plants, suggesting that they can stably share the host.

Under natural conditions, phytoplasma vectors can be exposed to mixed phytoplasma acquisition either by feeding on multiple-infected source plants or by feeding sequentially on different plants infected by different phytoplasmas. Interference between different phytoplasmas in the vector may have epidemiological consequences, i.e. the vector may 'select' a given phytoplasma from a plant and selectively transmit one strain, or the acquisition of one strain may provide cross-protection against the successive acquisition of another strain, or the insect may acquire and transmit more than one phytoplasma strain/species. In this latter case, the transmission pattern over time is determined by competition between the two (or more) strains and by the possibly different latent periods of the different phytoplasmas. Different phytoplasmas require different lengths of incubation in the same vector before being transmitted, and when sharing the same vector species the phytoplasma with the shorter latent period (LP) may be more efficiently transmitted early on after acquisition.

The length of the LP is not characteristic of the phytoplasma or of the vector but of the phytoplasma–vector association. The same phytoplasma may require different incubation times in different vector species: CYP is transmitted after an average latent period of 18 days by *M. quadripunctulatus* and 30 days by *E. variegatus* under the same experimental conditions (Bosco *et al.*, 2007).

Interactions of Multiple Pathogens in the Vector

Phytoplasma interactions in the vector insects may result in interactive or independent transmission. According to Purcell (1982), interference is generally most pronounced between closely related strains of the same pathogen; however, some suppressive interactions have been observed between phytoplasmas and other plant pathogens.

Interactive transmission

Aster yellows and oat blue dwarf virus (OBDV, Tymoviridae) in the vector *Macrosteles quadrilineatus* Forbes (= *fascifrons*) provide an example of interactive transmission between phytoplasmas and unrelated plant pathogens (Hsu and Banttari, 1979). Transmission rates of both AY and OBDV decreased when the leafhoppers were allowed to acquire both pathogens sequentially. Despite this negative interference, few leafhoppers simultaneously transmitted both phytoplasma and virus, as also demonstrated for the same pathogen–vector association by Frederiksen (1964). No explanation was provided for the interaction between AY and OBDV, but since both circulate and multiply in the vector they eventually compete for tissues and organs, and this competition results in the lower transmission efficiency of each pathogen in double-infected compared with single-infected leafhoppers.

An interesting example of interactive transmission between two mollicutes in the vector involves *D. maidis*, which may acquire both maize bushy stunt phytoplasma (MBSP) and corn stunt spiroplasma (CSS). Maramorosch (1958) described the competition between 'Mesa Central' and 'Rio Grande' maize-infecting mollicute strains. These pathogens are now thought to have been maize bushy stunt phytoplasma and corn stunt spiroplasma, respectively (Purcell, 1982). In this case, unilateral cross-protection in the vector *D. maidis* was found: prolonged acquisition of CSS suppressed the transmission of MBSP. When the vector fed for 1 day on MBSP-infected maize and 1 day on CSS-infected maize (or vice versa), both pathogens were transmitted in initial transmissions while only CSS was transmitted in later inoculations. This type of unilateral cross-protection suggests that CSS may multiply faster and/or move more rapidly in the insect body, thus suppressing transmission and/or multiplication of MBSP. As for AY and OBDV, some hoppers transmitted both pathogens to the same test plant (Maramorosch, 1958).

The interaction between related strains of phytoplasmas in the insect has been studied in detail by Freitag (1967). Competition between three aster yellows (AY) strains, Severe, Dwarf and Tulelake, was investigated in the vector *M. quadrilineatus* (=*fascifrons*). Leafhoppers were allowed to feed sequentially on plants infected with two different strains to investigate competition between pairs of strains in the combinations Severe–Dwarf and vice versa and Dwarf–Tulelake and vice versa. Unfortunately, this detailed study, as well as other studies on mollicute competition in vectors, was done well before molecular diagnostic tools were available, and thus only transmission results are provided; data on the differential acquisition of the different strains can only be inferred by transmission results and it is possible that phytoplasma strains were acquired but not transmitted. The most important result of Freitag's experiments is that more than 90% of the leafhoppers transmitted only one strain of the phytoplasma, although they were sequentially fed on two plants, each infected with a different strain. The majority of the leafhoppers transmitted only the first strain to which they had access. These findings clearly indicate an interaction in the vector between the different phytoplasmas, leading to the suppression of one of the strains or to a specific transmission pattern over time, in which the strain acquired first was the first to be transmitted and then the second strain was eventually transmitted. A high degree of cross-protection occurred between most of the combinations of phytoplasma strains, and only the Tulelake strain failed to provide cross-protection when challenged by the Dwarf strain. Few insects were able to transmit both strains that they had access to at the same time.

Cross-protection between related phytoplasma strains may be due to competition in the vector body for the tissues/cells that are infected. It can be speculated that competition between two equally virulent strains is mainly influenced by the time of acquisition. The first strain acquired is the most competitive because it starts to multiply first and it is probably the first to occupy the midgut epithelium, haemolymph and salivary glands. It is interesting to note that even when leafhoppers were allowed to acquire phytoplasmas for short, alternate 2-day feeding periods, instead of 2-week periods, cross-protection was evident. When a strain is less virulent in the insect, i.e. has slower multiplication and movement, it is transmitted only if it is acquired first, and then eventually displaced by the second, more aggressive strain. The production of allelo-chemicals by a strain which are active against another strain cannot be ruled out, but so far production of antibiotics by phytoplasmas has not been reported.

It is interesting to note that a similar competition pattern of AY strains was found in the host plants (Freitag, 1964). Therefore strain virulence was similar in both plant and vector hosts, suggesting that virulence could be related to speed of multiplication rather than to differential mobility in the insect.

The same phytoplasma–plant–vector associations were studied to investigate the ability of *M. quadrilineatus* to acquire phytoplasmas from dual-infected source plants (Freitag, 1964). Double-infected source plants showed symptoms of either one or the other strain or some symptoms of each. When fed on double-infected plants showing symptoms of one strain, leafhoppers usually acquired the strain that had induced the symptoms, but in a few cases they were able to acquire the other strain or both. In these experiments the vector probably did not 'select' the strain and simply acquired the most abundant one, and the strain that induced most symptoms was presumably the most abundant in the plant. In the plants showing both symptoms, we can assume that both strains were more or less equally represented and leafhoppers actually acquired the two strains with comparable efficiency.

Independent transmission

To our knowledge, independent transmission of phytoplasmas together with other plant pathogens has not been reported in the literature, although independent transmission of a spiroplasma with a virus has been reported (Gamez, 1973). *D. maidis* is the common vector of CSS and Rayado fino virus (RFV) to maize. When the vector acquired CSS and RFV sequentially, transmission of CSS did not appear to be influenced by the virus. After sequential acquisition of CSS and RFV, leafhoppers transmitted RFV first and CSS later. Interestingly, CSS has a longer latent period in the vector compared with RFV.

Case Study 1: FDP–CYP Interaction in the Vector

Interactions between different phytoplasmas in the same vector can now be studied in detail with molecular tools that allow detection of phytoplasmas and tracking of their path in the insect body (Bosco, 2006). In competition experiments, transmission assays provide information on the results of the interaction between/among phytoplasmas, while molecular detection in the whole body, in the dissected organs or in artificial feeding media provides details and information on the mechanisms of such competition.

We studied the competition between two genetically unrelated phytoplasmas in the common vector *E. variegatus* (D'Amelio *et al.*, 2007; D'Amelio, unpublished results). For this study FDP ('*Ca.* Phytoplasma vitis') and CYP ('*Ca.* Phytoplasma asteris') were chosen because they are both transmitted by the Deltocephalinae leafhopper *E. variegatus* to broad bean plants.

Nymphs of the leafhopper fed for 1 week on CYP-infected plants and for 1 week on FDP-infected plants or in the reverse sequence. Control leafhoppers were fed for 1 week on CYP or FDP source plants only. Leafhoppers were then assayed for the presence of the phytoplasmas by PCR and for transmission capability by feeding on test plants for the rest of their life. Phytoplasmas were also quantified by real-time PCR at different times post-acquisition, to investigate the competition pattern over time in the insect. Interestingly, the results showed a unilateral interaction: CYP interfered with FDP acquisition and transmission, regardless of the sequence of acquisition, while FDP did not interfere with CYP. Between 70 and 85% of the leafhoppers acquired CYP, in both single and serial acquisition with FDP. Similarly, *E. variegatus* transmitted CYP with high efficiency, starting from about 20 days post-acquisition and for the rest of its life, both when acquired alone or with FDP. More than 50% of the leafhoppers acquired FDP when feeding only on an FDP-infected source, but acquisition was severely reduced when leafhoppers also fed on CYP-infected plants. Moreover, FDP transmission was almost completely inhibited when leafhoppers had access to both CYP- and FDP-infected plants. To exclude the possibility that the failure of transmission of FDP could be due to the competition between phytoplasmas in the test plants after insect transmission rather than in the body of the vector, we checked the presence of phytoplasmas in the salivary glands and in the saliva (by analysing artificial feeding media) of double-infected *E. variegatus* by PCR. While CY phytoplasmas were consistently recovered from salivary glands and saliva, we detected FDP in very few salivary glands and failed to recover FDP from saliva, indicating that FDP is unable to colonize salivary glands efficiently and reach saliva (when co-infecting leafhoppers with CYP).

This unilateral cross-protection was confirmed by quantification of the two phytoplasma species in the insect body over time. When fed only on FDP source plants, *E. variegatus* hosted several hundred thousand cells of FDP per ng of insect DNA, but when fed on both CYP and FDP sources, regardless of the sequence, the titre was reduced to less than one hundred thousand. On the contrary, the titre of CYP in the vector was not affected by the simultaneous presence of FDP. In conclusion, *E. variegatus* transmits CYP with high efficiency following both single and double acquisition with FDP, and CYP actively colonizes salivary glands and is consistently present in the saliva. In contrast, *E. variegatus* is an efficient vector of FDP following single acquisition only, and, when competing with CYP in the vector, FDP is displaced, even if it persists for life.

Euscelidius variegatus is a natural vector of CYP and is a 'laboratory' vector of FDP. In nature, FDP infects grapevine and *E. variegatus* does not feed on grape. Therefore, this leafhopper and FDP have no history of evolutionary interaction and this may explain the less efficient colonization of the host insect and the very high level of phytoplasma multiplication in the vector, which also results in pathogenic effects (Bressan *et al.*, 2005). The possible interactions between the two different phytoplasmas in the vector insect are summarized in Fig. 16.2.

Conclusions

Even though we have attempted to define the concept of transmission specificity on a theoretical basis, the demonstration of transmission specificity is mostly empirical, and, since environmental conditions and plant- and insect-related factors can affect transmission efficiency, evaluation of different insect species under the same experimental conditions is needed to define vector and non-vector species. Very often these conditions cannot be reproduced for different species (e.g. an insect not feeding on plant A cannot be tested as a vector of phytoplasmas restricted to plant A), so we cannot prove whether a particular insect is not a vector under any circumstances. This should be taken into consideration when adopting strict categories, such as vectors and non-vectors, to identify factors regulating vector competency that are possibly present in vector species and absent in non-vector species. Interactive transmission of multiple phytoplasmas appears to be more common than independent transmission and, interestingly, cross-protection occurs between both genetically related and unrelated phytoplasma strains/species.

Transmission of phytoplasmas by insects is entering into a promising new era of exciting discoveries, especially because of the availability of molecular tools, of phytoplasma genome sequences and of new proteomic approaches, which provide the possibility of identifying transmission determinants in both insects and mollicutes and of explaining complex interactions, such as the ones between multiple phytoplasmas infecting single vectors or plants. Ecological investigations aimed at understanding the interactions among insects, phytoplasmas and host plants, as well as the search for new vectors, will also provide precious information that cannot be obtained in the laboratory. Experiments carried out with vectors, phytoplasmas and host plants under controlled conditions are also required to investigate the role of environmental factors and of plant- and insect-related factors that influence transmission competency. Finally, meta-analysis of published

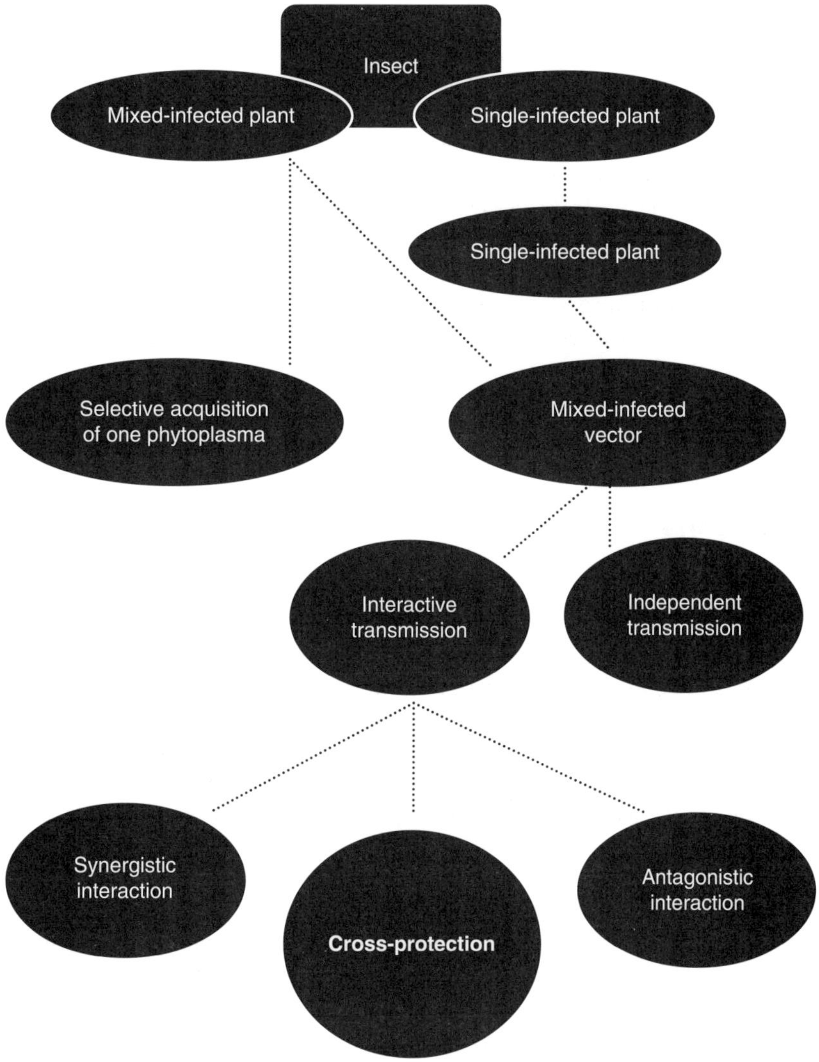

Fig. 16.2. Schematic diagram of the possible interactions of two different phytoplasmas in the vector insect. Cross-protection seems the most common type of interaction.

data on phytoplasma–vector associations may also provide some insights on transmission specificity.

Contributions from different levels of investigations will provide exhaustive explanations of phytoplasma–vector relationships under different environmental conditions and will help in the design of rational and advanced management of phytoplasma epidemics.

References

Alma, A., Marzachì, C., D'Aquilio, M. and Bosco, D. (2000) Cyclamen (*Cyclamen persicum* L.): a dead-end host species for 16Sr-IB and -IC subgroup phytoplasmas. *Annals of Applied Biology* 136, 173–178.

Barbara, D.J., Morton, A., Clark, M.F. and Davies, D.L. (2002) Immunodominant membrane proteins from two phytoplasmas in the aster yellows clade (chlorante aster yellows and clover phyllody) are highly divergent in the major hydrophilic region. *Microbiology* 148, 157–167.

Baton, L.A., Garver, L., Xi, Z.Y. and Dimopoulos, G. (2008) Functional genomics studies on the innate immunity of disease vectors. *Insect Science* 15, 15–27.

Berho, N., Duret, S. and Renaudin, J. (2006a) Absence of plasmids encoding adhesion-related proteins in non-insect-transmissible strains of *Spiroplasma citri*. *Microbiology* 152, 873–886.

Berho, N., Duret, S., Danet, J.L. and Renaudin, J. (2006b) Plasmid pSci6 from *Spiroplasma citri* GII-3 confers insect transmissibility to the non-transmissible strain *S. citri* 44. *Microbiology* 152, 2703–2716.

Bonfils, J. and Schvester, D. (1960) Les cicadelles (Homoptera Auchenorrhyncha) dans leurs rapports avec la vigne dans le sud-ouest de la France. *Annales des Epiphyties* 11, 325–336.

Bosco, D. (2006) Molecular detection of phytoplasmas in their insect vector: a basic tool to study the epidemiology of phytoplasma-associated diseases. *Trends in Entomology* 5, 19–27.

Bosco, D., Leoncini, P., Saracco, P., Raccah, B. and Marzachì, C. (2007) Interrelationships between '*Candidatus* Phytoplasma asteris' and its leafhopper vectors (Homoptera: Cicadellidae). *Journal of Economic Entomology* 100, 1504–1511.

Boudon-Padieu, E., Larrue, J. and Caudwell, A. (1989) ELISA and dot-blot detection of flavescence dorée MLO in individual leafhopper vectors during latency and inoculative state. *Current Microbiology* 19, 357–364.

Bressan, A., Clair, D., Sémétey, O. and Boudon-Padieu, E. (2005) Effect of two strains of flavescence dorée phytoplasma on the survival and fecundity of the experimental leafhopper vector *Euscelidius variegatus* Kirschbaum. *Journal of Invertebrate Pathology* 89, 144–149.

Bressan, A., Clair, D., Sémétey, O. and Boudon-Padieu, E. (2006) Insect injection and artificial feeding bioassays to test the vector specificity of flavescence dorée phytoplasma. *Phytopathology* 96, 790–796.

Bressan A., Turata, R., Maixner, M., Spiazzi, S., Boudon-Padieu, E. and Girolami, V. (2007) Vector activity of *Hyalesthes obsoletus* living on nettles and transmitting a stolbur phytoplasma to grapevines: a case study. *Annals of Applied Biology* 150, 331–339.

Cainelli, C., Forno, F., Mattedi, L. and Grando, M.S. (2007) Can apple aphids be vectors of '*Candidatus* Phytoplasma mali'? *IOPC/WBRS Bulletin* 30(4), 261–266.

D'Amelio, R., Marzachì, C. and Bosco, D. (2007) Double infection of '*Candidatus* phytoplasma asteris' and '*Candidatus* phytoplasma vitis' in the vector *Euscelidius variegatus* Kirschbaum. *Bulletin of Insectology* 60(2), 317–318.

Favali, M.A., Fossati, F., Toppi, L.S. and Musetti, R. (2008) *Catharanthus roseus* phytoplasmas. In: Harrison, N.A., Rao, G.P. and Marcone, C. (eds) *Characterization, Diagnosis and Management of Phytoplasmas*. Studium Press LLC, Houston, Texas, pp. 195–218.

Fletcher, J., Wayadande, A., Melcher, U. and Ye, F. (1998) The phytopathogenic mollicute–insect vector interface: a closer look. *Phytopathology* 88, 1351–1358.

Frederiksen, R.A. (1964) Simultaneous infection and transmission of two viruses in *Flax* by *Macrosteles fascifrons*. *Phytopathology* 54, 1028–1030.

Freitag, J.H. (1964) Interaction and mutual suppression among three strains of aster yellows virus. *Virology* 24, 401–413.

Freitag, J.H. (1967) Interaction between strains of aster yellows virus in the six-spotted leafhopper *Macrosteles fascifrons*. *Phytopathology* 57, 1016–1024.

Galetto, L., Bosco, D. and Marzachì, C. (2008a) Amp-mediated interactions between '*Candidatus* Phytoplasma asteris' chrysanthemum yellows (CY) and leafhopper vectors. In: Porta-Puglia, A. and Gonthier, P. (eds) Proceedings of the 9th International Congress of Plant Pathology. *Journal of Plant Pathology* 90(2, Suppl.), 461.

Galetto, L., Fletcher, J., Bosco, D., Turina, M., Wayadande, A. and Marzachì, C. (2008b) Characterization of putative membrane protein genes of the chrysanthemum yellows phytoplasma isolate (CY), '*Candidatus* Phytoplasma asteris'. *Canadian Journal of Microbiology* 54, 341–351.

Galetto, L., Nardi, M., Saracco, P., Bressan, A., Marzachì, C. and Bosco, D. (2009) Variation in vector competency depends on chrysanthemum yellows phytoplasma distribution within *Euscelidius variegatus*. *Entomologia Experimentalis et Applicata* 131, 200–207.

Gamez, R. (1973) Transmission of rayado fino virus of maize (*Zea mays*) by *Dalbulus maidis*. *Annals of Applied Biology* 73, 285–292.

Hsu, T.P. and Banttari, E.E. (1979) Dual transmission of the aster yellows mycoplasma-like organism and the oat blue dwarf virus and its effect on longevity and fecundity of the aster leafhopper vector. *Phytopathology* 69, 843–845.

Jensen, D.D. (1969) Comparative transmission of western X-disease virus by *Colladonus montanus*, *C. geminatus*, and a new leafhopper vector, *Euscelidius variegatus*. *Journal of Economic Entomology* 62, 1147–1150.

Kakizawa, S., Oshima, K., Jung, H., Suzuki, S., Nishigawa, H., Arashida, R., Miyata, S., Ugaki, M., Kishino, H. and Namba, S. (2006) Positive selection acting on a surface membrane protein of the plant-pathogenic phytoplasmas. *Journal of Bacteriology* 188, 3424–3428.

Killiny, N., Castroviejo, M. and Saillard, C. (2005) *Spiroplasma citri* spiralin acts *in vitro* as a lectin binding to glycoproteins from its insect vector *Circulifer haematoceps*. *Phytopathology* 95, 541–548.

Kwon, M., Wayadande, A.C. and Fletcher, J. (1999) *Spiroplasma citri* movement into the intestines and salivary glands of its leafhopper vector, *Circulifer tenellus*. *Phytopathology* 89, 1144–1151.

Larsen, M.L. and Walter, G.H. (2007) Intraspecific variation within *Orosius argentatus* Evans (Hemiptera: Cicadellidae): colour polymorphisms, morphometric analyses and host associations. *Australian Journal of Entomology* 46, 207–216.

Lee, I., Gundersen-Rindal, D.E. and Bertaccini, A. (1998) Phytoplasma: ecology and genomic diversity. *Phytopathology* 88, 1359–1366.

Lee, I., Davis, R.E. and Gundersen-Rindal, D.E. (2000) Phytoplasma: phytopathogenic mollicutes. *Annual Review of Microbiology* 54, 221–255.

Lefol, C., Caudwell, A., Lherminier, J. and Larrue, J. (1993) Attachment of the flavescence dorée pathogen (MLO) to leafhopper vectors and other insects. *Annals of Applied Biology* 123, 611–622.

Liu, H.Y., Gumpf, D.J., Oldfield, G.N. and Calavan, E.C. (1983) The relationship of *Spiroplasma citri* and *Circulifer tenellus*. *Phytopathology* 73, 585–590.

Maixner, M. (1994) Transmission of German grapevine yellows (Vergilbungskrankheit) by the planthopper *Hyalesthes obsoletus* (Auchenorrhyncha: Cixiidae). *Vitis* 33, 103–104.

Maixner, M., Ahrens, U. and Seemüller, E. (1995) Detection of the German grapevine yellows (Vergilbungskrankheit) MLO in grapevine, alternative hosts and a vector by a specific PCR procedure. *European Journal of Plant Pathology* 101, 241–250.

Maramorosch, K. (1958) Cross-protection between two strains of corn stunt virus in an insect vector. *Virology* 6, 448–459.

Marzachì, C., Milne, R.G. and Bosco, D. (2004) Phytoplasma–plant–vector relationships. In: Pandalai, S.G. (ed.) *Recent Research Development in Plant Pathology*, vol. 3. Research Signpost, Trivandrum, Kerala, India, pp. 211–241.

Özbek, E., Miller, S.A., Meulia, T. and Hogenhout, S.A. (2003) Infection and replication sites of *Spiroplasma kunkelii* (Class: Mollicutes) in midgut and Malpighian tubules of the leafhopper *Dalbulus maidis*. *Journal of Invertebrate Pathology* 82, 167–175.

Purcell, A.H. (1982) Insect vector relationships with prokaryotic plant pathogens. *Annual Review of Phytopathology* 20, 397–417.

Purcell, A.H., Richardson, J. and Finlay, A. (1981) Multiplication of the agent of X-disease in a non-vector leafhopper *Macrosteles fascifrons*. *Annals of Applied Biology* 99, 283–289.

Rosenberger, D.A. and Jones, A.L. (1978) Leafhopper vectors of the peach X-disease pathogen and its seasonal transmisssion from chokecherry. *Phytopathology* 68, 782–790.

Saracco, P., Marzachì, C. and Bosco, D. (2008) Activity of some insecticides in preventing transmission of chrysanthemum yellows phytoplasma ('*Candidatus* Phytoplasma asteris') by the leafhopper *Macrosteles quadripunctulatus* Kirschbaum. *Crop Protection* 27, 130–136.

Schvester, D., Carle, P. and Moutous, G. (1963) Transmission de la flavescence dorée de la vigne par *Scaphoideus littoralis* Ball. (Homopt., Jassidae). Expériences de 1961. *Annales des Epiphyties* 14, 175–198.

Suzuki, S., Oshima, K., Kakizawa, S., Arashida, R., Jung, H., Yamaji, Y., Nishigawa, H., Ugaki, M. and Namba, S. (2006) Interaction between the membrane protein of a pathogen and insect microfilament complex determines insect-vector specificity. *Proceedings of the National Academy of Sciences of the USA* 103, 4252–4257.

Tonkyn, D.W. and Whitcomb, R.F. (1987) Feeding strategies and the guild concept among vascular feeding insects and microorganisms. In: Harris, K.F. (ed.) *Current Topics in Vector Research*, vol. 4. Springer-Verlag, New York, pp. 179–200.

Vega, F.E., Davis, R.E., Barbosa, P., Dally, E.L., Purcell, A.H. and Lee, I.M. (1993) Detection of a plant pathogen in a non-vector insect species by the polymerase chain reaction. *Phytopathology* 83, 621–624.

Vidano, C. (1964) Scoperta in Italia di *S. littoralis* Ball cicalina americana collegata alla 'Flavescence dorée' della vite. *L'Italia Agricola* 101, 1031–1049.

Wayadande, A.C. (1994) Electronic monitoring of leafhoppers and planthoppers: feeding behavior and application in host-plant resistance studies. In: Ellsbury, M.M., Backus, E.A. and Ullman, D.L. (eds) *History, Development, and Application of AC Electronic Insect Feeding Monitors*. Thomas Say Publications in Entomology, Entomological Society of America, Lanham, Maryland, pp. 86–105.

Weintraub, P.G. and Beanland, L. (2006) Insect vectors of phytoplasmas. *Annual Review of Entomology* 51, 91–111.

17 Current and Possible Future Distributions of Phytoplasma Diseases and their Vectors

XAVIER FOISSAC[1] AND MICHAEL R. WILSON[2]

[1]INRA et Université Victor Ségalen, France; [2]National Museum of Wales, UK

Geographical Distribution and Impact of Phytoplasmas

Phytoplasmas are responsible for numerous crop diseases worldwide, out of which two kinds of epidemiological situations can be distinguished: epidemic and non-epidemic. Many diseases are not epidemic on the crop itself, meaning an infected plant is an epidemiological dead-end host for the phytoplasma. This is particularly true when the phytoplasma reservoir consists of wild plants and when the insect vector is living in the wild component of the ecosystem and not developing on the crop. In this case, the phytoplasma insect vector may occasionally feed on the cultivated plant, causing monocyclic epidemics. An example is the bois noir disease of grapevine in the Euro-Mediterranean basin, for which the stolbur phytoplasma is transmitted from bindweed and stinging nettle to grapevine (Langer and Maixner, 2004). The economic impact of such epidemics is directly linked to the abundance and infectivity of the insect populations and corresponds to the yield loss. Such diseases are generally not considered as quarantine diseases and control measures only rely on the prophylactic reduction of weeds identified as the main reservoirs for the phytoplasmas.

When the crop itself acts as the main reservoir and if the insect vector completes its life cycle on the infected crop, then the situation resulting from the transmission from plant to plant by the insect, into or between cultivated plots, corresponds to polycyclic epidemics. Most of the diseases that are spread this way are usually classified as quarantine diseases. This can occur on annual plants when there are no time gaps between cropping periods but occurs mostly in perennial, woody crops. This is, for instance, the case of flavescence dorée of grapevine in the southern European vineyards (Boudon-Padieu, 2002). In these cases, prophylactic management includes elimination of infected plants, certification of planting material and chemical control of the insect vector. As a consequence, the economic impact expands as the cost

 Phytoplasmas: Genomes, Plant Hosts and Vectors (eds P.G. Weintraub and P. Jones)

of control measures such as pesticide spray, field survey and nursery protection is added to the yield loss.

The geographical distribution and impact of phytoplasma diseases depends on the host range of the phytoplasma as well as the feeding behaviour of the insect vector. Some have a broad range of plant hosts and polyphagous vectors and therefore have a wide distribution. This is the case for '*Candidatus* (*Ca.*) Phytoplasma asteris', which has been reported in many crops worldwide. But many phytoplasmas have restricted host ranges and oligophagous or monophagous insect vectors, which restrict their geographical distribution. This review will focus on the main phytoplasma taxonomic groups and will describe the areas (Fig. 17.1) and the crops that are affected.

Taxonomic Phytoplasma Groups

16SrI

The taxonomic group 16SrI, also known as aster yellows (AY) group, is widely distributed in North America, where it induces 'yellows' and 'dwarfing' in vegetables such as lettuce, carrot, onion, cabbage, celery, potato and tomato; small fruits, such as blueberry and strawberry; and ornamentals, such as China aster, oenothera and periwinkle in the USA, as well as clover in both the USA and Canada (Lee *et al.*, 2004a). It is also responsible for some of the yellows of grapevine in the mid-Atlantic and north-eastern regions of the USA (Beanland *et al.*, 2006). Its occurrence in North and South America as the causal agent of maize bushy stunt has been demonstrated (Harrison *et al.*, 1996; Bedendo *et al.*, 1997). It has also been demonstrated to be one of the agents of sugarcane yellow leaf disease in Cuba (Arocha *et al.*, 2005). In Europe, phytoplasmas of the 16SrI group also affect lettuce, onion, potato and tomato, are responsible for clover phyllody and strawberry green petal, and are also the agent of various disorders of gladiolus, hydrangea, primula, anemone, ranunculus and chrysanthemum (Lee *et al.*, 2004a). Strains of '*Ca.* Phytoplasma asteris' have been associated with the decline of poplar in France, Germany and Croatia (Berges *et al.*, 1997; Seruga *et al.*, 2003). The involvement of '*Ca.* Phytoplasma asteris' in gramineous plant diseases has been described in Lithuania as well as in northern China (Gu *et al.*, 2005; Urbanaviciene *et al.*, 2007). In Asia, group 16SrI phytoplasmas are responsible for marguerite yellows and Paulownia witches'-broom. More recently, it was shown to be associated with the 'Al-Wijam' disease of date palm in Saudi Arabia (Alhudaib *et al.*, 2008). In Israel, some cases of grapevine yellows are due to '*Ca.* Phytoplasma asteris' infections.

16SrII

Witches'-broom disease of acid lime (*Citrus aurantifolia* L.) is a lethal disease that is caused by '*Ca.* Phytoplasma aurantifolia', a member of the

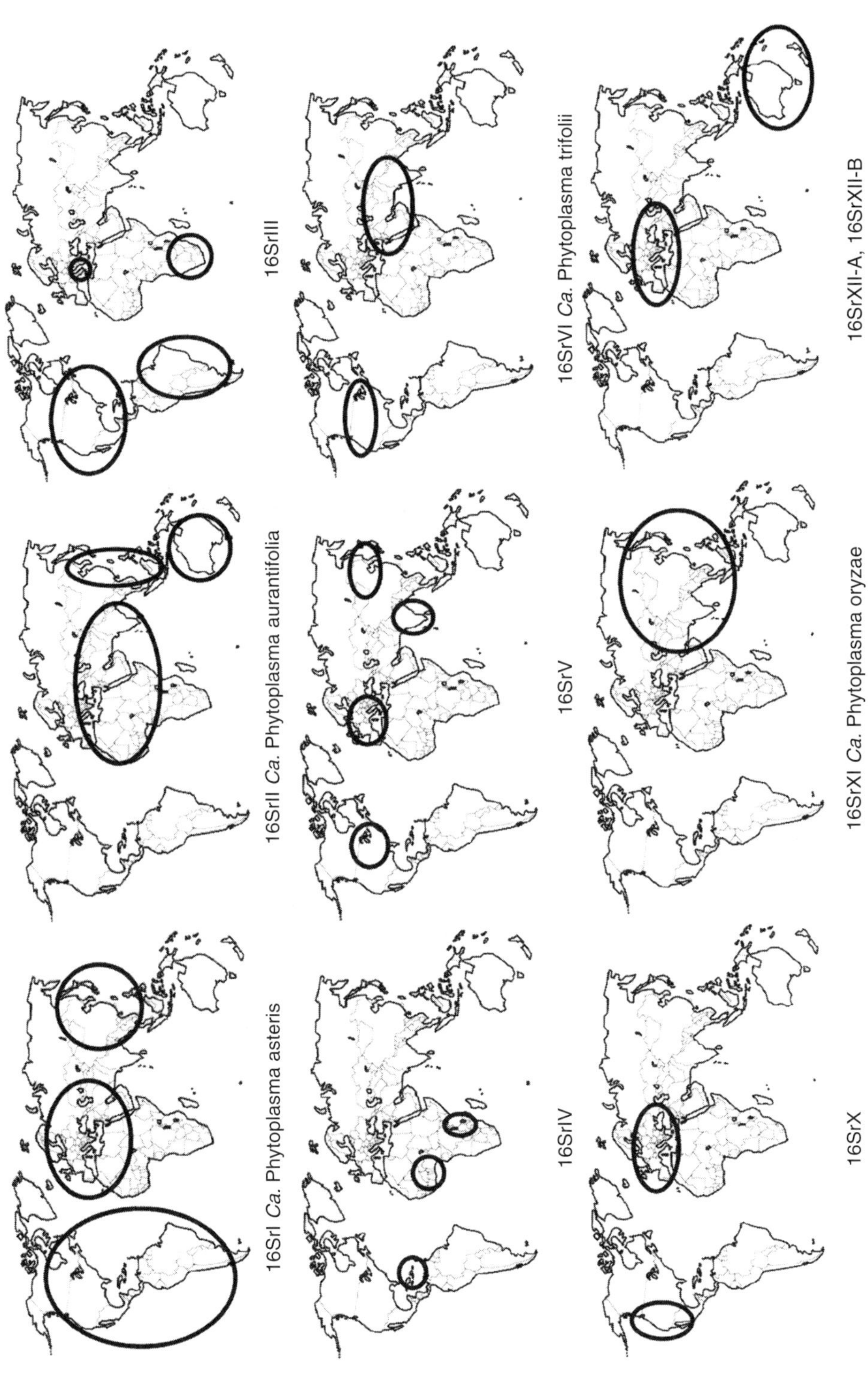

Fig. 17.1. Main geographical areas affected by the most important phytoplasma taxonomic groups.

16SrII-B taxonomic group. It appeared in the Sultanate of Oman in the late 1970s and in the neighbouring country, the United Arab Emirates, in 1989 (Garnier *et al.*, 1991). It is now reported in India and Iran (Ghosh *et al.*, 1999; Salehi *et al.*, 2005). Other strains of group 16SrII also induce witches'-broom in lucerne in Oman Sultanate (Khan *et al.*, 2002). Members of this taxonomic group are responsible for cotton phyllody in western Africa (Desmidts and Rassel, 1974) and faba bean phyllody in Sudan (Jones *et al.*, 1984). Phytoplasmas of the group 16SrII are present in Australia, where they cause tomato big bud and sweet potato little leaf (Schneider *et al.*, 1999). Phytoplasmas belonging to group 16SrII are known to cause peanut, sunn hemp and crotalaria witches'-broom in Asia (Yang, 1985; Sharma, 1990; Yang *et al.*, 2008).

16SrIII

In North America, western-X disease, caused by a phytoplasma of the 16SrIII taxonomic group, is the major phytoplasma disease threat to peach and cherry production (Granett and Gilmer, 1971) but can be found in other crops, such as walnut and pecan (Lee *et al.*, 2000). It has also occasionally been described in Italy. Other members of the 16SrIII phytoplasma clade are commonly present in Brazil on various plants (Montano *et al.*, 2007) and in South Africa, where they are responsible for sugarcane yellows (Cronje *et al.*, 1998). This group is present but has few hosts or impact in Asia (Lee *et al.*, 2000).

16SrIV

The members of the coconut lethal yellows group (16SrIV) and Nigerian coconut lethal decline group (16SrXXII-A) affect coconuts and date palms in Caribbean islands, Florida, Mexico, western Africa and Tanzania.

16SrV

Southern French, Spanish, northern Italian, Slovenian and Serbian vineyards are affected by the flavescence dorée (FD) phytoplasma, a quarantine pathogen of grapevine belonging to the 16SrV taxonomic group (Boudon-Padieu, 2002). This group consists of phytoplasmas with specific biological niches restricted to woody, perennial hosts. '*Ca.* Phytoplasma ulmi' is responsible for yellows of elm species in North America and Europe (Lee *et al.*, 2004b). In Europe, other phytoplasmas of group 16SrV mainly infect grapevine, alder, blackberry and *Spartium*. In Asia, '*Ca.* Phytoplasma ziziphi' is the agent of jujube witches'-broom in China, Korea and Japan, of cherry lethal yellows in China and of peach yellows in India (Zhu *et al.*, 1998; Jung *et al.*, 2003a; Lee *et al.*, 2004b).

16SrVI

Phytoplasmas in the 16SrVI group, taxonomically described as '*Ca.* Phytoplasma trifolii', were first described as being associated with a disease of alsike clover in Canada and later with phytoplasma diseases of tomato, as well as potato and elm, in North America (Chen and Hiruki, 1975; Lee *et al.*, 1991; Shaw *et al.*, 1993; Jacobs *et al.*, 2003). In Europe and the Middle East, this phytoplasma clade was recently associated with diseases of pepper and tomato in Spain, Jordan and Lebanon (Castro and Romero, 2002; Anfoka *et al.*, 2003; Choueiri *et al.*, 2007). Phytoplasmas of this group are also responsible for the little leaf disease of aubergine in India (Mitra, 1988; Schneider *et al.*, 1995).

16SrIX

The economic impact of the phylogenetic group 16SrIX is mostly restricted to the Middle East and is a major threat to almond production in Iran and Lebanon. The almond witches'-broom disease was first reported as almond brooming disease in Fars province of Iran (Salehi and Izadpanah, 1995). A similar epidemic disease of almond was reported in Lebanon (Choueiri *et al.*, 2001), and the phytoplasma agent was finally described as '*Ca.* Phytoplasma phoenicium' (Verdin *et al.*, 2003). A related phytoplasma was also associated with *Knautia arvensis* phyllody detected on field scabious in Italy (Marcone *et al.*, 2001). Production of almonds has been seriously affected since the 1990s in Lebanon and Iran, as the phytoplasma induces a lethal disease. Thousands of almond trees have died over the past 15 years in Lebanon, since the beginning of the first epidemic, which occurred in the south of the country in the early 1990s (Abou-Jawdah *et al.*, 2003). Other members of this clade cause pigeon pea witches'-broom in North America (McCoy *et al.*, 1983) and were recently described as being associated with a yellowing disease of citrus in Brazil (Teixeira *et al.*, 2008).

16SrX

Phytoplasmas of the 16SrX taxonomic group mostly affect temperate pome and stone fruit trees in Europe and the USA. They correspond to three different phytoplasmas, inducing European stone fruit yellows, pear decline and apple proliferation (Jarausch *et al.*, 1994; Lorenz *et al.*, 1994). The causal agents of these diseases have been respectively described as '*Ca.* Phytoplasma prunorum', '*Ca.* Phytoplasma pyri' and '*Ca.* Phytoplasma mali' (Seemüller and Schneider, 2004). '*Ca.* Phytoplasma mali', the agent of apple proliferation, affects only European and Turkish apple orchards, reducing the size and weight of the fruit of infected apple trees by half, which are therefore unmarketable. The disease is absent in North and South America as well as Asia. In Europe, young trees are more susceptible to the disease, but no tree mortality has been reported to date. The highest economic impact seems to occur in Germany and northern Italy. Damage to pear production caused by the pear decline phytoplasma ('*Ca.* Phytoplasma pyri') has been reported in Europe, North America and Asia Minor, and also in Taiwan. The disease kills trees in its quick decline form when varieties are grafted on sensitive rootstocks. Pear decline has been reported to reduce pear production by half in certain states of the USA and killed 50,000 trees in the late 1940s in Italy. In the severe form, dieback of the trees can occur within a few weeks. Economic damage to *Prunus* species is very common in Europe and Asia Minor because of European stone fruit yellows caused by '*Ca.* Phytoplasma prunorum', which kills, for example, about 5% of apricot trees per year in southern France due to apricot chlorotic leaf roll. '*Ca.* Phytoplasma prunorum' also induces plum leptonecrosis on *Prunus salicina* (Japanese plum) and yellows on most of the peach accessions in southern Europe. '*Ca.* Phytoplasma prunorum' does not affect sour and sweet cherry and the economic impact on European plum is restricted to some orchards in Italy planted with susceptible cultivars. This yellowing disease has also been reported on almond around the Mediterranean basin.

16SrXI

The taxonomic group 16SrXI is present in Asia, where it causes rice yellow dwarf disease. Rice infected by '*Ca.* Phytoplasma orizae' becomes pale yellow and then stunted, with no grain being produced. It is present in most rice-growing countries in Asia. Known phytoplasma relatives are the phytoplasmas associated with sugarcane white leaf and sugarcane grassy shoot found in sugarcane, annual bluegrass white leaf, bermuda grass white leaf and Brachiaria grass white leaf (Jung *et al.*, 2003b). This group seems to be absent from other rice-producing areas in the world but has been occasionally described in Europe on *Cirsium arvense* and on Napier grass (*Pennisetum purpureum*) in Kenya.

16SrXII

The stolbur phytoplasma (STOL), a member of the taxonomic group 16SrXII-A, infects a wide range of cultivated plants in Europe and the Mediterranean basin, such as the solanaceous crops, grapevine, celery, sugarbeet, strawberry and lavender (Garnier, 2000). STOL is classified as a quarantine pest on potato plants owing to the possible transmission through tubers. Symptoms of STOL disease are leaf discoloration, stunting and abnormal floral development, leading to sterility. In European vineyards, it causes a severe grapevine yellows, the bois noir disease. The main reservoirs of STOL in France, Germany and Italy are weeds such as bindweeds (*Convolvulus arvensis* and *Calystegia sepium*) or stinging nettles (*Urtica dioica*), from which it is transmitted by cixiid planthoppers to other weeds or cultivated plants

(Fos *et al.*, 1992; Langer and Maixner, 2004; Bressan *et al.*, 2007). The only crop on which STOL is epidemic is lavender (*Lavendula*), as the insect vector *Hyalesthes obsoletus* Signoret (Cixiidae) completes its life cycle on the crop, allowing young overwintering nymphs to acquire the pathogens early, thereby promoting an efficient epidemic transmission of the phytoplasma. Recently, this phytoplasma group has been associated with maize redness in Serbia, where another member of the Cixiidae, *Reptalus panzeri* (Löw), vectors the disease (Jovic *et al.*, 2007). '*Ca.* Phytoplasma australiense', representative of the group X16SrXII-B, is the agent of Australian grapevine yellows (Davis *et al.*, 1997). It also causes papaya dieback and Phormium yellows in Australia and New Zealand (Liefting *et al.*, 1998). Other subgroups in the 16SrXII group affect strawberries in Lithuania and *Hydrangea* in Japan (Sawayanagi *et al.*, 1999; Valiunas *et al.*, 2006).

Sixteen other taxonomic groups have recently been described or are currently being investigated, but it is too early to clearly describe their impact, host range and precise geographical distribution.

Geographical Distribution of Phytoplasma Insect Vectors

Number of species as vectors

Only a small proportion of phytoplasmas in plants are currently known and an even smaller number of their Auchenorrhyncha vectors. This number of insect species is a very small percentage of the described species – less than 1%. It is likely to remain a small number (and percentage), given the eventual rise in described species of leafhoppers – from perhaps 20,000 to 100,000 species.

The majority of known vectors, not surprisingly, are known as pests of cultivated food and industrial crops and ornamental plants. Even if we know host plants for only a relatively small number of species, we know enough to make generalizations about the patterns of host plant utilization in Auchenorrhyncha (e.g. Nickel, 2003). Most species are habitat specialists, but this is likely to be based on host plant specialization. Very few species are polyphagous. The xylem-feeding spittlebug *Philaenus spumarius* (L.) (Aphrophoridae) is recorded from a large number of species of dicotyledonous plants, but most other xylem-feeding hoppers seem much more restricted in their use of host plants. Among phloem-feeding Auchenorrhyncha, most are narrowly oligophagous at the plant genus level or monophagous to a narrow range of plant genera or strictly monophagous. A very few are known to alternate between hosts. Feeding hosts may be less restricted than those host plants on which eggs are laid.

Transport of species

Auchenorrhyncha are well suited for transport as alien species since the eggs are usually laid into plant tissue, where they may remain for considerable periods in diapause. Movements of species (with or without climate change) are going to be either as alien introductions from one region to another – most likely by transport of plants – or by changes of distribution within a region. This may also be made possible by transport of plants within a region (perhaps even by movement of originally introduced alien species). Climate change may increase the chances of recently introduced species surviving and increasing in distribution.

Alien species

Some of the most significant phytoplasma vector species are alien introduced species. In Europe, perhaps the most significant species is *Scaphoideus titanus* Ball, introduced from North America (Arzone *et al.*, 1987). *Fieberiella flori* (Stål) is a European species introduced into North America, as was *Circulifer tenellus* (Baker), introduced to western USA from the Mediterranean area. Lastly, *Orientus ishidae* Matsumura, introduced into North America and Europe, appears to be an Asian species. It is significant that all of these species (and the planthopper *Metcalfa pruinosa* (Say)) have limited economic impact in their native range.

Monitoring of species movements and introductions due to climate change is only going to be possible when there is a network of observers and also where the fauna is well known taxonomically. It is not surprising that the increasingly large literature on alien species (e.g. Rabitsch, 2008) is centred on northern Europe, where there is a concentration of specialists and the availability of identification literature.

There is an increasing interest in both the extent of arrival of alien (i.e. non-indigenous) species and the impact they might have on the native flora and fauna. Inevitably such data for invertebrates are most available in those countries that have an efficient interception and quarantine system but also in those countries or regions in which the fauna is well enough known for new arrivals to be distinguished from native species. Roques *et al.* (2009) have summarized information on alien terrestrial invertebrate species in Europe. Kenis *et al.* (2009) reviewed the ecological effects of invasive alien insects. They have also reviewed the data on the invasion processes and invasive alien insect species management in central Europe from two databases: a compilation of two inventories of alien insects in Austria and Switzerland and a list of interceptions of non-indigenous plant pests in Europe, gathered by the European and Mediterranean Plant Protection Organization (EPPO) for the period 1995–2004 (Kenis *et al.*, 2007). For one-third of the insects established in Switzerland and Austria, the region of origin is unclear. Others come mainly from North America, Asia and the Mediterranean region. Among the intercepted insects, 40% were associated with commodities from Asia, 32% from Europe and only 2% from North America. Sternorrhyncha, Coleoptera and Psocoptera were particularly well represented in the alien fauna compared with the native fauna. In the interception database, Sternorrhyncha were also well represented, but Diptera accounted for the highest number of records. Sap feeders and detritivores were the dominant feeding niches in the alien insect fauna. In contrast, external defoliators, stem borers, gall makers, root feeders, predators and parasitoids were under-represented. Nearly 40% of the alien insects in Switzerland and Austria live only indoors. Another 15% live outdoors but exclusively or predominantly on exotic plants. Less than 20% are found mainly in 'natural' environments. The majority of introductions of alien insects in Europe are associated with the international trade in ornamental plants. An economic impact was found for 40% of the alien insects in Switzerland and Austria, whereas none is known to have an ecological impact. Rabitsch (2008) reviewed the alien (non-native, non-indigenous, exotic) true bug (Heteroptera) species in Europe. Forty-two established alien Heteroptera are recognized, of which 12 species are alien to Europe (originating outside Europe: eight from North America, three from the Eastern Palaearctic, one from New Zealand), 24 species are translocated within Europe and six cryptogenic species are of unknown origin. Since 1990, an approximate arrival rate of seven species per decade has been observed. A recent trend of increased introductions from North America to Europe is suggested. The most important pathway of alien Heteroptera is translocation as contaminants (49%), usually with ornamental plants, followed by unintentional introduction through natural dispersal (unaided) across political borders within Europe (28%), and translocation as stowaways within a transport (21%).

It is likely that climate change will allow alien species whose distribution has been limited by climate to expand their distribution to new regions. The ability to tolerate the environmental extremes characteristic of their invaded range is important in their distribution. Among Auchenorrhyncha, the North American flatid planthopper *Metcalfa pruinosa* was first found in Italy in 1980 (Arzone *et al.*, 1987) and for some years was confined to a small area in north-east Italy. The warm summers of recent years have allowed the species' rapid expansion into southern and central European countries. It is possible that warmer winters are not an important factor for this species but that a requirement for warm summers is more important (R. Remane, 2008, personal communication).

Sometimes the period between arrival and expansion may be quite long. Preisseravier *et al.* (2008) experimented with the invasive elongate hemlock scale *Fiorinia externa* Ferris, which feeds on eastern hemlock, *Tsuga canadensis*, on the east coast of North America. Following its 1908 arrival, it remained localized until entering a period of rapid northward range expansion in the 1970s. Experiments showed that northern populations were more tolerant of experimental exposure to cold temperatures than were southern populations. The results provide evidence for local adaptation to extreme temperatures in *F. externa* and provide one possible explanation for the lag period between the arrival of this species and its eventual northward range expansion.

Climate change effects

There have been few studies reporting the effects of potential climate change on Auchenorrhyncha and even less on those species known to be phytoplasma vectors. Masters *et al.* (1998) used manipulations of local climate to investigate how warmer winters (with either wetter or drier summers) would affect a range of grassland Auchenorrhyncha. They found both direct and indirect effects. Supplemented summer rainfall led to an increase in vegetation cover, leading to an increase in the abundance of Auchenorrhyncha. Summer drought, however, caused a decrease in vegetation cover, but this did not lead to a corresponding decrease in the abundance of Auchenorrhyncha. Egg hatch and the termination of nymphal hibernation occurred earlier in winter-warmed plots; however, the rate of nymphal development was unaffected.

Insufficient information is presently available to predict the widespread effects of a warming climate on interactions between plant hosts and disease vectors and the diseases they transmit. Some studies have already reported some predictions. Yamamura and Yokozawa (2002) examined the relationship among the prevalence of rice stripe virus disease transmitted in Japan by the delphacid planthopper *Laodelphax striatellus* (Fallén). The susceptible stage for virus transmission is within several weeks of transplanting rice seedlings. Any changes in synchronization between planthopper and host plants will alter the area vulnerable to rice stripe virus.

Changes in distribution of species in the same region

Although little is known yet about climate change effects on phytoplasma vectors, there are rapidly accumulating data on many other insect species. Northward shifts of many species have now been reported, based on distributional data accumulated over the past 25 years (e.g. Hickling *et al.*, 2005). Many of these data are somewhat anecdotal but the number of published studies with more quantitative data is increasing. There have been several recent reported studies on Heteroptera (true bugs) species, e.g. Musolin and Fujisaki (2006) and Musolin (2007). They concluded that Heteroptera species respond to climate change by shifting distribution ranges and changing abundance, phenology, voltinism, physiology, behaviour and community structure. These comments are also likely to apply to plant disease vectors.

The pentatomid *Nezara viridula* (L.) (Heteroptera: Pentatomidae) has been extensively studied in Japan. Musolin (2007), Tougou *et al.* (2009) and Yukawa *et al.* (2007) have studied various aspects of the comparative past and current limits of the distribution range in central Japan. In the early 1960s, the northern limit of the range was in Wakayama Prefecture and was limited by a +5°C isothermal line for the mean January temperature. In 2006–2007, a new survey demonstrated that this northern limit had shifted northwards by 85 km, i.e. at a mean rate of 19.0 km/decade. The shift was most probably promoted by milder winter conditions. *N. viridula* was mostly found close to those locations where: (i) the mean January temperature exceeded +5°C; (ii) the mean number of cold days did not exceed 26 in January–February; and (iii) the mean annual lowest temperature did not drop below −3.0°C. The mean January temperature and number of cold days are the most important factors controlling the northern limit of distribution of *N. viridula*. All the climatic data suggest that, over the last 45 years, environmental conditions have become more favourable for overwintering of *N. viridula* at many locations in central Japan. This has probably promoted the northward spread of the species, representing a direct response to climate warming

Prediction of the Influence of Climate Change on Phytoplasma Diseases

To date, no studies have linked changes of phytoplasma disease impact or geographical distribution to changes in climatic conditions. Many biological parameters influencing phytoplasma epidemiology can theoretically be affected by climate change. As a result of global warming, local increase in mean temperature can act at the level of insect vector population dynamics, biology and fitness, but also at the level of the interaction between the phytoplasma and its two hosts: the plant and the insect vector. Events such as a storm or change in wind conditions can affect insect vector dispersal.

The predicted increase of mean temperature over the planet will increase the phytoplasma multiplication rate early in the season, when temperature is suboptimal. For example, in a temperate climate, it may be surmised that an increase in mean temperature during spring will result in a higher multiplication of phytoplasmas in plants and insects. On the infected plant side, it will result in an earlier development of symptoms, which might also be more severe, as a higher number of phytoplasmas in the plant may result in an increased disease severity. This reduction of the incubation period in the plant should also reduce the acquisition access period: the time necessary for acquisition of the pathogen by the insect vector feeding on the infected plant. Phytoplasmas will also multiply faster in the insect vector, thus decreasing the latency period necessary for the insect colonization by the phytoplasma. One can therefore predict a shortening of the duration of the epidemiological cycle of phytoplasmas where temperature conditions are suboptimal. Earlier stages of insect development should be able to transmit the disease when eggs are the overwintering stage of the insect, with a necessary acquisition of the phytoplasma by the nymphs. In life cycles where adults or nymphs are the overwintering stages, earlier transmission of the phytoplasma should be accomplished by a larger population of infected adults that survived the milder winter or by nymphs that survived the winter on infected plants. In both systems, more insects containing higher titres of phytoplasmas will perform earlier transmission of the phytoplasmas. With earlier transmission, the time period in which plants are infected should therefore be longer, increasing the possibility of subsequent acquisition by the insect vectors. All these parameters should promote enhanced spread and expression of the disease.

Conversely, the detrimental influence of phytoplasma infection on insect fitness can reduce the opportunities of phytoplasma disease propagation. It is known that the maize bushy stunt phytoplasma (group 16SrI) and the flavescence dorée phytoplasma (group 16SrV) reduce the lifespan of their respective insect vectors, *Dalbulus longulus* DeLong and *S. titanus* (Nault *et al.*, 1984; Bressan *et al.*, 2005). Interestingly, the pathogenicity of the western-X phytoplasma (group 16SrIII) to its vector *Paraphlepsius irroratus* (Say) is temperature dependent (Garcia-Salazar *et al.*, 1991).

In conclusion, it is difficult, due to the lack of long-term monitoring, to predict accurately the influence of global warming on the distribution and incidence of phytoplasma diseases, but key parameters such as the influence of increased temperature on insect vector population dynamics and the temperature optimum for phytoplasma multiplication could be monitored and determined. Risk assessment of phytoplasma disease spread in changing climate conditions would also benefit from building models integrating different temperature-dependent parameters influencing the phytoplasma life cycle.

References

Abou-Jawdah, Y., Dakhil, H., El-Mehtar, S. and Lee, I.M. (2003) Almond witches'-broom phytoplasma: a potential threat to almond, peach, and nectarine. *Canadian Journal of Plant Pathology – Review of Canadian Phytopathology* 25, 28–32.

Alhudaib, K., Arocha, Y., Wilson, M. and Jones, P. (2008) First report of a 16SrI, *Candidatus* Phytoplasma asteris group phytoplasma associated with a date palm disease in Saudi Arabia. *Plant Pathology* 57, 366.

Anfoka, G.H.A., Khalil, A.B. and Fattash, I. (2003) Detection and molecular characterization of a phytoplasma associated with big bud disease of tomatoes in Jordan. *Journal of Phytopathology* 151, 223–227.

Arocha, Y., Lopez, M., Fernandez, M. *et al.* (2005) Transmission of a sugarcane yellow leaf phytoplasma by the delphacid planthopper *Saccharosydne saccharivora*, a new vector of sugarcane yellow leaf syndrome. *Plant Pathology* 54, 634–642.

Arzone, A., Vidano, C. and Alma, A. (1987) Auchenorrhyncha introduced into Europe from the Nearctic region: taxonomic and phytopathological problems. In: Wilson, M.R and Nault, L.R. (eds) *Proceedings 2nd International Workshop on Leafhoppers and Planthoppers of Economic Importance*. CIE, London, pp. 3–17.

Beanland, L., Noble, R. and Wolf, T.K. (2006) Spatial and temporal distribution of North American grapevine yellows disease and of potential vectors of the causal phytoplasmas in Virginia. *Environmental Entomology* 35, 332–344.

Bedendo, I.P., Davis, R.E. and Dally, E.L. (1997) Molecular evidence for the presence of maize bushy stunt phytoplasma in corn in Brazil. *Plant Disease* 81, 957.

Berges, R., Cousin, M.T., Roux, J., Maurer, R. and Seemüller, E. (1997) Detection of phytoplasma infections in declining *Populus nigra* 'Italica' trees and molecular differentiation of the aster yellows phytoplasmas identified in various *Populus* species. *European Journal of Forest Pathology* 27, 33–43.

Boudon-Padieu, E. (2002) Flavescence dorée of the grapevine: knowledge and new developments in epidemiology, etiology and diagnosis. *ATTI Giornate Fitopatologiche* 1, 15–34.

Bressan, A., Girolami, V. and Boudon-Padieu, E. (2005) Reduced fitness of the leafhopper vector *Scaphoideus titanus* exposed to flavescence doree phytoplasma. *Entomologia Experimentalis et Applicata* 115, 283–290.

Bressan, A., Turata, R., Maixner, M., Spiazzi, S., Boudon-Padieu, E. and Girolami, V. (2007) Vector activity of *Hyalesthes obsoletus* living on nettles and transmitting a stolbur phytoplasma to grapevines: a case study. *Annals of Applied Biology* 150, 331–339.

Castro, S. and Romero, J. (2002) The association of clover proliferation phytoplasma with stolbur disease of pepper in Spain. *Journal of Phytopathology* 150, 25–29.

Chen, M.H. and Hiruki, C. (1975) Electron microscopy of mycoplasma-like bodies associated with clover proliferation disease. *Proceedings of the American Phytopathological Society* 2, 52.

Choueiri, E., Jreijiri, F., Issa, S., Verdin, E., Bové, J. and Garnier, M. (2001) First report of a phytoplasma disease of almond (*Prunus amygdalus*) in Lebanon. *Plant Disease* 85, 802.

Choueiri, E., Salar, P., Jreijiri, F., El-Zammar, S., Massaad, R., Abdul-Nour, H., Bové, J. M., Danet, J.L. and Foissac, X. (2007) Occurrence and distribution of 'Candidatus Phytoplasma trifolii' associated with diseases of solanaceous crops in Lebanon. *European Journal of Plant Pathology* 118, 411–416.

Cronje, C.P.R., Tymon, A.M., Jones, P. and Bailey, R.A. (1998) Association of a phytoplasma with a yellow leaf syndrome of sugarcane in Africa. *Annals of Applied Biology* 133, 177–186.

Davis, R.E., Dally, E.L., Gundersen, D.E., Lee, I.M. and Habili, N. (1997) '*Candidatus* phytoplasma australiense', a new phytoplasma taxon associated with Australian grapevine yellows. *International Journal of Systematic Bacteriology* 47, 262–269.

Desmidts, M. and Rassel, A. (1974) Presence of mycoplasma-like organisms in plant species affected by yellows symptoms in Upper Volta – relationships with cotton phyllody. *FAO Plant Protection Bulletin* 22, 138–141.

Fos, A., Danet, J.L., Zreik, L., Garnier, M. and Bové, J.M. (1992) Use of a monoclonal-antibody to detect the stolbur mycoplasma-like organism in plants and insects and to identify a vector in France. *Plant Disease* 76, 1092–1096.

Garcia-Salazar, C., Whalon, M.E. and Rahardja, U. (1991) Temperature-dependent pathogenicity of the X-disease mycoplasma-like organism to its vector, *Paraphlepsius irroratus* (*Homoptera, Cicadellidae*). *Environmental Entomology* 20, 179–184.

Garnier, M. (2000) The stolbur phytoplasma: a ubiquitous agent. *Comptes Rendus de l'Académie d'Agriculture de France* 86, 27–33.

Garnier, M., Zreik, L. and Bové, J.M. (1991) Witches-broom, a lethal mycoplasmal disease of lime trees in the Sultanate of Oman and the United Arab Emirates. *Plant Disease* 75, 546–551.

Ghosh, D.K., Das, A.K., Singh, S., Singh, S.J. and Ahlawat, Y.S. (1999) Association of a phytoplasma with witches' broom, a new disease of acid lime (*Citrus aurantifolia*). *Current Science* 77, 174–177.

Granett, A.L. and Gilmer, R.M. (1971) Mycoplasmas associated with X-diseases in various *Prunus* species. *Phytopathology* 61, 1036–1037.

Gu, P., An, F., Wu, Y., Yang, D., Luo, Z., Xiang, J. and Yang, Y. (2005) Comparison and analysis of 16S rDNA fragment of phytoplasma of wheat blue dwarf. *Acta Phytopathologica Sinica* 35, 403–409.

Harrison, N.A., Richardson, P.A. and Tsai, J.H. (1996) PCR assay for detection of the phytoplasma associated with maize bushy stunt disease. *Plant Disease* 80, 263–269.

Hickling, R., Roy, D.B., Hill, J.K., Chris, D. and Thomas, C.D. (2005) A northward shift of range margins in British Odonata. *Global Change Biology* 11, 502–506.

Jacobs, K.A., Lee, L.M., Griffiths, H.M., Miller, F.D. and Bottner, K.D. (2003) A new member of the clover proliferation phytoplasma group (16SrVI) associated with elm yellows in Illinois. *Plant Disease* 87, 241–246.

Jarausch, W., Saillard, C., Dosba, F. and Bové, J.M. (1994) Differentiation of mycoplasmalike organisms (MLOs) in European fruit trees by PCR using specific primers derived from the sequence of a chromosomal fragment of the apple proliferation MLO. *Applied and Environmental Microbiology* 60, 2916–2923.

Jones, P., Cockbain, A.J. and Freigoun, S.O. (1984) Association of a mycoplasma-like organism with broad bean phyllody in the Sudan. *Plant Pathology* 33, 599–602.

Jovic, J., Cvrkovic, T., Mitrovic, M., Krnjajic, S., Redinbaugh, M.G., Pratt, R.C., Gingery, R.E., Hogenhout, S.A. and Toševski, I. (2007) Roles of stolbur phytoplasma and *Reptalus panzeri* (*Cixiinae, Auchenorrhyncha*) in the epidemiology of maize redness in Serbia. *European Journal of Plant Pathology* 118, 85–89.

Jung, H.Y., Sawayanagi, T., Kakizawa, S. *et al.* (2003a) '*Candidatus* Phytoplasma ziziphi', a novel phytoplasma taxon associated with jujube witches'-broom disease. *International Journal of Systematic and Evolutionary Microbiology* 53, 1037–1041.

Jung, H.Y., Sawayanagi, T., Wongkaew, P. *et al.* (2003b) '*Candidatus* Phytoplasma oryzae', a novel phytoplasma taxon associated with rice yellow dwarf disease. *International Journal of Systematic and Evolutionary Microbiology* 53, 1925–1929.

Kenis, M., Rabitsch, W., Auger-Rozenberg, M.A. and Roques, A. (2007) How can alien species inventories and interception data help us prevent insect invasions? *Bulletin of Entomological Research* 97, 489–502.

Kenis, M., Auger-Rozenberg, M.-A., Roques, A., Timms, L., Péré, C., Cock, M.J.W., Settele, J., Augustin, S. and Lopez-Vaamonde, C. (2009) Ecological effects of invasive alien insects. *Biological Invasions* 11, 21–45.

Khan, A.J., Botti, S., Al-Subhi, A.M., Gundersen-Rindal, D.E. and Bertaccini, A.F. (2002) Molecular identification of a new phytoplasma associated with alfalfa witches'-broom in Oman. *Phytopathology* 92, 1038–1047.

Langer, M. and Maixner, M. (2004) Molecular characterisation of grapevine yellows associated phytoplasmas of the stolbur-group based on RFLP-analysis of non-ribosomal DNA. *Vitis* 43, 191–200.

Lee, I.M., Davis, R.E. and Hiruki, C. (1991) Genetic interrelatedness among clover proliferation mycoplasmalike organisms (MLOs) and other MLOs investigated by nucleic acid hybridization and restriction fragment length polymorphism analyses. *Applied and Environmental Microbiology* 57, 3565–3569.

Lee, I.M., Davis, R.E. and Gundersen-Rindal, D.E. (2000) Phytoplasma: phytopathogenic mollicutes. *Annual Review of Microbiology* 54, 221–255.

Lee, I.M., Gundersen-Rindal, D.E., Davis, R.E., Bottner, K.D., Marcone, C. and Seemüller, E. (2004a) '*Candidatus* Phytoplasma asteris', a novel phytoplasma taxon associated with aster yellows and related diseases. *International Journal of Systematic and Evolutionary Microbiology* 54, 1037–1048.

Lee, I.M., Martini, M., Marcone, C. and Zhu, S.F. (2004b) Classification of phytoplasma strains in the elm yellows group (16SrV) and proposal of '*Candidatus* Phytoplasma ulmi' for the phytoplasma associated with elm yellows. *International Journal of Systematic and Evolutionary Microbiology* 54, 337–347.

Liefting, L.W., Padovan, A.C., Gibb, K.S., Beever, R.E., Andersen, M.T., Newcomb, R.D., Beck, D.L. and Forster, R.L.S. (1998) '*Candidatus* Phytoplasma australiense' is the phytoplasma associated with Australian grapevine yellows, papaya dieback and Phormium yellow leaf diseases. *European Journal of Plant Pathology* 104, 619–623.

Lorenz, K.H., Dosba, F., Poggipollini, C., Llacer, G. and Seemüller, E. (1994) Phytoplasma diseases of *Prunus* species in Europe are caused by genetically similar organisms. *Journal of Plant Disease Protection* 101, 567–575.

Marcone, C., Ragozzino, A., Camele, I., Rana, G.L. and Seemüller, E. (2001) Updating and extending genetic characterization and classification of phytoplasmas from wild and cultivated plants in southern Italy. *Journal of Plant Pathology* 83, 133–138.

Masters, G.J., Brown, V.K, Clarke, I.P, Whittaker, J.B. and Hollier, J. (1998) Direct and indirect effects of climate change on insect herbivores: Auchenorrhyncha (Homoptera). *Ecological Entomology* 23, 45–52.

McCoy, R., Tsai, J.H., Norris, R.C. and Gwin, G.H. (1983) Pigeon pea witches'-broom in Florida. *Plant Disease* 67, 443–445.

Mitra, D.K. (1988) Little leaf disease of eggplant. In: Maramorosch, K. and Raychaudhuri, S.P. (eds) *Mycoplasma Disease of Crops*. Springer-Verlag, New York, pp. 343–348.

Montano, H.G., Brioso, P.S. and Pimentel, J.P. (2007) List of phytoplasma hosts in Brazil. *Bulletin of Insectology* 60, 129–130.

Musolin, D.L. (2007) Insects in a warmer world: ecological, physiological and life-history responses of true bugs (Heteroptera) to climate change. *Global Change Biology* 13, 1565–1585.

Musolin, D.L. and Fujisaki, K. (2006) Changes in ranges: trends in distribution of true bugs (Heteroptera) under conditions of the current climate warming. *Russian Entomological Journal* 15, 175–179.

Nault, L.R., Madden, L.V., Styer, W.E., Triplehorn, B.W., Shambaugh, G.F. and Heady, S.E. (1984) Pathogenicity of corn stunt spiroplasma and maize bushy stunt mycoplasma to their vector, *Dalbulus longulus*. *Phytopathology* 74, 977–979.

Nickel, H. (2003) *The Leafhoppers and Planthoppers of Germany (Hemiptera, Auchenorrhyncha): Patterns and Strategies in a Highly Diverse Group of Phytophagous Insects*. Pensoft, Sofia.

Preisseravier, E.L., Elkington, J.K. and Abell, K. (2008) Evolution of increased cold tolerance during range expansion of the elongate hemlock scale *Fiorinia externa* Ferris (Hemiptera: Diaspididae). *Ecological Entomology* 33(6), 709–715.

Rabitsch, W. (2008) Alien true bugs of Europe (Insecta: Hemiptera: Heteroptera). *Zootaxa* 1827, 1– 44.

Roques, A., Rabitsch, W., Rasplus, J.-Y., Lopez-Vaamonde, C., Nentwig, W. and Kenis, M. (2009) Alien terrestrial invertebrates of Europe. In: DAISIE, *Handbook of Alien Species in Europe*. Springer Science, Berlin, pp. 63–79.

Salehi, M. and Izadpanah, K. (1995) Almond brooming. In: *12th Iranian Plant Protection Congress*, Karaj, Iran, Entomological Society of Iran, Iran, p. 238.

Salehi, M., Heydarnejad, J. and Izadpanah, K. (2005) Molecular characterization and grouping of 35 phytoplasmas from central and southern provinces in Iran. *Iranian Journal of Plant Pathology* 41, 150–154.

Sawayanagi, T., Horikoshi, N., Kanehira, T., Shinohara, M., Bertaccini, A., Cousin, M.T., Hiruki, C. and Namba, S. (1999) '*Candidatus* Phytoplasma japonicum', a new phytoplasma taxon associated with Japanese hydrangea phyllody. *International Journal of Systematic Bacteriology* 49, 1275–1285.

Schneider, B., Cousins, M.T., Klinkong, S. and Seemüller, E. (1995) Taxonomic relatedness and phylogenetic positions of phytoplasmas associated with diseases of faba bean, sunnhemp, sesame, soybean, and eggplant. *Journal of Plant Disease Protection* 102, 225–232.

Schneider, B., Gibb, K.S., Padovan, A., Davis, R.I. and De La Rue, S. (1999) Comparison and characterization of tomato big bud and sweet potato little leaf group phytoplasmas. *Journal of Phytopathology* 147, 31–40.

Seemüller, E. and Schneider, B. (2004) '*Candidatus* Phytoplasma mali', '*Candidatus* Phytoplasma pyri' and '*Candidatus* Phytoplasma prunorum', the causal agents of apple proliferation, pear decline and European stone fruit yellows, respectively. *International Journal of Systematic and Evolutionary Microbiology* 54, 1217–1226.

Seruga, M., Skoric, D., Botti, S., Paltrinieri, S., Juretic, N. and Bertaccini, A.F. (2003) Molecular characterization of a phytoplasma from the aster yellows (16SrI) group naturally infecting *Populus nigra* L. 'Italica' trees in Croatia. *Forest Pathology* 33, 113–125.

Sharma, O.P. (1990). Witches'-broom disease of sunn hemp (*Crotalaria juncea* L.) by MLO. *Indian Journal of Mycology and Plant Pathology* 20, 264–265.

Shaw, M.E., Kirkpatrick, B.C. and Golino, D.A. (1993) The beet leafhopper-transmitted virescence agent causes tomato big bud disease in California. *Plant Disease* 77, 290–295.

Teixeira, D.C., Wulff, N.A., Martins, E.C., Kitajima, E.W., Bassanezi, R., Ayres, A.J., Eveillard, S., Saillard, C. and Bove, J.M. (2008) Phytoplasma closely related to the pigeon pea witches'-broom phytoplasma (16Sr IX) is associated with citrus huanglongbing symptoms in the state of São Paulo, Brazil. *Phytopathology* 98, 977–984.

Tougou, D., Musolin, D.L. and Fujisaki, K. (2009) Some like it hot! Rapid climate change promotes changes in distribution ranges of *Nezara viridula* and *Nezara antennata* in Japan. *Entomologia Experimentalis et Applicata* 130, 249–258.

Urbanaviciene, L., Jomantiene, R., Valiunas, D. and Davis, R.E. (2007) Molecular identification of 16Srl-A, 16Srl-B, 16Srl-C, and 16Srl-L subgroups of phytoplasmas in gramineous plants in Lithuania. *Bulletin of Insectology* 60, 127–128.

Valiunas, D., Staniulis, J. and Davis, R.E. (2006) 'Candidatus Phytoplasma fragariae', a novel phytoplasma taxon discovered in yellows diseased strawberry, *Fragaria × ananassa*. *International Journal of Systematic and Evolutionary Microbiology* 56, 277–281.

Verdin, E., Salar, P., Danet, J.L., Choueiri, E., Jreijiri, F., El Zammar, S., Gélie, B., Bové, J.M. and Garnier, M. (2003) '*Candidatus* Phytoplasma phoenicium' sp. nov., a novel phytoplasma associated with an emerging lethal disease of almond trees in Lebanon and Iran. *International Journal of Systematic and Evolutionary Microbiology* 53, 833–838.

Yamamura, K. and Yokozawa, M. (2002) Prediction of a geographical shift in the prevalence of rice stripe virus disease transmitted by the small brown planthopper, *Laodelphax striatellus* (Fallén) (Hemiptera: Delphacidae), under global warming. *Applied Entomology and Zoology* 37, 181–190.

Yang, L.L. (1985) Host responses of peanut witches' broom disease. *Journal of Agricultural Research of China* 34, 464–468.

Yang, Z.H., Chen, Q.B., Yang, L.F., Li, H.C. and Bai, C.J. (2008) Occurrence of a 16SrII group phytoplasma associated with crotalaria witches' broom in Hainan, China. *Plant Pathology* 57, 364.

Yukawa, J., Kiritani, K., Gyoutoku, N., Uechi, N., Yamaguchi, D. and Kamitani, S. (2007) Distribution range shift of two allied species, *Nezara viridula* and *N. antennata* (Hemiptera: Pentatomidae), in Japan, possibly due to global warming. *Applied Entomology and Zoology* 42, 205–215.

Zhu, S.F., Hadidi, A., Gundersen, D.E., Lee, I.M. and Zhang, C.L. (1998) Characterization of the phytoplasmas associated with cherry lethal yellows and jujube witches'-broom diseases in China. *Acta Horticulturae* 472, 701–714.

Index